AF896566

Theory of Mountainurbanology

Guangyu Huang

Theory of Mountainurbanology

Springer

Guangyu Huang
School of Architecture and Urban Planning
Chongqing University
Chongqing, China

ISBN 978-981-16-0818-6 ISBN 978-981-16-0819-3 (eBook)
https://doi.org/10.1007/978-981-16-0819-3

© The Editor(s) (if applicable) and The Author(s), under exclusive license to Springer Nature Singapore Pte Ltd. 2021
This work is subject to copyright. All rights are solely and exclusively licensed by the Publisher, whether the whole or part of the material is concerned, specifically the rights of translation, reprinting, reuse of illustrations, recitation, broadcasting, reproduction on microfilms or in any other physical way, and transmission or information storage and retrieval, electronic adaptation, computer software, or by similar or dissimilar methodology now known or hereafter developed.
The use of general descriptive names, registered names, trademarks, service marks, etc. in this publication does not imply, even in the absence of a specific statement, that such names are exempt from the relevant protective laws and regulations and therefore free for general use.
The publisher, the authors and the editors are safe to assume that the advice and information in this book are believed to be true and accurate at the date of publication. Neither the publisher nor the authors or the editors give a warranty, expressed or implied, with respect to the material contained herein or for any errors or omissions that may have been made. The publisher remains neutral with regard to jurisdictional claims in published maps and institutional affiliations.

This Springer imprint is published by the registered company Springer Nature Singapore Pte Ltd.
The registered company address is: 152 Beach Road, #21-01/04 Gateway East, Singapore 189721, Singapore

Foreword

With the growth of the population and advancement of science and technology, the process of urbanization is rapidly expanding worldwide. Urbanization not only plays a positive role in social development but also brings a series of negative problems, such as environmental pollution, ecological degradation, biodiversity decline, heat island effect, and traffic congestion. Modern architecture is breaking through its original traditional fields, guiding the construction and development of cities with the scientific theory of human settlements. Architects have changed from the narrow field in the past, which mainly pursued the objective of material substance and external landscape of architectures to the integration of architecture, landscape, and urban planning. Furthermore, combined with engineering with ecology and culture, this is the concrete embodiment of the people-oriented, all-round, coordinated, and sustainable development in urban planning that the Scientific Outlook on Development adheres to.

Although urban planning is becoming increasingly ecological, ecological disciplines are also undergoing reform. The present ecology has got rid of the tendency of pure naturalism in the early stage of development, and this is increasingly oriented to the society to solve the urgent ecological and environmental problems faced by the society, which would inevitably result in the hybridization and integration of adjacent disciplines, the differentiation of disciplines, and the emergence of new ones. Among these, urban ecology is a new research direction generated in the process of transformation, which shows great vitality. It is also in this context where ecologists, scientists, and workers engaged in architecture and urban planning encounter are acquainted, and associate with each other on the big stage of urban development and planning. Approximately 10 years ago, I met Mr. Huang Guangyu at an academic conference. I was touched by his modest and simple academic attitude and professional dedication. Furthermore, I resonated with his idea of introducing the concept of ecology into urban planning. In particular, his tireless work in mountainurban planning and design aroused my great interest and admiration. This is not only because China is a mountainous country, where the distribution and quantity of mountain cities account for a large proportion, but also because I have conducted scientific surveys in mountain areas for many years, especially in the southwest mountain areas, and know the characteristics of mountains and mountain cities, as well as the unique problems these face. The common interests and personalities made us feel like old friends at the first meeting. Hence, we began a long-term friendship and close contact. Mr. Huang Guangyu often participated in symposiums on urban ecology and planning organized by the Ecological Society of China (ESC), and appraisal meetings on relevant achievements, and presented his profound insights. After that, I received his articles and academic monographs, and also attended the academic appraisal meetings of his research results. After several communications, I learned that Mr. Huang and his students carried out a lot of research and practical work in mountainurban planning and design. He organically combined hardware and software planning with ecological concepts, which demonstrated the characteristics of mountain cities.

In this book, the author made a systematic exposition, in terms of the analysis on the advantages and disadvantages of mountainurban construction, the necessity of ecological construction in mountain cities, the definition, research objects and contents, research methods and trends of mountainurbanology, the characteristics and service functions of mountainurban ecosystems, specific survey techniques and methods, site selection, structure, eco-environment, transportation, architecture, disasters, aesthetics, and governance of mountain cities, etc. This book, with the essence of Mr. Huang's work over the years, is a systematic and in-depth monograph on the planning and design of mountain cities based on the optimization of human settlements, which has laid a solid foundation for the future research and practice of ecological mountainurban planning and construction. Mr. Huang asked me to write a preface for his book, which I esteem as a great honor. I would like to make this preface to express my support, learning, and appreciation, and hope that this book would play an active and exemplary role in the residential environment in mountain areas and the ecological construction of cities.

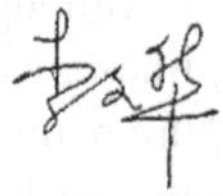

July 2006

Li Wenhua
Academician of Chinese Academy of Engineering
Beijing, China

Preface

From latter last century, due to the impact of the global environmental crisis and the tide of economic globalization, sustainable development has become a common concern. Therefore, urban planning is becoming more and more important, and apparently, including the tendency of multi-discipline in urban study and the involvement of ecology.

As the largest developing country, China's urbanism development and economic growth is having a distinctive influence on the world. Mountain cities are widely distributed in west and south China. However, the unique condition and inherent character of mountain cities have not been given enough concern. The two major problems in the development of mountain cities are losing the unique city character and environmental disasters.

Therefore, the purpose of this book is to emphasize the multi-discipline study on mountain cities, promote the ecological theory, and innovative thoughts on urban science research, in order to better direct mountain human settlement, balance human beings, and nature, and avoid the "plainization" in mountain cities.

The book was based on the author's teaching and practice experience on mountainurban study.

Chongqing, China
Spring 2006

Guangyu Huang

Acknowledgments

The *Theory of Mountainurbanology*, which has been prepared for many years, is finally finished today, which makes me feel relieved. I would like to thank my seniors and colleagues for completing this book.

Mr. Wu Liangyong, with his strategic vision of a commanding height in the academic field, has promoted the construction of Chinese architecture, urban planning disciplines, and educational development, which constantly inspires me. He led us to compile the first textbook *Urban and Rural Planning* during the hard times in the beginning of 1960s, recommended me to undertake the chief review of textbook *Principles of Urban Planning* in the early 1980s, and accepted me as a part-time professor of Tsinghua University to recruit doctoral students in the early 1990s. I have been receiving enthusiastic support and help from him and other leaders of the School of Architecture of Tsinghua University such as Mr. Zhao Bingshi and Mr. Li Daozeng. Mr. Wu's profound insights on "Habitat Environmental Science", interdisciplinary and interdisciplinary interaction, and methodology of scientific research are all precious teachings that will benefit me for life.

Mr. Li Wenhua, who often gives me advice on ecology, makes me see the light and hope in the combination of interdisciplinary subjects. He has fully affirmed my efforts in integrating ecology and mountainurbanology, and wrote an enthusiastic and encouraging preface for the publication of this book.

Mr. Qi Kang has devoted his whole life to the architectural discipline construction and the development of urban planning education in China. I deeply admire his passion for and attention to the development of architectural discipline in the vast areas of China, especially in the west and northeast, and I am also moved by his concern and support for me in the research of regional architectural culture and mountainurbanology.

Mr. Tang Pu, Mr. Ye Qishen, Mr. Liu Chaoyang, Mr. Liao Zuyi, Mr. Chen Qigao, Mr. Yu Zhuoqun, Mr. Chen Shujin, Mr. Wang Yuechang, Mr. Wu Yingwei, Mr. Lu Zhenwei, Mr. Chen Zhihong, Mr. Lu Xiaodi, Mr. Chen Dexiang, and Mr. Ren Zhouyu are all beloved mentors who taught me professional courses. Their pragmatic and rigorous academic attitude and style of teaching by verbal and deed instructions have made me unforgettable and deeply grateful.

Mr. Tang Pu had always been concerned about and supporting my research of mountainurbanology and the publication of series of books, who set an example by publishing his research achievements of many years in 1995, *Mountain Residential Architecture*. With his tragic passing away, I have lost a respected mentor whom I will miss forever.

I would like to express my great gratitude to Mr. Zhang Xingguo, Dean of the School of Architecture and Urban Planning of Chongqing University, and other leaders for warm care, help, and support.

I would like to express my deep appreciation to Xing Zhong, Yang Liu, Ying Wen, Li Hao, Ye Lin, Wu Yong, Pei Wen, Qiao Xin, and Wei Xiaofang for their hard work in collating manuscripts, illustrations, literature review, and proofreading.

I would also like to thank my beloved wife, Ms. Yuan Wenqiong, and our daughter, Huang Jian, for their tedious work in printing the manuscripts, the photography, and the translation.

Finally, I would also like to thank the staff of China Architecture & Building Press for everything they have done for the publication of this book.

As a new interdisciplinary subject, mountainurbanology is gradually taking shape with continuous improvement, which is far from being complete. Therefore, in case of any deficiency and mistake, I would like to implore the seniors, colleagues, and readers of the academic circle for criticism and correction.

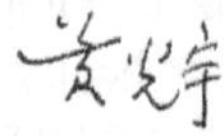

Guangyu Huang
Written in the spring of 2006 in Mountainurban Chongqing.

Brief Introduction

This book was first published by China Construction Industry Press in Beijing in September 2006. It is about mountainurbanology grounded in China. From a dynamic, multi-discipline perspective in a changing context, a comprehensive framework, which includes the location of mountain city, planning, design, building, transportation, disaster control, esthetics, and governance was established based on the investigation of natural, social, and economic studies. This book is recommended as a reference book for planning and design professionals, decision-makers, college faculties, and students.

Contents

About the Author

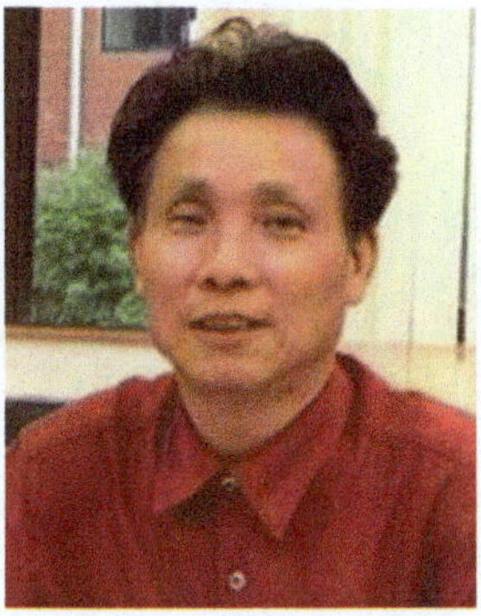

Prof. Guangyu Huang (1935–2006) was born in Yueqing, Zhejiang Province, China. He was a renowned Chinese urban planning scholar, senior professor at the School of Architecture and Urban Planning of Chongqing University, and a pioneer of urban planning major in Chongqing University.

Professor Huang laid the ground for China's mountainurbanology and took the lead in the theory and design approaches of eco-city. In 1992, he founded the Research Center for Mountain City and Regional Environment and established extensive contacts with other research institutes in the United States, Germany, Japan, India Nepal, etc., by which he has greatly prompted the research of mountain cities in China. Under his advocacy, the Urban Planning Society of China established the Urban Ecology Committee in November 2001 and engaged Prof. Huang as the first chairman. In March 2004, he was invited to deliver a keynote speech "Ecological Patterns on Spatial Structure of Mountain Cities" at Harvard University, where his argument that the most marvelous eco-cities of future would appear in mountain-water blending regions drawn attention of international scholars.

Professor Huang researched and practiced urban planning in the mountain regions of southwestern China for almost 40 years. During this time, he established the theory of mountainurbanology and developed a framework for this subject. By combining urbanology with ecology, he put forward the theoretical and methodological system for the ecological construction in mountain cities in China, which integrated modern theories of urban planning with mountainology and ecology.

In 2005, the research project "Key Technologies for the Ecological Planning and Construction in Mountain Cities" with him as the leading expert won the second prize of China National Science and Technology Progress Award, which is so far the highest prize on urban planning in China.

Dedicated to the education of urban planning, Prof. Huang has students all over the country, who are now the backbones of the field. In 2006, the Professional Steering Committee of Urban Planning Discipline, an affiliation of the Ministry of Housing and Urban-Rural Development of China, granted him the "Lifetime Achievement Award for Urban Planning Education".

List of Figures

Chapter 4

Chapter 5

Chapter 6

Chapter 7

Chapter 8

Chapter 9

Chapter 10

List of Tables

Chapter 1
Introduction

The earth surface, which is the geographical environment on which human beings live, comprise of the two largest geomorphic units, the continent and the ocean, with a total area of 510 million km^2. The land area is 148.9 million km^2, which account for 29.2% of the total surface area of the earth, while the ocean area is 361.1 million km^2, which account for 70.8% of the total surface area of the earth, and is 2.4 times of the total land area. The land area is surrounded by the ocean.

On the continent, the mountains (at altitudes of 500–8,000 m) cover the largest area, which account for 47.82% of the land area, hills (at altitudes of 200–500 m) are the second largest area, which account for 26.8%, and the plains (at altitudes of 0–200 m) account for 24.85%. The area of depression with an altitude of less than 0 m is the smallest, accounting for only 0.53%. The area above the mountains and hills total to 111.1 million km^2, accounting for 74.62% of the land area (Liu 2001).

China covers a land area of 9.6 million km^2, accounting for 6.4% of the world's land area, and ranks third in the world. The mountain area is 6.62 million km^2, accounting for 69% of the country's total land area, and approximately 6% of the world's mountain area (including hills). In terms of both absolute area and proportion, this is very prominent in all countries of the world.

The landforms of China's land are diverse in development, with high terrains in the west and low terrains in the east. The world's roof Qinghai-Tibet Plateau has an average altitude of above 4,000 m, while the highest peak in the world, Mount Qomolangma, has an altitude of 8,844.43 m. Furthermore, high mountains rear into the clouds, and there are also basins of different sizes and shapes. Moreover, there are broad and slightly undulating plateaus at the top, as well as low and flat plains. From the Turpan Basin at an altitude of -155 m to the offshore plains at below 50 m above sea level, the mountains and rivers are interlaced with varied weather conditions (Fig. 1).

In the long history of crustal development, the tectonic movement of the Earth's crust has laid great undulations on China's topography and the basic contour of its landform, constituting three major steps of the country's landform, descending from west to east (Fig. 2).

© The Author(s), under exclusive license to Springer Nature Singapore Pte Ltd. 2021
G. Huang, *Theory of Mountainurbanology*,
https://doi.org/10.1007/978-981-16-0819-3_1

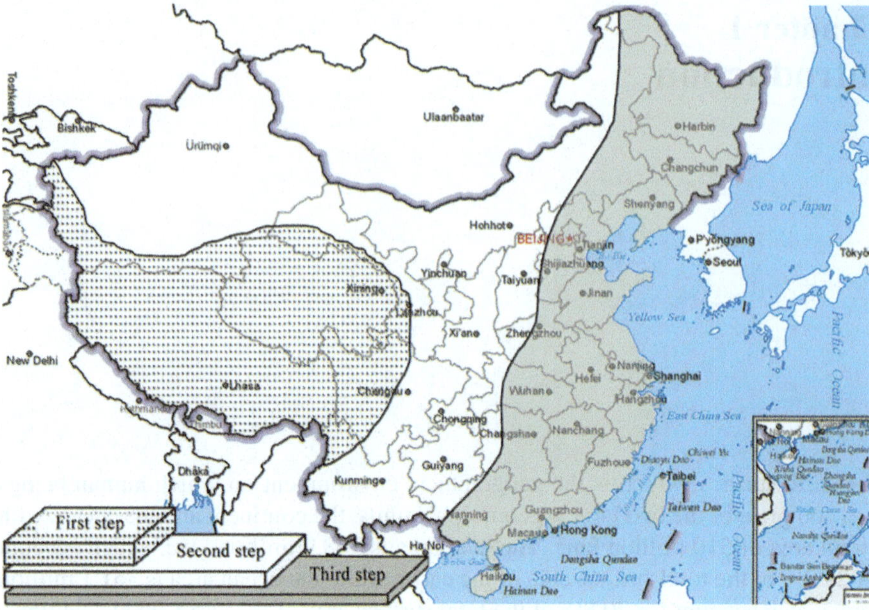

Fig. 1 Distribution of Major Mountain Systems in China. *Source* Zhao Ji, *Natural Geography of China*, Beijing: Higher Education Press. May 1995, p. 6

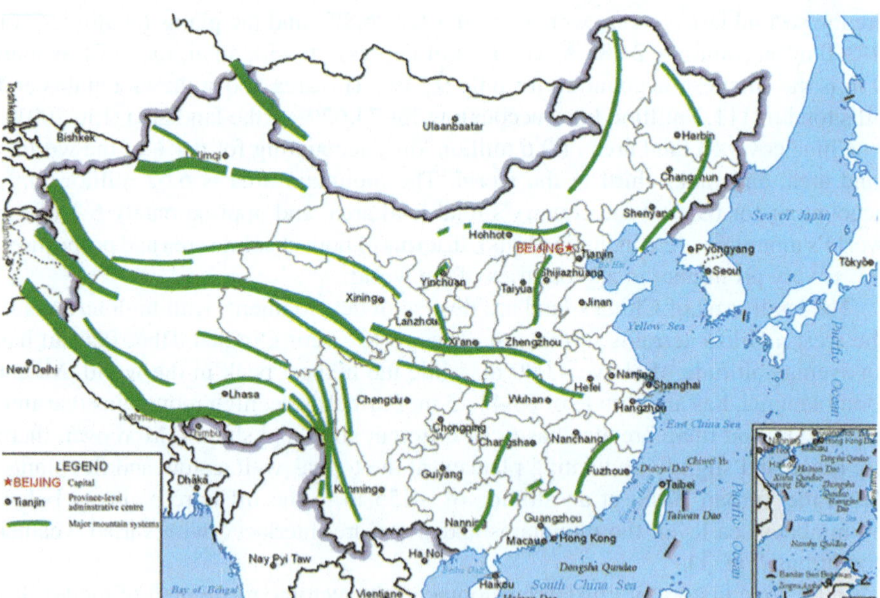

Fig. 2 Three Steps of China's Geomorphology. *Source* Cheng Jianjun, Kong Shangpu. *Geomancy and Architecture*. Jiangxi Science and Technology Press. October 1992, p. 19

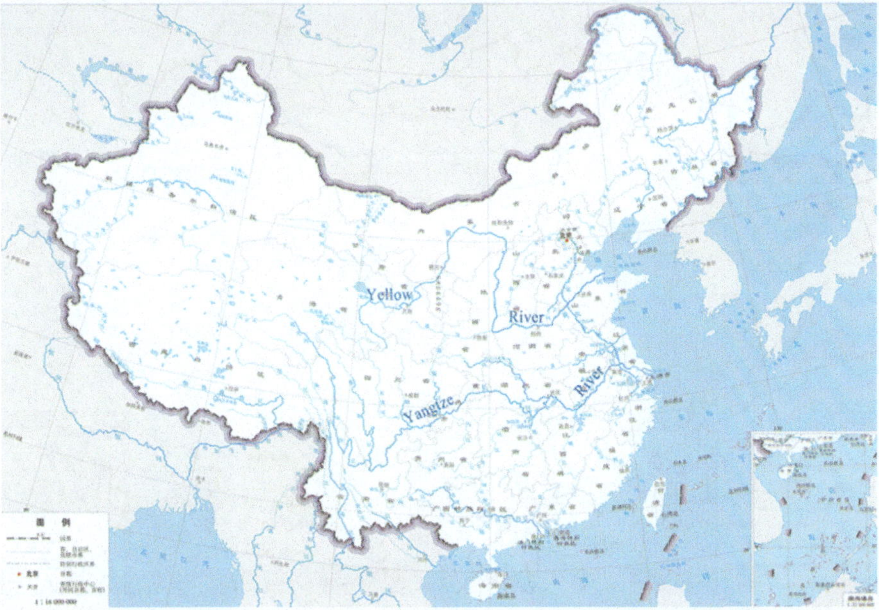

Fig. 3 Major River Systems in China. *Source* Website of China Meteorological Administration http://www.cma.gov.cn/

The first step is the Qinghai-Tibet Plateau at an altitude of more than 4,000 m, covering approximately a quarter of China's land area. Known as the roof of the world, its existence constitutes the inherent characteristics of China's natural geography and the main cause of global climate change.

For the second step, with an altitude of 3,000–1,000 m, this includes the plateaus and the Sichuan Basin in the central and western regions.

The third step, which is below 1,000 m above sea level, is the hilly areas of the middle and lower reaches of the Yangtze River and eastern coastal plain

The overall topographical features and the spatial distribution of mountain ranges in China determine the origin and trend of the water system, and the spatial distribution characteristics of rivers and lakes. The first river source zone is the first step of the southern Qinghai-Tibet Plateau, which is the birthplace of important rivers, such as the Yangtze River, Yellow River, Yarlung Zangbo River, Lancang River and Nu River. The second river source zone is the uplift zone on the edge of the second step, namely, the Greater Khingan Mountains, plateaus in Hebei and Shanxi provinces, and the Yunnan-Guizhou Plateau, from which the Amur River, Liao River, Haihe River, Huai River, and Xi River originate. The third source zone is the Changbai Mountains, the hills in Shandong Province, and the southeastern coastal hills. The rivers that originate from this zone include the Tumen River, Yalu River, Qiantang River, Ou River, Min River, Jiulong River, Gan River, and the tributaries of the Pearl River, Bei River and Dong River (Ji 2002) (Fig. 3).

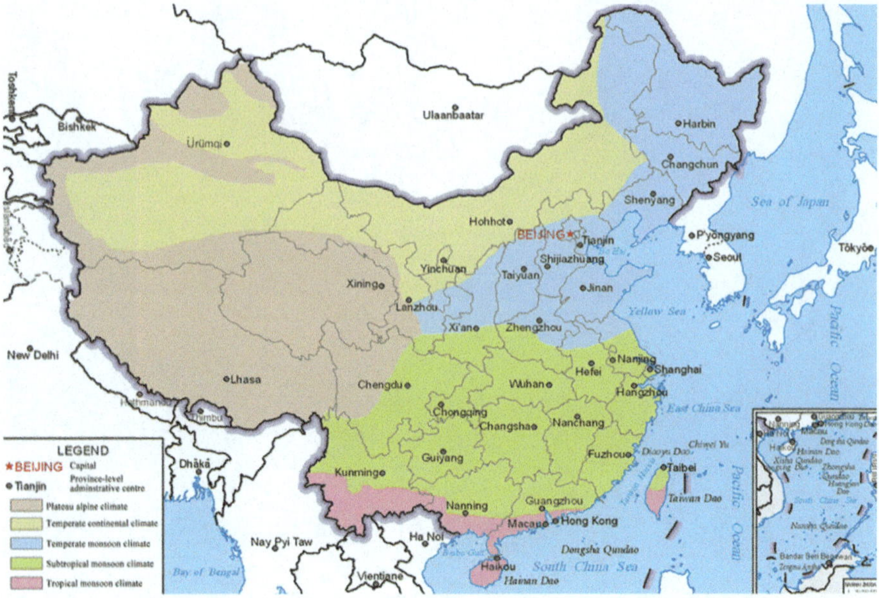

Fig. 4 Climate Zoning Map of China. *Source* Website of China Meteorological Administration http://www.cma.gov.cn/

The general trend of the landform, which is high in the west and low in the east, determines the flow direction of the water system and the overall distribution of the climate change in China (Fig. 4). This spans a latitude of 50° from north to south, passes through cold temperate zone, warm temperate zone, subtropical zone, tropical zone and the equator, respectively, and mainly consists of the warm temperate zone and subtropical zone. This crosses a longitude of 61° from east to west, and transits from humid and semi-humid zones to semi-arid and arid zones. This overall natural landscape pattern determines the overall ecological pattern, which has a profound impact on China's population distribution, the overall social and economic structure and development pattern, and the distribution and layout of mountain cities in China.

Under the overall natural landscape structure, the Chinese civilization of 5,000 years has brought about profound changes in China's natural and cultural environment. The crisscross plain water network and cultivated land, overlapping hills, mountains and terraces (Fig. 5), vast plateaus, grasslands of the Gobi Desert and oasis constitute the distinctive characteristics of the three natural landscape areas of Chinese civilization.

The climate zone can also be divided into three major zones: eastern monsoon zone dominated by plains, low mountains and hills, arid zone of the northwest plateau and paramo zone of the Qinghai-Tibet Plateau. The eastern monsoon zone extends from the bank of Amur River to the Nansha Islands. The northern part is a temperate zone and a warm temperate zone. This is a concentration area of agriculture, coal, oil

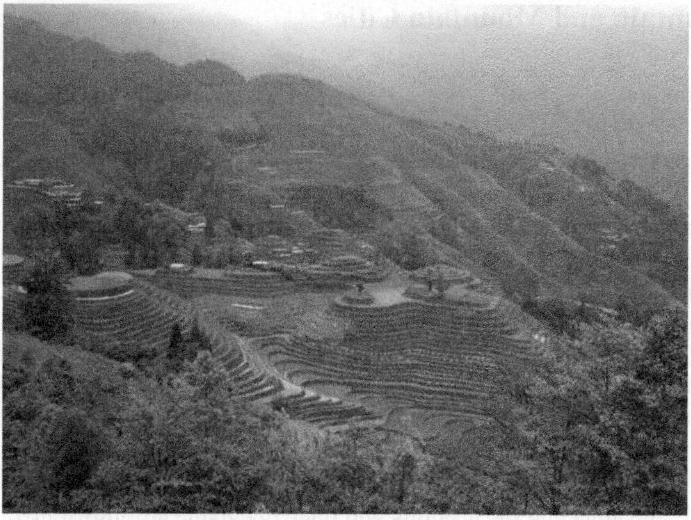

Fig. 5 Terraces in Guilin City, Guangxi Zhuang Autonomous Region. *Source* Photograph by the author in 2005

and iron ore resources, which is an old industrial base of Northeast China. Furthermore, this is a concentrated zone of coastal, marine and port industrial cities. In the south, there are mainly subtropical agricultural and non-ferrous metal industrial areas, and most of these are costal hilly cities, and densely populated cities and populations. The arid zone of the Northwest Plateau comprises of the arid basin, grassland, Gobi, desert, and inter-boundary mountains, which is an intensive agriculture, animal husbandry, and oil and gas base in the oasis, with dry climate, very low population and urban density. The Qinghai-Tibet Plateau, as the roof of the world, is characterized by strong neotectonics, with plateau permafrost and modern glaciers (frequent earthquakes and natural disasters), complex construction conditions, inconvenient transportation, distinct landscape features of altitudinal zonation of marginal mountain systems, abundant resources, vast land, and sparse population. With the implementation of the Western Development Strategy, ecological reconstruction has been vigorously carried out, and new progress has been made in regional infrastructure construction. The construction conditions are undergoing major changes. The Qinghai-Tibet Plateau Railway with the highest altitude and the most complicated construction conditions in the world is about to be open to traffic. The traffic conditions between China and its neighboring countries in the west would be further improved, and economic and cultural exchanges would also be further strengthened.

1 Mountain and Mountain Cities

(1) Mountains

A mountain is a kind of geomorphic type with a certain altitude and slope. Mountain areas can be divided into a broad sense and a narrow sense, which include low mountains, middle mountains, high mountains, and extremely high mountains in a narrow sense, as well as mountains, hills, and plateaus in a broad sense. The so-called "mountain" in this book is a generalized concept, which generally refers to landforms with high altitude and undulating terrain, including mountains, hills, and rugged plateaus in natural geography. The generalized mountains in China covers an area of approximately 6.5 million km^2 (Zhong 2000). Among these, the mountain area (including extremely high mountains, high mountains, middle mountains and low mountains) accounts for 33%, the hill area (including high hills and low hills) accounts for 10%, and the plateau area accounts for 26%. The classification of topography in China and the absolute and relative heights are shown in Table 1.

(2) Mountain distribution

Mountain areas are widely distributed in China, with a large area in most provinces and regions. There is a popular saying: "70% of mountains, 10% of waters and 20% of cultivated land" nationwide. For example, Guizhou Province, Yunnan Province, Sichuan Province, Chongqing Municipality, Fujian Province, Guangxi Zhuang Autonomous Region, Shaanxi Province and Hunan Province have more than 80% of the mountain area. There is a saying: "30% of hills, 70% of mountains, and almost no flat land" in Chongqing Municipality, while "90% of mountains and 10% of waters and cultivated land" and "the terrain is too rugged to find a flat within one meter" in Guizhou Province. The basic situation of mountain areas in China is shown in Table 2. Even the coastal areas and islands of the Yangtze River Delta and

Table 1 Classification of topography in China

Name		Absolute height (m)	Relative height (m)
Extremely high mountain		>5,000	>1,000
High mountain	Deep cut	3,500–5,000	>1,000
	Medium cut		500–1,000
	Shadow cut		100–500
Middle mountain	Deep cut	1,000–3,500	>1,000
	Medium cut		500–1,000
	Shadow cut		100–500
Low mountain	Medium cut	500–1,000	500–1,000
	Shadow cut		500–1,000
Hill		Generally <500	<200

Table 2 Distribution of mountain areas in provinces, municipalities and special administrative regions in China

Provinces, cities and districts	Area (10,000 km^2)	Mountain proportion (%) (Including hills)	Land framework
Guizhou Province	17.6	95.1	80% of mountains, 10% of waters and 10% of cultivated land
Yunnan Province	39.4	95	90% of mountains and 10% of waters and cultivated land
Sichuan Province	48.5	90.1	Mountainous province in China
Chongqing Municipality	8.24	86.9	30% of hills, 70% of mountains, and almost no flat land
Fujian Province	12.14	86.9	80% of mountains, 10% of waters and 10% of cultivated land
Guangxi Zhuang Autonomous Region	23.6	86	80% of mountains, 10% of waters and 10% of cultivated land
Shaanxi Province	20.56	84.6	
Hunan Province	21.18	83.4	70% of mountains, 10% of waters and 20% of cultivated land
Shanxi Province	15.63	79.4	70% of mountains, 20% of waters and 20% of cultivated land
Jiangxi Province	16.69	78.8	30% of mountains, 30% of grasslands, 20% of gobi desert, 10% of forests and 10% of cultivated land
Gansu Province	45.4	77.8	"Spine of the World"
Tibet Autonomous Region	122.84	76.5	
Taiwan Province	3.6	73.9	Plateau provinces and regions
Qinghai Province	72.12	69.9	70% of mountains, 10% of waters and 20% of cultivated land 60% of mountains, 20% of waters and 20% of cultivated land
Zhejiang Province	10.18	67.1	

(continued)

Table 2 (continued)

Provinces, cities and districts	Area (10,000 km^2)	Mountain proportion (%) (Including hills)	Land framework
Guangdong Province	17.79	66.2	70% of mountains, 10% of waters and 20% of cultivated land 60% of mountains, 10% of waters and 30% of cultivated land
Beijing Municipality	1.68	65.4	60% of mountains, 5% of waters and 35% of cultivated land
Hubei Province	18.74	65.2	Two basins surrounded by three mountains
Liaoning Province	14.59	63	
Jilin Province	18.74	56.6	
Xinjiang Uygur Autonomous Region	165	56	
Shandong Province	15.67	54.7	50% of mountains, 10% of waters, 10% of grasslands and 30% of cultivated land
Ningxia Hui Autonomous Region	5.18	54.1	60% of mountains, 10% of waters and 20% of cultivated land
Heilongjiang Province	45.46	52.4	Plateau provinces and regions
Hebei Province	18.77	56.6	"Three large proportions and five small proportions"
Inner Mongolia	118.3	48.5	
Jiangsu Province	10.15	4.9	40% of mountains, 10% of waters and 50% of cultivated land
Henan Province	21.18	38.8	60% of mountains, 10% of waters and 30% of cultivated land
Hainan Province	3.54	38.7	–
Anhui Province	13.96	36.8	
Shanghai municipality	0.63	–	
Tianjini municipality	1.17	4.9	
Hong Kong special administrative region	0.11		

(continued)

Table 2 (continued)

Provinces, cities and districts	Area (10,000 km^2)	Mountain proportion (%) (Including hills)	Land framework
Macao special administrative region	0.0024		

[a]*Source Comprehensive Atlas of China*, Editorial Board of *China Geography Series*. Beijing: SinoMaps Press, Dec. 1990

[b]The data in the table was mainly calculated equivalently according to the statistical surface area of the national topographic area in the *Handbook of Natural Resources in China (1990)* edited by Cheng Hong, with the corresponding supplements made on this basis

[c]In the table, the "three large proportions and five small proportions" in Jiangsu Province refers to the large proportions of cultivated land, water area and urban settlements (including land for mining and industry), accounting for 54.91, 24.75 and 10.08% of the total rural land area of the province, respectively, while the small proportions of garden land, forest land, pasture land, transportation land and unused land account for 1.97, 3.13, 0.57, 2.53 and 1.95% of the total land area of the province, respectively

[d]The data on the proportion of mountains in Hong Kong and Macao special administrative regions is not available

Pearl River Delta Plains also have a large proportion of hills and mountains. For example, there is a saying: "70% of mountains, 10% of waters and 20% of cultivated land" in Zhejiang Province, and "60% of mountains, 20% of waters and 20% of cultivated land" in Guangdong Province. Merely the Shanghai Municipality has no mountain area. Jiangsu Province has a small mountain area, accounting for only 4.9%. Furthermore, the mountain area of Beijing, which is located in the North China Plain, accounts for 65.4%, while Tianjin's mountain area is very small, accounting for only 4.9%.

(3) Mountain cities

There has been no strict definition on a mountain city. Mountain cities, called slide cities in some foreign countries, such as Japan, or hillside cities in Europe and the United States, refer to cities built on sloping hillsides. Building a city on slopes and flats will have different changes and influences on the safety, practicality and economic efficiency of urban planning, and urban design, construction and use, which should be given special attention and particularly studied. However, the previous definition is not enough to fully understand the concept of mountain cities. Sinnce this only considers the basic characteristics and effects of the "slope", and ignores other important features of mountain cities, such as changes in vertical gradients, landforms around cities, and differences in the environment, this will have a significant impact on mountainurban planning and construction. Therefore, the natural characteristics of mountain cities should be defined from the following two aspects:

(1) Mountain cities are built on undulating slopes, with a gradient of more than five degrees, which are different from flat cities, regardless of the altitude, such as

Chongqing Municipality, Lanzhou City, Panzhihua City, Hong Kong Special Administrative Region, Qingdao City, Yan'an City, Zunyi City, etc.;

(2) Although mountain cities are built on flat land, the complex surrounding topography and natural environment have a major impact on the layout structure, development direction, and eco-environment, such as Guiyang City, Kunming City, Guilin City, Hangzhou City, Yantai City, etc.

For example, many cities are initially built on flat land, which have continued to expand to the surrounding hills and hillsides with economic development and population expansion, and formed completely different spatial structures from the original flat cities, both in the scale of land use and the form of development.

Since a city is closely connected with its surrounding environment, we cannot just talk about it in isolation in our understanding and planning. Furthermore, the trend of urban regionalization and regional urbanization has been very obvious in the process of rapid urbanization. Therefore, the mountainurban mentioned in the present study is also a relative and generalized concept.

(4) Distribution of mountain cities (towns)

China's mountain cities (towns) are widely distributed, which are located in the western region, central region, southeast coastal hilly areas, and coastal islands. As of the end of 2005, there are 661 designated cities nationwide Among these, 231 cities are mountain cities, accounting for about 35% of the total. Among more than 1,900 counties in the country, there are approximately 960 mountain towns, which is more than half of the total number of counties (Guangyu 2002).

(5) Advantages and disadvantages of mountainurban construction

(1) **Favorable conditions for mountainurban construction**

In the process of mountainurban construction, the following factors are favorable conditions.

① **Topography and landform enrich the urban landscape**

Mountain cities are rich in topography and landforms, with the integration of mountains and waters, and this is conducive to the formation of urban landscapes, such as Chongqing Municipality, Yibin City, Leshan City, Luzhou City, Wuhan City, Nanjing City in the Yangtze River Basin, and Guiyang City, Kunming City, Dali City, and Lijiang City on the Yunnan-Guizhou Plateau.

② **Abundant resources**

Mountain cities boast of abundant natural animal and plant resources, mineral resources, and hydraulic resources, with superior conditions for the development of diversified businesses, which serves as a driving force for China's modernization drive.

③ **Multi-ethnic cultural integration**

Multi-ethnic cultures exchange, infiltrate and merge with each other, in which ethnic minorities are widely distributed, forming a rich and colorful ethnic culture and cultural landscape in the Chinese nation.

④ **Varied changes in spatial and temporal landscape**

Due to the changes in terrain, climate, vertical vegetation and undulating landscape, a city is endowed with dynamic beauty. With four seasons in one mountain and different weather within 10 miles, the city will be very attractive when well planned and constructed.

⑤ **Rich tourism resources and cultural landscapes**

Rich tourism resources and cultural landscapes are conducive to the development of tourism, particularly natural ecotourism. Mountains, springs, water, waterfalls, forests, animals and rich ethnic customs are favorable places for the development of smokeless industry, ecological tourism, traditional Chinese medicine and green food.

China is a country with the most complex and diverse natural and artificial eco-environments in the world. China's cities (towns) are not only numerous in number and type, but also have great regional differences. Southern and northern cities, coastal and inland cities, mountain and plain cities, upstream and downstream cities, plateau and hilly cities, large and small cities, industrial and mining cities, and historical and cultural cities differ greatly in natural conditions, economic conditions, and social and eco-environments. Therefore, urban planning must be carried out on the basis of the comprehensive analysis of the specific conditions and characteristics of the cities mentioned above.

(2) **Unfavorable factors for mountainurban construction**

The following are the unfavorable factors in mountainurban construction:

① **Complex terrain and difficult construction**

The main characteristics that mountain cities differ from plain cities are complex terrain, and the various changes caused by height, slope and microclimate.

② **High cost of infrastructure investment**

The investment cost of mountainurban construction is generally higher than that in flat cities. For example, the investment in infrastructure construction generally increases by 15–30%.

③ **Inconvenient transportation and communication**

In mountain areas, the obstruction of mountains and waters and terrain undulations increase the complexity and difficulty of traffic connections.

④ **Prominent contradiction of cultivated lands**

There are few flat lands suitable for mountainurban construction, which is prominently contradictory to agricultural cultivated land.

⑤ **Fragile eco-environment**

With a fragile eco-environment, mountain cities are vulnerable to environmental disasters, such as soil erosion, mountain torrents, floods, landslides, mudslides, collapse, and earthquakes, which easily induce secondary disasters, endangering personal and urban safety.

⑥ **Large energy consumption**

Mountain cities need to overcome the obstacles of gravity and terrain, and increase energy consumption.

⑦ **Backward social, economic and cultural development**

Most areas in mountain cities have relatively congested traffic, underdeveloped economy and education, and low level of urban culture, where the development lags behind, when compared with coastal areas.

(6) Mountainurban ecological planning and construction

A mountain city is an artificial natural mountain environment, a product of natural transformation, and an integral part of natural mountain environment. The specific natural environment and socio-economic conditions of mountain areas and cities determine that we must carefully choose development patterns and construction methods, optimize the allocation of natural ecological resources, and reduce the damage to the natural eco-environment in the process of development and construction. The natural eco-environment in mountain areas influences and limits the form and expansion direction of urban space. The planning and layout of various material elements, such as urban plane organization, spatial structure, urban green space system and road traffic, and the layout and construction of various buildings must be coordinated with the natural geographical environment. Practical experience has repeatedly shown that disasters, such as the landslides, collapse, mudslides, erosion and mountain torrents that occur in mountain areas and cities, are often caused by improper planning and construction, or poor management. Therefore, we must change all the planning, design and management concepts that do not conform to the principles of ecologicalization, promote sustainable ecological planning and design, implement the ecologicalization of transportation networks, promote environmentally friendly ecological buildings and technologies, and accelerate the pace of urban and rural ecological development.

On a global scale, the energy crisis and environmental movements in the second half of the last century have prompted the intersection of natural and social sciences (Shang 2002). Although urban science is expanding towards ecology, environmental science and decision management science, this has also prompted for the extension of ecology from traditional biology to human ecology, social ecology, urban ecology and landscape ecology. Ecology has become the theoretical basis for solving environmental and urban problems. The ecologicalization of urban planning and design has also become one of the important basic theories for the development of urban planning discipline.

The 21st century is a new century, in which mankind move towards ecological civilization. This is also a century of sustainable development for mankind to pursue the harmonious development between man and nature, and between man and society. Under the guidance of the Scientific Outlook on Development and the Thought on Building a Harmonious Society, China's urban and rural ecological planning and construction will certainly provide a new experience for the harmonious development of the world.

2 Mountainurbanology

(1) Purpose and significance of mountainurbanology

The continuous differentiation and integration of disciplines is a common phenomenon and law in the development of modern science. With the rapid development of China's national economy and urban and rural construction, great progress has been made in science, technology and discipline construction, and new areas of scientific research and discipline are continuously being explored, in which research on the development and conservation of mountain areas, and the planning and construction of mountain cities is attracting more and more attention from academia and planning, and construction workers. The reasons are, as follows:

(1) Mountain areas and cities not only occupy a large proportion of China's land area and the number of cities, but also take a particularly important strategic position. China is a mountainous country, with a mountain area accounting for two-thirds of its land area and a population, accounting for approximately half of its total population. Mountain cities account for approximately half of the country's total, which is even larger in the western region. The vast majority of national minorities live in mountain areas, where most of the main resources, such as mineral resources, hydroenergy and forests, are distributed. Therefore, China's mountain areas have a great potential for development and very important strategic position. Mountain cities are an important part of China's urban system, and this is the core of the development of the mountain economy, and an important base for the development and revitalization of mountain areas, and the building of prosperous and culturally-advanced new mountain areas and countryside. At the same time, mountain cities generally have the characteristics of rich landscape diversity and biodiversity. As long as they are planned

in a scientific way, it is possible to build charming and dynamic cities (towns) with the coordinated development of artificial and natural environments.

(2) With the acceleration of China's modernization and urbanization, the urban population has increased rapidly, the suburban farmland and arable land have reduced constantly, and the development and utilization of mountain areas. Furthermore, the number and scale of mountainurban construction have also been steadily increasing.

(3) The characteristics of mountain eco-environment and its sensitivity and vulnerability determine the importance of mountain conservation, the complexity of mountain development and utilization, and the arduousness of engineering technology. The development and utilization of mountain areas will inevitably bring about eco-environmental problems. If improperly developed and utilized, these will cause many man-made mountain disasters, and bring serious losses and impacts on the national economy and people's life.

(4) As many mountain areas and cities are located in the middle and upper reaches of river basins and lake catchments, the occurrence of mountain disasters will affect the downstream areas, with an extraordinarily extensive range of impact and serious damage.

(5) In recent years, natural disasters have frequently occurred. With the development of mountain areas, the mountainurban scale expansion and the lack of reasonability in construction, the natural eco-environment of mountain basins cannot be effectively protected. Natural disasters, such as mountain torrents, soil erosion, landslides, collapse, mudslides and ground subsidence, have a tendency to intensify.

(6) The complex natural geographical conditions of mountain areas increase the difficulty of its development and construction. Mountain areas and cities are far inferior to plains and coastal areas, in terms of transportation and communication, with a relatively backward and imbalanced economic and cultural development. Therefore, it is necessary to provide active technological, economic and policy support, and more importantly, the guidance of scientific construction theory, in order for the mountainurban eco-environment to be well protected, and allow the mountainurban economy to be rapidly revitalized, thereby further promoting the fundamental improvement of national eco-environmental conditions, and the sustainable and coordinated development of the society and economy.

According to the statistical bulletin of the National Bureau of Statistics, the total population of China was 1,284.53 million at the end of 2004, in which the urban population was 502.12 million, and the rural population was 782.41 million, with the level of urbanization reaching 39.1%. According to statistics of the Ministry of Land and Resources, the land area occupied by cities and towns in the country expanded from nearly 13,000 km^2 in 1990 to nearly 34,000 km^2 in 2004, with an area of 130 m^2 per capita. According to the forecast of China Population Information Research Center on China's population development, the total population of China will reach 1,472.2 million by 2020 based on the increase of the urbanization level by an average of one percentage point every year, with 840.36 million in urban areas and 631.64 million in rural areas, and the urbanization level will reach 57.1%. Calculated on the basis of the comprehensive urban land use of 100 m^2 for each urban resident, 23,824 km^2 of urban construction land needs to be increased (360 million mu on contract), while China's existing cultivated land area is only 1.831 billion mu, which is only 1.4 mu per capita. Obviously, such a huge demand for land is unbearable to provide on flat land. Hence, the demand for hillside land in urban construction is bound to further increase. As the consumption of mountain resources and the pressure on the mountain environment continue to increase, the contradiction between man and land will become more prominent in mountain areas and cities. The "destructive construction" carried out in many areas in disregard of ecological conditions has resulted in the imbalance of environmental security, and gradual loss of biodiversity, landscape diversity, mountain folk cultural diversity and architectural style diversity in mountain settlements. If this situation continues, it will not only lead to the unsustainable development of living environment in mountain areas, but also cause the loss of mountain cultural heritage, resulting in the substantial decline in the quality of life of people of all ethnic groups in mountain areas.

"Science and technology are the primary productive forces". The construction of mountain areas and cities relatively lags behind, foremost, in terms of science and technology. In recent years, China has conducted a lot of work in scientific research on mountain areas. For example, in 1984, the Geographical Society of China established the Mountain Research Committee, the China Society of Natural Resources established the Professional Committee for Mountains, and the professional journal of *Mountain Research* (now revised as the *Journal of Mountain Science*) was published. In 1988, the Chengdu Institute of Geography, Chinese Academy of Sciences was renamed the "Institute of Mountain Hazards and Environment" to strengthen scientific research on mountain hazards and environment. However, in general, the above research work mostly concentrated in the field of natural science, which still lags far behind developed countries.[1] Professor Ding Xizhi, a famous Chinese geographer, pointed out that "the potential for China's development lies in mountains" (*Collected Works of Ding Xizhi*, p. 289). Mountain cities are an important base for the development of urban and rural economy in mountain areas and the building of modern, prosperous and culturally-advanced new mountain areas, with a profound impact on the natural eco- environment. Therefore, carrying out systematic and scientific theoretical mountainurban research, fundamentally improving the understanding of formation, the development and evolution of mountain cities, enhancing the scientific nature of mountainurban planning and construction, reducing the blindness in the development and construction of mountain areas, and accelerating the process of mountainurban modernization are the glorious and arduous tasks for China's urban science researchers and planners, and the establishment of mountainurbanology is undoubtedly a strategic task in the scientific research on mountain cities.

---

[1]Huang Guangyu. "Attaching Importance to the Development and Protection of Mountain Areas". Beijing: *Outlook*. Issue No.21,1989.

(2) Definition, content and disciplinary framework of mountainurbanology

(1) **Definition and content of mountainurbanology**

Mountainurbanology is an emerging discipline that takes mountain cities as its research subject, which is a comprehensive inter-discipline that combines urban science with mountain science. mountainurbanology is the technology and art of studying the characteristics of urbanization in mountain areas, the law of formation, development and evolution of mountainurban system, as well as the planning, mountainurban construction and management under the comprehensive action of specific natural, economic and social conditions. Therefore, mountainurbanology is a branch of urban science, or a branch of urbanology, which serves as the theoretical and technical basis for guiding the scientific construction and reasonable development of mountain cities.[2]

The task of mountainurbanology is to understand and master the internal mechanisms and external conditions of the development and changes of mountain cities, as well as the basic characteristics and laws of mountainurban development and construction through the comprehensive investigation and study of mountain cities and its regions, thereby achieving the purpose of scientific construction and reasonable development. Scientific construction aims to meet the natural and economic laws and ecological balance requirements for mountain cities. Reasonable development means that the mountainurban system and the scale can coordinate with the structure, function and land layout of each city, which not only needs to meet the requirements of creating a good living and working environment for mountain cities, but also leaves room for its long-term development in the future, in order to meet the requirements of sustainable development, and achieve the unity of economic, social and environmental benefits. Obviously, it is not a discipline that can undertake such a complex task, but a group of disciplines that comprises of two related categories, natural science and social science, including the basic disciplines, basic disciplines of technology, and disciplines of professional technology.

(2) **Disciplinary framework of mountainurbanology**

Mountainurbanology generally includes the following related disciplines:

① **Basic disciplines of mountainurbanology**: Urban philosophy, urban history, urban aesthetics, urban geography, urban economics, urban sociology, urban ecology, urban climatology, urban disasters, geomancy theory, etc.

② **Basic disciplines of technology of mountainurbanology**: Urban planning, urban architecture, landscape architecture, regional planning, mountainurban catastrophology, etc.

③ **Disciplines of professional technology of mountainurbanology**: Mountainurban planning, mountainurban design, mountain architecture design, mountain garden landscape engineering design, mountainurban road engineering,

[2]Huang Guangyu. *Planning, Construction and Ecological Environment of Mountain Cities*. Beijing: Science Press, 1994.

Table 3 Disciplinary framework of mountainurbanology

Mounturbanology		
Montology	Urbanology	Ecology
Mounturbanology Montology Urbanology Ecology Mountain Environmentology Mountain Geomorphology Mountain Climatology Mountain Hydrology Mountain Biology Science of Mountain Disasters Mountain Ecology	Urban Philosophy Geomancy Science of Urban History Urban Aesthetics Regional Ekistics Urban Ekistics Architecture Urban Geography Urban Sociology Urban Economics Urban Management Science Urban Public Health	Regional Ecology Urban Ecology Landscape Ecology Industrial Ecology Pollution Ecology Engineering Ecology

mountainurban traffic engineering, mountainurban municipal utilities engineering (including water supply and drainage engineering, electric power, telecommunications, thermal power, gas and municipal public health), development and utilization of mountainurban underground space, mountainurban construction machinery and technology, mountainurban disaster prevention engineering, mountainurban outdoor engineering and vertical design, mountainurban environmental protection, mountainurban construction and management, etc.

Many of the above disciplines constitute the discipline system of mountainurbanology. It can be observed that mountainurbanology encompasses a wide range of disciplines and a wealth of specialized contents. From this perspective, it can be noted that mountainurbanology is the general term of multidisciplinary research on mountain cities. The disciplinary framework of mountainurbanology is shown in Table 3.

(3) Research methods of mountainurbanology

According to the definition, tasks and contents of mountainurbanology, there must be scientific and reasonable methods for the relevant research, which mainly include:

(1) **Systematic analysis and comprehensive approach**

A mountain city is a huge natural, social and economic system, and each of which comprises of many subsystems. Each system influences and interacts with each other. The change in each subsystem will cause the change of large systems. Therefore, the approach of systematic analysis must be applied to decompose intricate mountain cities into various organic systems for analysis and research. However, the existence of each system is not isolated, and has an inseparable connection with other systems.

Therefore, mountain cities must be taken as an organic whole in mountainurbanology for comprehensive research, in order to draw a correct conclusion.

(2) **Analytical approach of ecological factors**

Mountainurban is a social, economic and natural complex ecosystem, with urban society as the main body and mountain regional space and various facilities as the environment. The analytical approach of ecological factors is to apply ecological principles to analyze, study and evaluate various natural ecological factors and social and economic ecological factors of mountain cities, and on this basis, carry out macro strategic development planning, regional planning, land use planning, overall layout planning, architectural environment planning and landscape design, engineering facility planning and design, etc. Due to the complex natural conditions and land use characteristics of mountain cities, the urban system distribution in mountain areas, the overall development and layout structure of mountain cities, the construction of buildings and various municipal facilities, as well as people's daily life and activities, are more restricted and influenced by natural conditions. Therefore, it has become an important approach of mountainurbanology research to focus on the analysis of ecological factors in mountain cities, particularly in the analysis of natural ecological factors.

Through the single factor classification and grading analysis of natural ecological factors, a comprehensive conclusion is drawn based on the superposition, synthesis and comprehensive analysis of social and economic factors, which can be used as a basis for the development planning and construction decision-making of mountain cities, and make research results more scientific. The application of remote sensing technology, geographic information system and computer-aided design can greatly improve the quality and effectiveness of analysis. The analytical approach of natural ecological factors aims to implement the guiding principle of giving priority to the natural eco-environment, and initially delineate environmental conservation areas that are not allowed to be developed from the perspective of environmental control, in order to establish a reasonable development pattern for the sustainable and coordinated development of the natural environment, economy and society in this area.

(3) **Dialectical approach**

The social and cultural system and the spatial environment system of mountain cities are the unity of contradiction and opposition, which are both contradictory and unified in the organic mountainurban unity. This dialectical relationship, which is both contradictory and unified, commonly exists in all phenomena of mountain areas and cities. Therefore, it is necessary to use the dialectical view of unity of opposites to understand the law of the formation and development of mountain cities, such as the relationship between the formation and development of mountain cities and the large mountain areas or river basin environments, and the relationship between the reasonable division of labor and cooperation between mountainurban systems, the relationship between the centralized and compact layout and organic dispersion of mountain cities, the relationship between the protection and transformation of mountain terrain, the relationship between the mountain development and utilization, and the maintenance of mountain ecology and environmental protection, as well as to study and deal with various contradictions in mountainurban construction, such as the contradiction between the development of agriculture, forestry, animal husbandry and sideline industries and the land for urban construction and development, the contradiction between flat land and sloping land and wasteland for urban construction, the contradiction between the land for infrastructure, such as road traffic and construction land, the contradiction between urban construction land and non-development land and reserve land, the contradiction between terrain slope, slope direction and striving for sunshine and ventilation in buildings, and the contradiction between the elevation selection of construction site and the economic efficiency of construction investment. The correct handling of these relationships and contradictions directly affects the success or failure of construction and the economic efficiency of construction investment. Merely by applying dialectical views and approaches can we grasp the main contradictions and essentials without attending to one thing and losing sight of the other.

China's national economy and urban and rural construction are undergoing a new and important period of development. The establishment of mountainurbanology, and the implementation and in-depth development of scientific mountainurban research will greatly benefit the mountain development and conservation, and bring mountainurban construction and management into a scientific and reasonable track for development.

J. D. Vanvers, the former president of the International Mountain Society and the famous mountain ecologist, once pointed out that "The relationship between mankind and mountains has never been more important than it has been in the last quarter of the century. The future survival of human beings depends on the development and protection of mountains".

(4) Research and development trend of mountain cities

(1) **Concern of international organizations on mountain issues**

As the global environmental changes have intensified, mountain issues have attracted the attention of the international community since the 1970s. In 1971, the United Nations Educational, Scientific, and Cultural Organization (UNESCO) launched the "Man and Biosphere Program" (MAB), which gave special attention to urban and eco-environmental issues, and listed tropics, drought, mountain and urban areas as one of the key research subjects. Since then, the study of mountain and urban ecosystems has gradually drawn the attention of countries all over the world. In 1974, the International Development Foundation and the Mountain Geography Committee of the National Geographic Society held the International Conference on Mountain Environment and Development in Munich, and issued the *Munich Declaration*, which revealed the serious situation of population growth, forest destruction, aggravated erosion and environmental deterioration in many parts of the world, and called on the people of the world to raise the awareness of protecting mountain eco-environment and strengthen cooperation on the relevant research. In 1976, an international symposium on mountain environment was held again in Cambridge, Massachusetts, USA, which emphasized the necessity to strengthen collaborative research among mountain research institutions of the countries, and issued the *Cambridge Declaration.* In 1980, the UNESCO formally established the International Mountain Society (IMS), and the purpose was to "strive to seek a good balance between the human and mountain environment and resource development, and published the journal of *Mountain Research and Development* in 1981. In 1983, under the impetus of the MAB, the International Center of Integrated Mountain Development (ICIMD) was established in the capital of Nepal to carry out the development and research on the utilization and management of mountain natural resources, mountain enterprises and infrastructure construction.

The *Agenda 21* and four other important documents (the *Rio Declaration*, the *Convention on the Protection of Biological Diversity*, the *Framework Convention on Climate Change*, and the *Policy Statement on Forest Issues*) adopted at the United Nations Conference on Environment and Development in 1992 are closely correlated to mountain issues. In particular, Chapter 13 of the *Agenda 21*, "Management of Fragile Ecosystems—Sustainable Development of Mountains", gives high priority to the mountain environment and development as a global environmental issue. Since then, many international organizations have strengthened the research on sustainable mountain development, and carried out relevant activities. From 1994 to 1998, the European Community Environmental and Climate Plan took the mountain ecosystem as a special and important global ecosystem, and increased financial support for research. In 1998, the UN Secretary General Kofi Annan announced that the year of 2002 would be designated as the "International Year of Mountains", indicating that sustainable mountain development will be a major event in the global environment and development in the 21st century, and one of the important tasks in the global implementation of the *Agenda 21.*

(2) **Domestic mountainurban research trends**

Mountain human settlements are an important part of the global human settlements, and mountain habitat science is an important component of habitat science, especially for China, which is a mountainous country that occupies an important position in the natural ecosystem of the global mountain areas. Since the 1960s, the older generation of architects, such as Ye Zhongji, Gu Qiyi, Tang Pu, Ye Qishen, Huang Zhongshu, Zhao Changgeng, and Yu Zhuoqun from Chongqing University (the former College of Architectural Engineering), have developed a keen interest in the study of regional architectural culture in Southwest China, successively carried out the investigation and research on Chongqing folk houses, Qiang Minority's architecture, Tibetan architecture and garden architecture, and successively published monographs. In 1958, the college opened the specialized program of urban and rural planning. In the second year, the major of urban and rural planning was officially enrolled. The research office of architectural design and urban and rural planning was established. The education and scientific research of mountain urban planning was gradually carried out. From the 1980s to the 1990s, mountainurban planning, township planning, and rural architecture became the main contents of teaching and research. In 1987, the Sino-Canada Joint Postgraduate Program was held, and in 1988, the Committee of Foreign Urban Planning Specialty and the Committee of Small Town Specialty of China Association of City Planning were established, and the academic seminars were held, which promoted the exchange of academic research. In 1992, as approved by the Ministry of Housing and Urban-Rural Development and the Chinese Academy of Sciences, the Research Center for mountainurban and Regional Environment was established in Chongqing University. Its mission was to "carry out research on mountainurban and regional development strategies, and scientific research on mountainurban planning and design, allowing this to gradually become a basis for the development of research talents of the mountainurban and regional environment in China, and academic exchanges at home and abroad". In the same year, the First National Symposium on Mountain Cities and Construction was held, at which academic exchanges were held on a wide range of issues such as mountainurban planning and design, eco-environment, and disaster prevention and control, and the concept of "mountainurbanology" was discussed. In 1997, the First International Symposium on Sustainable Development of Mountain Habitat Environment was held in Chongqing, and the *Declaration of Mountain Human Settlements* was issued. Subsequently, young and middle-aged scholars, such as He Zhiya, Dai Zhizhong, Zhao Wanmin, Long Bin, Mao Gang and Yang Yuzhen, diligently worked and published many relevant works, exerting a good influence. In 1998 and 2002, the national and international academic seminars on mountain cities were successively held in Xi'an and Kunming, and the sub-centers for mountainurban and regional environment research were established in Yunnan and Guizhou. Both the academic research on mountain human settlements and the theory and practice of mountainurban planning and design have undergone gratifying progress. The establishment of sub-centers for mountainurban and regional environment research has created a new situation in the study of mountain cities (towns) in China. Hence, scientific research and academic exchange activities on mountains have been carried out in an in-depth way from the perspective of mountainurban ecosystem and human settlements through interdisciplinary intersection and integration. With the approach of "inviting overseas and going global", we have successively participated in the international academic conferences held by Hanover University, Waseda University, University of Tokyo, University of Manitoba, McGill University, Harvard University, and Massachusetts Institute of Technology, as well as those held in cities such as Frankfurt, France, London, the UK, Biborg, Spain and Niigata, Japan for academic exchanges, and invited experts from countries, such as Germany, the United States, Canada, Japan, Britain, France, Belgium, Malaysia, South Korea, and North Korea, to give lectures and conduct academic exchanges.

The study of mountainurban habitat environment has also received the concerns, support and enthusiastic guidance from the older generation of experts, such as Tang Pu, Wu Liangyong, Qi Kang, Zhong Xunzheng, Zhou Ganzhi, Li Wenhua, Li Dehua, Dong Jianhong, Tao Songling, Peng Yigang, He Jingtang, Zou Deci and Zhang Jinqiu. Mr. Wu Liangyong encouraged in the opening paper of the First International

Academic Conference on the Sustainable Development of Human Settlements in 1997 that "I hope later scholars will carry forward the arduous and practical style of study, and take improving the living environment in mountain areas and promoting the science and art of mountain construction as their own responsibility. Any relevant achievement will add luster to the architectural culture of mountains. When the mountain flowers of Southwest China are in full bloom and the fine works are emerging, it will be the time when the architecture school of Southwest China stands proudly in the architectures of the world". The ardent hope of the older generation of scholars is not only a spur on but also an encouragement to the later scholars who are dedicated in the research on mountain habitat environment and construction.

3 Mountainurban Ecosystem

(1) Characteristics of mountainurban ecosystem

The mountainurban ecosystem is a combination of the mountain natural ecosystem and the mountain artificial ecosystem. In order to correctly understand the mountainurban ecosystem, we must first know the characteristics of these two ecosystems.

(1) **Characteristics of the mountain natural ecosystem**

The mountain natural ecosystem is an ecosystem characterized by distinct regional differentiation and diversity. Due to the change of the gradient law of the three major steps in China's topography, the trend characteristics of mountains and water systems, and the different combination features of latitude, longitude and vertical geographical locations, the regional differences and biodiversity of complex hydrothermal conditions are formed, which have a profound impact on the mountain ecosystem.

① **Influence of geographical location and climate**

China is located in the southeast of Eurasia, bordering on the west coast of the Pacific Ocean and the South Asian continent and facing the Indian Ocean in the southwest. It spans 50 latitudes from north to south and 61 longitudes from east to west, creating favorable geographical conditions for the formation of diverse ecosystem types. Due to strong monsoon airflows formed by the huge differences in thermodynamic conditions between land and sea, the southwest and southeast monsoon prevails in summer, making most of Southeast China, and the southwest and northwest regions experience abundant rainfall at the same latitude in the world. This forms a rainy and hot climate during the same period, which is beneficial to the mountains, human settlements and the growth of plants, animals and crops in these areas. However, in winter, due to the influence of cold airflow over Siberia and Mongolian Plateau, the dry and cold climate appears in most parts of China, which adversely affects the mountain human settlements and the development and growth of animal and plant life systems.

② **Influence of meridional variation**

From the eastern coast to the hinterland of inland, the distribution of precipitation shows a decreasing trend, which has a significant impact on the structure of the mountain ecosystem. From east to west, it forms a natural ecosystem of humid, semi-humid, semi-arid and arid mountains. The humid mountain ecosystem is mainly distributed in the south of the first and second steps for the development of forest vegetation, which is the area with the most intense human activities, the highest population density, and the most exploitation and utilization. The baseband vegetation of the semi-humid mountain ecosystem is mainly forest steppes, meadows, grasslands and partially dry forests. The baseband vegetation of the semi-arid mountain ecosystem is mainly arid grassland. The baseband vegetation of the arid mountain ecosystem is mainly desert grassland and desert.

③ **Influence of latitudinal variation**

From Hainan Island in the south to Heilongjiang in the north, the latitudinal changes in the energy and functions of the mountain natural ecosystem in tropical, subtropical, temperate and cold zones, with the distribution of different heat zones, and these are formed due to the difference in solar radiation heat. Among these, the subtropical mountain ecosystem has the most extensive distribution area, including the southeast coastal mountain system and the mountain system of Wumeng Mountain-Dalou Mountain-Wuling Mountain, as well as the Liang Mountain and Ailao Mountain in the south of the Hengduan Mountain System.

④ **Influence of vertical variation**

The mountain ecosystem shows the feature of obvious vertical variation. Due to the differences of mountains in altitude, relative height and geographical location, the corresponding degree of variation and vertical structure of the mountain ecosystem are different. Hence, the creatures and vegetation demonstrate different changes in the vertical band spectrum. For example, Mount Emei (Leshan Giant Buddha), the world cultural and natural heritage, has a relative height difference of 2,600 m and an altitude of up to 3,300 m, with the vertical band spectrum differentiation in subtropical, temperate, and cold zones. Therefore, it has the climate characteristics of "four seasons in one mountain and different weather within ten miles", and abundant natural resources and biodiversity, with the species of animals and plants accounting for one-tenth of the country.

(2) **Characteristics of the mountainurban ecosystem**

In the natural ecosystem, the relationship between creatures and the environment is mostly passive and adaptive. However, in the urban ecosystem with human beings as the main body, the relationship between humans and the environment is often manifested as a relationship of active adaptation and transformation. The mountainurban ecosystem is a mountainurban eco-environment system that combines the natural ecosystem and the artificial ecosystem.

① **Artificialization of the mountain natural ecosystem**

The mountainurban spatial environment is composed of two parts: the urban natural spatial environment where a mountainurban itself is located, and the regional environment of its surrounding area, including mountain terrain and landform, regional climate, hydrology, soil and biological vegetation, which directly affects the mountainurban ecosystem. In turn, human activities in mountain cities will affect and change the mountainurban natural environmental conditions and the surrounding areas. In the mountainurban natural ecosystem, the natural environmental elements such as land, atmosphere and creatures are severely interfered by human activities, which brings significant changes in the natural physical environment of urban areas, and result in artificial landforms (buildings), artificial climate (heat island and greenhouse effect, etc.), artificial ground (concrete or asphalt pavement, etc.), artificial "water system" (water supply and drainage pipe network, etc.), and many other artificial facilities, thereby forming an ecosystem with an artificial environment. When the changes in human activities in the mountain natural ecosystem exceeds the allowable carrying capacity of the natural environment, it is likely to cause an imbalance in the mountainurban ecosystem, and bring damage to the survival and development of humans in mountain cities.

Although such artificial construction in mountain cities is inevitable, for the fragile and sensitive mountain natural ecosystem, determing how to minimize the negative impact of the interference of such human activities, control it within the allowable bearing capacity of the environment, and achieve the harmonious coexistence of human activities and the natural environment are the primary tasks that must be faced in mountainurban planning and construction.

② **Complexity and diversity of the mountainurban ecosystem**

In mountain cities, the regional differences formed by the particularity of the natural ecosystem enable the structure and function of the mountainurban ecosystem to clearly present its complexity and diversity, reflecting the diversity of the mountainurban spatial structure, the multidimensional nature of traffic structure, the diversity of industrial structure, the regional nature of cultural structure, and the rich diversity of landscape structure. At the same time, the service function, and the transmission and conversion of matter and energy of the mountainurban ecosystem also have its own characteristics.

③ **Vulnerability and sensitivity of the mountainurban ecosystem**

Due to its own vulnerability, the mountainurban eco-environment system is not a "self-sufficient" system. In the mountainurban eco-environment system, energy and matter depend on the input of external systems. A mountainurban, such as an organ of a giant, constantly absorbs all kinds of matter and energy and discharges various wastes. The larger a city is, the greater the disturbance to the natural eco-environment system, the greater its dependence on the surrounding environment, the more sensitive and complex the eco-environment problems, and the greater the interference and influence on the self-regulation function of the mountain natural ecosystem will be, which further exacerbates the vulnerability and sensitivity of the mountainurban ecosystem.

A dynamic mountainurban must have a healthy urban ecosystem that must be composed of two ecological subsystems, that is, the natural eco-environment subsystem and the artificial eco-environment subsystem. Merely when both the systems are healthy and well coupled can a healthy urban ecosystem be formed. The so-called healthy natural eco-environment system is able to enduringly maintain the ecological process of natural evolution without man-made sabotage and interruption, while a healthy artificial eco-environment system can continuously maintain the coordinated operation of urban social and economic activities without being stopped or interrupted. The natural eco-environment system and the artificial ecological system are mutually conditional, symbiotic and co-prosperous. Without any one, the city will cease to exist. In an urban ecosystem, the city will also decline or even die once the natural eco-environment system is destroyed, which is the same for the artificial ecological system.

(2) Mountain landscape ecosystem

(1) **Characteristics of the constituent elements of the mountain landscape ecosystem**

Landscape ecology and ecosystem is a branch of ecology and landscape ecology. In general, landscape is a comprehensive geomorphic feature that reflects the topography or scenery of a geographical area, and landscape ecology is the ecological interpretation of landscape. The concept of "landscape ecology" was put forward by the German geographer A • C • Troll in 1939, who defined landscape ecology as the science of studying the intricate causal feedback relationship among biological communities in a landscape, emphasizing that landscape ecology is a comprehensive study that combines aerial photogrammetry, geography and plant ecology. Patch, corridor and matrix are the three basic structural units of landscape ecology. With the in-depth development of applied ecology, landscape ecology has also developed rapidly in many countries around the world, in which a landscape ecosystem with frequent human activities is the main research subject. The three basic component units of the mountain landscape ecology, namely, the patch, corridor and matrix, have distinctive characteristics:

① **Patch**: The complex and multi-dimensional geomorphic characteristics of mountain areas often help form natural geomorphic units with different shapes and obvious vertical gradient changes, as well as irregular patch shapes and boundaries, with obvious edge effects.

② **Corridor**: The characteritics of mountain areas and cities vary with topography, which generally change along river systems and ravines with the composiiton of rich and diverse landscape corridors, and often form intersecting network structures such as mountain corridor, river and water corridor, gully and valley corridor.

③ **Matrix**: Matrix presents the change of vertical gradient with the characteristics of complex watershed distribution and geomorphic changes. For example, the Yangtze River Basin in the Sichuan Basin and the Yunnan-Guizhou Plateau forms a V-shaped valley in its upper reaches, and the terraces along both sides of the river are extended along the valleys and narrowly distributed, featuring narrow river bed, large slope decline and rapid water flow. In the U-shaped valley area of the middle reaches, the river surface is gradually widened, and the flow velocity slows down with the slope drop. The valleys along the banks of the river are wide and the terraces, are obviously vertically distributed. In the downstream alluvial plain area, the river bed is broad, and the floodplain extends along the horizontal direction with a wide field of vision. In the Loess Plateau and gully areas of the middle and upper reaches of the Yellow River Basin, the loess layers on both banks of the river are densely gullied, with obvious cutting and serious soil erosion, and the loess tableland, ridge, hills and plains are distributed in a vertical gradient way.

The different combinations of the above three basic structural units with different spatial scales and characteristics constitute the different landscape ecological patterns and network systems in mountain areas and cities.

(2) **Service function of the mountain landscape ecosystem**

A city is an artificially dominated ecosystem, in which the natural and semi-natural ecosystem is fragmented due to serious interference of the artificial ecosystem. Therefore, it is particularly important to protect the natural environment and strengthen the ecological service function of the natural environment system in regional and urban planning. However, in the long-term utilization and transformation of nature, people often only pay attention to the direct consumption value and market value of natural resources, and turn a blind eye to the important ecological service value of natural environment. For example, in the mountainurban development and construction, people often only pay attention to the consumption value and real estate development value of land as important natural resources, but overlook its ecological service value of landform with rich natural changes. Therefore, people often adopt the development approach of large-scale excavation and filling without respect for the natural landform to "plainize" the mountain landscape. This not only causes serious soil erosion, and increases the investment in building infrastructure and municipal construction, but also makes the natural landscape features of mountain changes completely disappear.

Ecological service function is the basis of human survival and modern civilization. Due to the lack of understanding of the ecosystem service function and its importance of the natural environment, humans have led to the unreasonable use of natural resources and the destruction of the eco-environment. Countless facts tell us that unreasonable exploitation and utilization of natural resources can sometimes bring disastrous consequences to a society and a nation. The grim history and reality, such as the ecological degradation of Western China and the Yellow River Basin, the demise of the ancient city of Loulan, and the decline of the ancient city of Gaochang, are thought-provoking, and provide important warnings to future generations.

In 2005, the author investigated the ancient city of Gaochang, located at the south foot of the "Huoyan Shan (Flaming Mountains)", 45 km east of Turpan City. Built in the Western Han Dynasty in the first century BC, it used to be the capital city of the Qu's Gaochang Kingdom and the Uighur Gaochang Kingdom successively. In history, it was once a prosperous capital city, a key point of communication between China and the West, and an important gateway to the western region of the ancient Silk Road where Xuanzang passed by when he took a journey to India for Buddhist scriptures, but was destroyed in the war at the end of the 13th century. Due to long-term vegetation destruction, ecological degradation and severe drought, the city has been abandoned, and most of the ground buildings have disappeared. The ancient ruins are walls of rammed loess, and the contour of tall, thick, and deeply sunken moat remains. According to historical records, the highest part of the city wall is 11.5 m, the base is 12 m wide, the plane is slightly square, and the circumference is approximately 5.4 km. This is divided into three parts: outer city, inner city and palace city. The inner city is in the middle of the outer city, and the palace city is located in the northernmost end, and there is a temple ruin in the southwest corner of the outer city. With the vicissitudes of the times, we can no longer see the prosperous scene of the former capital in the ancient city of Gaochang, leaving only dilapidated walls and loess. In 1961, the ancient city of Gaochang was listed as one of the first batch of key cultural heritage sites under state protection (Fig. 6).

(3) Service function of the mountainurban ecosystem

In addition to the functions of life and production, a city also has the function of ecological service that refers to the ecological service with special value played by

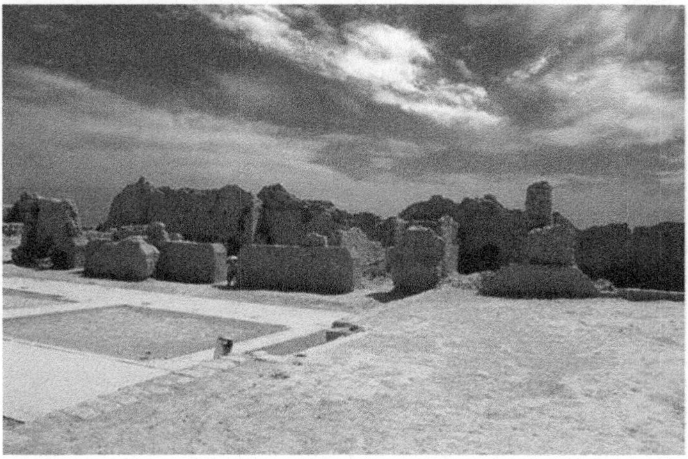

Fig. 6 Ancient Ruins of Gaochang. *Source* Photograph by the author in 2005

the environment system composed of a relatively stable and orderly structure, which provides beneficial use and corresponding value for the survival and development of humans and other living organisms (National Environmental Protection Agency 1994). For example, the green eco-environment patches composed of mountainurban forests and green space can save water, conserve water resources, improve and adjust microclimate, preserve soil and water, prevent soil erosion, release oxygen, absorb dust and carbon dioxide, serve as a breeding habitat for a variety of birds, insects and other wild animals and plants, and provide citizens with functions, such as leisure, recreation and entertainment activities. The belt-shaped ecological corridors, such as river valleys, gullies, wetlands and river systems in mountain cities, can improve the microclimate of urban areas, increase air convection, provide places for biological migration, reproduction and habitation, improve biodiversity and landscape diversity of urban areas, and provide citizens with cultural and aesthetic functions, such as leisure, sightseeing and tourism. Various small animals, birds, butterflies, bees and other insects living in these ecological patch corridors can also prey on pests, spread plant seeds and pollen, and provide plant nutrients, information, energy transmission and regulation.

A city's ecological service function is manifested in various aspects, with many classification methods. The representative classification is shown in the Table 4 (Zhiyun and Tongqian 2004).

From the above analysis, it can be seen that the service function of urban ecosystem is mainly the direct and indirect ecological service function that the natural ecosystem of urban areas performs in humans and other living organisms.

The strengthening of urban artificial ecosystem, the expansion of population, and the construction of road traffic system make the natural landscape ecosystem in urban areas be severed and seriously fragmented, which greatly weakens the service function of urban natural ecosystem. Therefore, it is necessary to strengthen

Table 4 Ecosystem service function

Ecosystem service function

Function of providing products	Regulating function	Cultural function	Support function
Food and fiber Wood Fuel Medicine Ornamental and environmental plants Genetic pool Fresh water Hydroenergy,etc.	Air quality control Climate regulation Water resources regulation Erosion control Water purification Waste disposal Human disease control Biological control Pollination Storm control, etc.	Cultural diversity Spiritual and religious values Knowledge system Educational value Inspiration Aesthetic value Social relations Perception Cultural heritage value Leisure tourism, etc.	Primary production Oxygen production Soil formation Nitrogen cycle Water cycle Habitat provision, etc.

the concept of ecological service function in the development and urban planning of mountain cities, increase the value of ecosystem service function as much as possible, and consciously mitigate this contradiction.

References

Agency NEP (1994) Agenda of the 21st century. China, Beijing
Guangyu H (2002) Mountainurbanology. China Architecture & Building Press, Beijing
Ji Z (2002) Natural geography of China. Higher Education Press, Beijing
Liu N (2001) Natural geography. Science Press
Shang Y (2002) General ecology, 2nd edn. Peking University Press, Beijing
Zhiyun O, Tongqian Z (2004) Regional ecology and sustainable development. China Meteorological Press, Beijing
Zhong X (2000) Introduction to mountain science and mountain research in China. Sichuan Science and Technology Press, Chengdu

Chapter 2
Investigation

1 Objective and Importance

Planning is a kind of prediction, arrangement and planning for the future. Therefore, planning must be well-founded. In other words, planning must be scientific. Where does the scientific nature of urban planning come from? In addition to the professional ethics and rigorous work attitude of urban planners, scientific planning theories and approaches, it is more important to have a down-to-earth and realistic attitude and style of investigation and research.

Exploration refers to detailed investigation and review, and inspection means careful observation, investigation and research. Investigation means to carry out survey and research carefully, which is the fundamental attitude and working method of planners to do planning work well. In the design process, the relationship between the investigation and research on mountain cities and the formation of planning and design results is the same as the relationship between "ten-month pregnancy" and "one-day delivery", in which investigation and research is similar to "ten-month pregnancy", and the planning and design results are similar to "one-day delivery".

(1) Important means of understanding cities and its regions

"No investigation, no right to speak". Investigation is an important means to understand the mountainurban ecosystem.

Mountain areas and cities often have complex natural ecosystems, such as topography, geomorphology and climate. In order to understand these, we must conduct in-depth and comprehensive investigations and analysis of mountain areas and cities under the guidance of disciplines, such as ecology, geography and engineering geology. This would also help to understand the geography, landform, natural climate of mountain areas and the compositional characteristics and interrelations of the complex natural, social, and economic ecosystem of mountain cities, which is an important means to fully understand the main natural elements and social and economic conditions of mountain areas and cities, and the basis for mountainurban planning.

© The Author(s), under exclusive license to Springer Nature Singapore Pte Ltd. 2021
G. Huang, *Theory of Mountainurbanology*,
https://doi.org/10.1007/978-981-16-0819-3_2

(2) Basic basis for planning and design

Urban planning and design is an important part of urban planning work. The results of high-quality urban planning preparation are an important prerequisite and basis for building and managing cities at a high level. The compilation of urban planning, namely, planning and design, involves many constraints, such as politics, economy, society, natural geography and the environment (Wenhua and Jingzhu 2004). Therefore, it is necessary to conduct scientific and efficient surveys to investigate the status quo and characteristics of the planning objects on the spot, and conduct comprehensive analysis and demonstration, in order to provide a scientific basis for reasonable planning and design. Without in-depth and extensive investigation as a guarantee, the results of urban planning and design can only be a mere formality and cannot effectively guide construction, which ultimately leads to the failure of planning.

The traditional approach of urban planning investigation is the most intuitive first-hand information obtained through field investigation in person. In the ancient times when science and technology were not yet developed, people mainly relied on this intuitive approach. Ancient Chinese people vividly called mountains "dragons". Ancient geomancers were very meticulous about the situation of mountains and rivers and site investigation on contexts. For example, Huang Miaoying's *Bo Shan Pian* has a record of the introduction of phase and ground approach: "Theoretically, anyone who wants to investigate the topography of a mountain must visit the mountain site in person. But at the mountain field, the first thing to observe is not its topography, but the flowing water in the mountain. The flood water will flow into rivers, and short water current will flow into mountain streams and creeks. Be sure to check where the water comes from and where it flows. How to observe and analyze the topography of a mountain? Specifically, if the range gathers upwards, you should ascend to observe. If the range gathers downwards, you should observe from below. If the range height is medium, you should observe in an opposite position. If the mountain has both incoming and outgoing water, you should observe on the side. If there are other hills surrounding the mountain, you should observe step by step. If there are corresponding mountains before and after the caves, you should observe from the front two angles". (*Vastu· Art Collection·Collection of Ancient and Modern Books* compiled in the Qing Dynasty). Shen Kuo, a famous scholar and geographer in the Northern Song Dynasty, conducted an in-depth investigation and research on the changes of natural geography in ancient China in his masterwork *Dream Pool Essays*. For example, after careful investigation of the landform of Yandang Mountain and its causes, he made an incisive scientific explanation: "The reason for the formation of the landform should be the result of the scouring of flood water in the valley. The sand and soil were all washed away, leaving only towering and huge rocks. The large and small Longqiu waterfalls, drencher curtains and Chuyue Valley are the caves formed by current scour. If you look up from below, the mountains are all high rocky cliffs. If you observe from above, the peaks in the valleys are just as high as the surrounding mountains, and even lower than the ground level. All the places in the world where the water runs through the valleys have towering and upright soil layers and shrine-like rocks, with the same causes as above". (Volume 24 of *Dream Pool Essays*).

The famous geographers and historians in Chinese history, such as Sima Qian, Li Daoyuan, Guo Pu, Hsuan-tsang, Guanxiu, Guo Shoujing and Xu Xiake, have traveled through famous mountains, rivers, cities and towns to learn about the geographical and climatic characteristics, social and conditions, ethnic cultures and local customs, mastered detailed first-hand information, wrote immortal historical masterpieces, or made decisions about city location and planning that have been passed down through the ages.

In the feudal society, cities were surrounded by walls and in a long-term closed state, where the development and changes were also very slow. In the modern times after the Industrial Revolution, the development of industrialization and the progress of science and technology brought about profound changes in the nature, scale and structure of cities. A city is not only a political and economic center, but also a center of culture, transportation and information. At the same time, the social, economic, cultural and natural environmental issues of cities are becoming more and more complex. Therefore, the theory of modern urban science and planning came into being, and the problems that needed to be investigated and studied in urban planning also became increasingly extensive and complicated. For example, in 1915, the British sociologist Patrick Geddes emphasized the importance of "investigation-analysis-planning" in urban planning in his book *Cities in the Evolution*. In urban planning and research, the application of remote sensing, geographic information, and digital technology systems also provided powerful means and tools for urban scientific investigation and planning decision-making, in addition to traditional investigation techniques and approaches.

2 Natural Conditions

The natural environment composition of mountain cities is very complicated. The combination of different geographic locations, climates, topography, geology, hydrology, plants and animals forms the unique landscape and environment of mountain cities in different regions, and exerts a profound impact on human activities. Due to its inflexibility, the natural environment can become an advantageous condition or a limiting factor for the development of mountain cities. Therefore, in order to study the characteristics of mountain cities and carry out relevant planning and construction, it is necessary to first conduct a comprehensive investigation and analysis on the above ecological factors of natural environment (Yongchang 2000). At the same time, in view of the trend of more and more frequent occurrence of various natural disasters in recent years, it is more necessary to conduct a comprehensive investigation and understanding of the natural environmental conditions in mountain areas, to understand the changes of natural conditions, and the causes of disasters and the measures that should be taken for disaster mitigation and prevention.

Here, we must clearly realize that the natural environment is a very complex organic integrity, in which various material phenomena and ecological processes are not isolated or accidental, but interrelated, mutually restricted and influenced.

Therefore, although we investigate and collect data on various physical phenomena under natural conditions, we must understand the interrelation and internal connection among various physical phenomena from a systematic and holistic perspective, and recognize the basic rules and characteristics of material exchange and energy conversion among these.

(1) Terrain

Terrain is an important influencing factor to display the morphology of mountain cities, which is the base of mountainurban morphology, with different effects on the local climate and eco-environment of mountain areas and cities. The terrain is closely correlated to the selection of land use, the layout of municipal facilities such as road engineering, water supply and drainage, the layout and construction of buildings, and the applicability and economic efficiency of development and construction of mountain cities.

Due to the complex changes of terrain, the environmental conditions of underground and ground are very different. According to the specific location, light, moisture, heat, and air flow all differ significantly, creating various regional microclimates that affect the living conditions of people and creatures. As the difference in solar radiation has a significant impact on the temperature and humidity of the south and north slopes, the south slope of mountain receives more solar energy, while the north slope receives less solar energy, and the solar energy received by other slopes will be between the two, depending on the specific direction of the slope. Therefore, the south of buildings is always warmer and drier than the north, and the north side of the tree trunk is cooler and more humid than the south side. Furthermore, bryophyte ombrophytes usually grow on the north side of the trunk, and on the north slope. The flowers on the south side of the crown usually bloom earlier and are brighter than those on the north side. In sub-tropical areas such as Chongqing and Guangzhou, the aerial roots of some arbors also grow more luxuriantly.

There are also obvious microclimate changes in the valley, such as low temperature at night (especially in winter), high daytime temperature (especially in summer), high relative humidity, relatively stagnant air, and weak wind. The cold air from high up at night sinks to the bottom of the valley, and the water vapor in the air condenses into mist and frost.

Mountains have the effect of blocking airflow. When an air mass meets the mountains, it rises, cools, becomes saturated with water (because the saturated water content of cold air is much less than that of hot air), and releases water on the windward side. When the cooled dry air sinks on the leeward side, the air warms up again and absorbs moisture. As a result, the vegetation on the windward side of a mountain grows more luxuriantly than that on the leeward side, and there are more species of animals.

The terrain includes different factors such as gradient, slope shape, slope surface, slope length, slope direction and the syntagmatic relations, and all of which will have an impact on the modes of planning, design and construction.

(1) **Gradient**

The gradient directly affects the layout of land, the route selection of roads and the layout of buildings.

In the analysis of mountainurban land use, the gradient of land used can be classified according to the requirements of land adaptability. In general, the gradient can be divided into six levels: slopes with a gradient less than 5% (<2.86°) has basically no or little impact on the urban layout, buildings and road layout. Slopes with a gradient between 5–15% (2.86°–8.53°) affects the organization of land use, the layout of buildings and the selection of road routes. Special treatment are required for land use organization, building layout and road traffic for slopes with a gradient of 15–30% (8.53°–16.70°), 30–45% (16.70°–24.3°) and 45–55% (24.3°–28.8°). Land with a gradient of more than 55% (>28.8°) is prohibited for development and construction.

(2) **Slope shape**

The terrain in nature is various, with different shapes such as convex slope, concave slope and flat and straight slope. W. Penck, a German geomorphologist, believed that the shape of a mountain slope depended on the contrast between the velocity of crustal movement and the rate of denudation. If the amount of crustal uplift is greater than that of denudation, the slope shape is convex. When the amount of rise is equal to that of denudation, the slope shape is flat and straight. When the rising amount is less than that of denudation, the slope shape is concave.

(3) **Slope orientation**

Slope orientation refers to the orientation of a slope, including the slope with a sunny exposure and the down-sun slope. According to the different slope orientation, it can be divided into four orientations of east, south, west and north, or eight orientations of east, south, west, north, southeast, northeast, southwest and northwest. In the slope land development, in order to strive for good orientation and energy saving requirements, most of the areas in China are more particular on the orientation requirements of the building layout.

(4) **Slope surface**

Slope surface refers to the size of the ground area of the slope. A large slope area is conducive to the unified consideration of land use planning on the slope, while a small slope area will increase the difficulty in the use and economic efficiency of development and investment.

(5) **Slope length**

Slope length refers to the length of a slope surface at the same gradient. If a slope surface exceeds a certain length, it will affect the landscape effect of road traffic organization and building layout.

(6) **Elevation**

Vertical zonality is one of the main structural features of mountain landforms, which is the rule that the natural geographical elements extend along the contour line and change vertically with the terrain height. From the perspective of the vast geographical environment, the three-dimensional spatial distribution formed by the interaction of vertical zonality, latitudinal zonality and meridional zonality is called "three-dimensional zonality".

Tectonic uplift and mountain terrain are the preconditions for the generation of vertical zones. The direct reason for vertical replacement is that the mountain heat and its combination with water vary with the change of the terrain height, and the general trend is that the mountain temperature decreases with the increase of altitude. As the altitude increases, the thickness and density of the atmosphere around the mountains decrease, and the contents of water vapor, CO_2 and dust in the atmosphere also decrease sharply, resulting in a significant increase in solar radiation intensity with the rise of altitude. However, with the increase of altitude, the effective surface radiation increases significantly, and the dissipation of heat increases faster than the heat received on the surface, which causes the mountain surface temperature to decrease with the rise of altitude. In general, when the altitude rises by 100 m, the temperature drops by 0.6 °C. In this way, the energy budget changes of mountain with the change of altitude are shown in the form of mountain temperature changes (Qiaoli 1994). Changes in altitude and temperature affect the vertical distribution of vegetation and creatures.

Elevation not only affects the distribution of vegetation and creatures, but also has a great impact on the mountainurban spatial layout, landscape characteristics and economic efficiency of construction. In mountain areas, cities are often impossible to build on the same elevation, but on different terraces, forming a mountainurban landscape with overlapping layers strewn at random. For example, the built-up area of Chongqing is built from the riverside with the lowest elevation of approximately 185 m to the hilly area with an altitude of below 350 m, forming a rich and multi-step three-dimensional mountainurban contour line.

(2) Geology

(1) **Geological structure**

For a construction base for mountain development, the stratum is generally in a state of three-dimensional spatial structure, which is represented by the stratum orientation and gradient that affect the layout of construction engineering. Considering the requirements for the safety of foundation, the building construction should be arranged on a dip slope with a small stratigraphic and a gradient consistent with the direction of stratum.

(2) **Rock types**

In mountain areas, the rock types are divided into igneous rock, sedimentary rock and metamorphic rock according to their genesis and nature. The lithologic strength of each type of rock is different, which has an important influence on the density and economic efficiency of development and construction.

(3) **Carrying capacity**

The tighter the lithology is, the greater the bearing capacity of the base rock stratums will be.

(3) Climate

A special local climate is formed in mountain areas due to the influence of altitude and terrain fluctuation. The difference of local climate is determined by the geographical location, altitude, slope orientation and mountain distribution. The vertical difference of climate is especially obvious when the mountain's vertical elevation difference fluctuates by several kilometers. Due to the different characteristics of "adaptation" of animals and plants to climate, there are different biological population distributions at different vertical heights.

Sunlight and air temperature are changes in the surface environment caused by solar radiation. The influencing factors mainly include sunshine and heat. The growth of human beings and any other creatures and plants is inseparable from the two basic ecological elements of solar radiation, namely, light and heat.

(1) **Sunlight**

Sunlight refers to the state of direct sunlight on the surface, which is composed of such factors as height angle of incidence of the sun, hours of sunshine, sunshine rate and solar shadow curve. These important natural ecological factors are all affected by the geographical location, altitude, topographic slope, slope direction of a mountainurban.

(2) **Temperature**

Air temperature is the average temperature of air mixture, which is an important ecological factor that plays a role everywhere. Both human beings and creatures live in an environment with a certain temperature and are affected by its changes. The surface temperature conditions are constantly changing. In space, it varies with the geographical latitude, the vertical height of the ecosystem and the change of various niches. In time, it has the changes of four seasons and day and night. In order to investigate the temperature changes in a region, it is also necessary to understand its annual and daily changes, absolute maximum and minimum temperatures, and daily temperature difference, in addition to the average temperature.

(3) **Humidity**

Humidity indicates the amount of water vapor contained in the air. Water vapor condenses into water and releases a lot of heat energy. Water evaporates into water vapor, which absorbs a lot of heat energy and affects the change of atmospheric temperature. The condensation of water vapor is the main cause of precipitation. The continuous evaporation and condensation of water cause continuous hydrological cycles.

The expression of humidity includes absolute humidity, relative humidity and specific humidity, in which relative humidity is the most commonly used. Relative humidity is the ratio of the actual partial pressure of water vapor to the weight of saturated water vapor in the air, which is greatly affected by temperature variation. In general, the relative humidity is low due to the high temperature in the daytime, and high due to the low temperature at night.

(4) **Precipitation**

Precipitation and temperature are the most important ecological factors that affect the distribution of humans and creatures on the Earth, whose combined effects determine the overall pattern of the distribution of humans and biological communities on the Earth. In mountain areas and cities, concentrated precipitation often causes floods and mountain torrent disasters, which should be given special attention.

When the warm air steam containing water vapor rises to a certain height, the temperature drops, cools and condenses into liquid or solid form, forming rain, hail and snow. The ways of expression of precipitation generally include average precipitation, daily maximum precipitation (also known as precipitation intensity), and one-hour and ten-minute maximum precipitation.

① **Forms of precipitation**

Water falling on the ground, whether liquid or solid, is collectively called precipitation.

Precipitation includes rain, snow, frost, dew, ice, hail, etc. In different forms of precipitation, rain is the most, followed by snow.

② **Factors affecting precipitation**

In addition to the main influence of water vapor in the atmosphere, precipitation is also related to comprehensive factors such as airflow, air pressure, wind direction, air temperature, terrain and elevation. In mountain areas, geographic location, terrain, temperature, altitude and humidity have the greatest impact. For example, the higher the altitude, the more rainfall there is. The rainfall on the windward slope is more than that on the leeward slope. In mountain areas, the temperature at night is lower than at daytime, and there is often night rain, such as Ya'an, which is known as the "rain city". There is also frequent night rain in Chongqing, just like an old saying goes "Overnight rain filled the autumn pool from Ba Mountain", which reflects the typical climate characteristics of the mountainurban.

③ Modes of precipitation

a. Convective rainfall

Convective rainfall is the convective action of the earth. When the ground is exposed to the sun for a long time, the ground temperature rises due to heat conduction, and the airflow rises to form convective rainfall that belongs to local rainfall featuring short time, large intensity and small range. Convective rainfall often occurs in summer, also known as thunderstorms, such as the ancient poetry of "the west is veiled in rain, the east enjoys sunshine. My gallant is as deep in love as day is fine" by the famous Chinese poet Liu Yuxi and "the black clouds in the sky were tossed by strong wind, and the big bean raindrops tilted down in an instant, which swept from east Zhejiang by the wind, as if the sea water was rolled up and fell from the sky" by the famous Chinese Ci-poet Su Dongpo, both of which describe the scene of thunderstorms in southern China in summer.

b. Ridge rainfall

Ridge rainfall is mainly affected by mountain terrain. When the moving direction of the moist flowing air mass is blocked by mountain ridge, it is forced to rise and condense when it is cold, forming ridge rainfall, also known as topographic rain, which is characterized by its long duration, but not high intensity. For example, Chongqing is located in the paralleled ridge valley of the East Sichuan Basin, with four mountains of Zhongliang Mountain, Jinyun Mountain, Zhenwu Mountain, and Mingyue Mountain distributed in parallel from north to south, where ridge rainfall is often formed.

c. Cyclonic rainfall

When rainfall and cyclone coexist, it is called cyclonic rainfall, which is formed by the intersection of cold air masses and warm air masses, or typhoon rain caused by typhoon. The rainfall intensity is high, and the rainfall area distribution is small, which has a great impact on the precipitation of small watersheds.

d. Impact of precipitation on the ground

Due to the high-speed impact of raindrops, the slope ground is scoured, and the gradient and the length of a slope affect the velocity and flow of precipitation, respectively. Theoretically, the larger the gradient is, the higher the flow velocity and the stronger the erosivity will be. When the gradient is between 40° and 50°, the amount of erosion is the largest, and when the gradient is exceeded, the erosion amount decreases. The reason is that the larger the gradient is, the smaller the actual rain area and the corresponding flow will be. The compactness of the ground structure of the and the roughness of its surface also affect the resistance of the ground to scouring.

Vegetation has a protective effect on ground erosion. Crowns, trunks, fallen leaves and turfs can all intercept dripping rainwater, slow down the impact of raindrops, and avoid direct impact on the ground. Tall and dense crowns can retain 15–80% of rainfall. Falling leaves and other litter can not only store water, but also block the erosion of the flowing water on the slope, which can also improve soil properties, increase soil permeability, and reduce the erosion of flowing water after decomposition. The rhizomes of plants can also consolidate soil layers on slopes and prevent erosion. Therefore, it is necessary to cover the urban open spaces with green vegetation as much as possible in mountain cities, in order to reduce the surface erosion and improve the permeability of the ground. The construction of roads, squares and parking lots should be as permeable as possible.

(5) **Wind**

Due to the different changes in landform and atmospheric temperature, an atmospheric pressure difference is formed, which makes the air flow from high to low pressure and generates wind. Wind has a great impact on the human settlements in mountain cities. The influence of wind must be considered in the overall urban layout, the organization of the green space system, the direction of roads, and the design and layout of buildings. For example, in the cold areas of northern China, priority should be given to the impact of wind, ice and snow protection. In the hot and humid areas of southern China, natural ventilation should be used as much as possible to improve comfort, and in the coastal areas, the impact of typhoon attack should also be considered.

In mountain areas and cities, the local climate and ventilation effects are changed due to the influence of terrain, resulting in so-called climate effects of positive and negative landforms.

① **Climate effects of positive landform include**:

Climate effects of isolated peak and its climate

Compared with the surrounding lowlands, the space from the part above the isolated peak piedmont to the top of the mountain has a low temperature, much cloud and rain, little sunshine and radiation, strong wind force and low vapor pressure, and also generates multi-day rain-type precipitation due to the influence of the slope wind in the daytime and the mountain wind at night.

Climatic effects of plateau and its climate

Compared with a plain of the same latitude, the typical plateau climate features low and uniform temperature, thin air, little precipitation, much sunshine and high solar radiation intensity, low absolute humidity, high wind speed and much high wind, which are more obvious with the increase of the plateau height.

② **Climate effects of negative terrain include**:

Climate effects of valley and its climate:

Compared with the valley slopes on both sides of the valley, the valley bottom features high temperature, little precipitation, much sunshine, strong solar radiation, large amount of evaporation, high dryness and low wind speed, with valley wind circulation, valley wind in the daytime and mountain wind at night. In the valley climate, the river valley climate is the most typical, such as that in the Three Gorges Reservoir on the upper reaches of the Yangtze River. Among these, the east-west river valleys have higher temperature due to the shielding of the cold air by the east-west mountain ranges on both sides. For example, the "three stove cities" in the Yangtze River Valley of China, that is, Nanjing, Wuhan, and Chongqing, are all in the east-west valley. The average temperature in July is 28.2 °C in Nanjing, 29 °C in Wuhan, and 28.6 °C in Chongqing.

Climate effect of basin and its climate:

Basins have a similar influence on light, heat, and water elements but different climatic characteristics compared with valleys, which are manifested, as follows: when the other conditions are the same, the basin bottom is a "heat island" (for example, the Turpan Basin in northwest China, with an average temperature of 32.7 °C in July, is 9.3 °C higher than the northeast Siping at the same latitude), "dry island" (little rainfall and dryness) and "sunlight island" (much sunshine and strong radiation), which are not only relative to the basin wall, but also to the non-basin with the same latitude and height. The main reason for its formation is good sealing, which acts as a barrier to cold, dry, or warm and humid airflows, and plays a role of climbing over mountains (Baopu 1995).

(4) Urban climate

Urban climate is a local climate formed by the strong influence of human activities in the process of urbanization. Due to the variation in the urban underlying surface, the climate changes in urban areas, such as temperature field, humidity field, wind field, and air visibility, have helped formed a heat island effect, a dry island effect, a wet island effect, and a turbid island effect, presenting local climate characteristics significantly different from those of the surrounding non-urbanized areas.

(1) **Urban heat island effect**

Due to the high concentration of population and buildings in urban areas, dense distribution of infrastructure and transportation, a large amount of anthropogenic heat and greenhouse gas emissions, the temperature in urban areas is often higher than that in non-urbanized areas. In the spatial distribution of air temperature, there is a "heat island" phenomenon, known as the heat island effect, which is one of the typical characteristics of urban climate. The heat island effect is usually expressed by the intensity of urban heat island, that is, the temperature difference between the highest temperature zone in a city and suburban countryside in the same period. The intensity of a city's heat island is closely correlated to its population size, population density, building density, and spatial structure. The larger the city scale, the more concentrated the distribution of population and buildings, the more greenhouse gas emissions, the more man-made heat, and the greater the intensity of the urban heat island. Under the control of the high pressure system, the intensity of the urban heat island is high when the weather is fine, the stratification is stable, the wind is weak, or the wind is calm. In sunny and stable weather conditions, the intensity of cities is mostly strong from night to early morning, and weak at noon. The annual variation in the intensity of urban heat island is also significant, which is usually high in winter and low in summer in China.

(2) **Urban dry island and wet island effect**

Due to the change of urban underlying surface and heat island effect, the relative humidity of cities is lower than that of suburbs, showing an obvious "dry island effect". Under the sun irradiation in the daytime, the amount of water vapor in the low-level air of urban areas is less than that of suburbs. Especially in the summer season, the crops in suburbs grow vigorously, and the difference in natural evapotranspiration among the suburbs is larger. Due to the roughness of the underlying surface and the heat island effect in urban areas, its mechanical and thermal turbulences are stronger than those in suburbs. Through the vertical exchange of turbulence, more water vapor is transported from the low layer of air in urban areas to the upper layer than in suburbs, both of which lead to the formation of "urban dry island" due to the reason that the water pressure on the ground in urban areas is lower than that in suburbs. At night, the wind speed decreases, the air stratification is stable, the temperature in suburbs drops faster than that in urban areas, the saturated water pressure reduces, a large amount of water vapor condenses into dew on the ground, with a small amount of water vapor remaining in the lower air. Hence, the water pressure rapidly decreases.

Due to the heat island effect, the condensation amount in urban areas is much smaller than that in suburbs, the nighttime turbulence is weak, the water vapor exchange with the upper air is small, and the water vapor pressure near the ground in urban areas is higher than that in suburbs, forming "urban wet islands". The urban wet island formed due to different amounts of condensation in suburbs is called "condensate wet island", which is mostly formed within a few hours after sunset and maintained at night. After sunrise, the temperature in suburbs rises, dew evaporates, and soon the water pressure in suburbs is higher than that in cities. Hence, it is transformed into an urban dry island. Therefore, the emergence of urban dry islands and urban wet islands must be accompanied by urban heat islands, because urban dry islands are one of the reasons for the formation of urban heat islands, and the formation of urban wet islands must first have the existence of urban heat islands. As the average water pressure in urban areas is lower than that in suburbs, coupled with the heat island effect, its relative humidity is lower than that in suburbs. Even when the distribution of water vapor pressure shows the features of urban wet islands, the distribution of relative humidity is still smaller in urban areas than in the suburbs.

(3) **Urban turbid island effect**

The urban turbid island effect mainly manifests as:

① There are far more pollutants in the air of cities than in the suburbs;
② The number of cloudy days in cities is more than that in the suburbs. There are more condensation nuclei in the air of cities, stronger low-altitude thermal and mechanical turbulence, and far more low cloud amounts and the number of

cloudy days with low cloud amount as the standard (the number of days with low cloud amount ≥8) than those in suburbs.

③ The turbidity in the air of cities is higher than that in suburbs. Due to the large amount of pollutants and low cloud cover in the urban atmosphere, the number of sunshine hours is reduced, and the solar direct radiation (S) is greatly weakened, while as there are many scattered particles, the diffuse solar radiation (D) is stronger than that in clean air. The atmospheric turbidity (or turbidity factor) is expressed by the D/S ratio, which is significantly higher in urban areas than in suburbs;

④ The number of foggy days in cities is more than that in suburbs, which is because there are more particulate pollutants in the urban atmosphere, with the effects of scattering and absorbing light and reducing visibility. At the same time, the number of foggy days in cities also increases due to the increase of condensation nuclei.

(4) **Urban wind**

On the whole, due to the uneven artificial underlying surface of a city, the average urban wind speed is lower than that of open suburbs with the same height, and the local wind direction and speed within the city vary greatly. Due to the difference in the angle of the dominant wind direction between the street and the city, or the difference in the width of streets and alleys and the layout of buildings, the narrow tube effect will be generated, with different wind forces and wind speed changes, such as "alley wind", "puerto wind" and "vortex wind". In a clear and stable climate with a small gradient of atmospheric pressure over a wide range, especially on a clear night, a weak low-pressure center is formed in urban areas due to the existence of urban heat island effect, and an updraft appears. The air near the ground in suburbs supplements the urban heat island center, and the hot air that rising to a certain height in the center flows to the suburb and sinks, forming a slow heat island circulation, which is called the urban wind system. The phenomenon of urban wind system causes pollutants to accumulate in urban areas and form dust cover, which is conducive to the formation of low clouds and local convective rain in urban areas (Nanwei 2001).

(5) Hydrology

(1) **Surface water**

Water is indispensable for human production, life and life maintenance. An old saying goes in the famous Chinese classics *Guanzi • Chapter of Water and Geography* that "What is water? It is the origin of all things and the root of all life…"

Li Daoyuan said in the "preface" to the *Commentary on the Waterways Classic* that "water occupies almost most of the Earth surface area. Without rivers and oceans, there would be no Earth. Without water, human beings and all the other creatures would not survive", which not only shows the importance of water for all things, but also points out the different effects and results of different qualities of water on human quality. Water is an indispensable component of creatures and human body. For example, 60–70% of the human body is water. It not only circulates in the body constantly to maintain its vitality, but also continuously exchanges matter and energy with the external environment. All metabolic activities of life must be mediated by water, which not only constantly circulates among the atmosphere, land and ocean, but also continually exchanges and transmits matter and energy between creatures and their environment. In fact, this kind of water circulation is like the "blood circulation" of the natural geographical environment, which communicates the exchanges of substances among the basic circles and promotes the occurrence of various connections. The process of water cycle also plays the role of hydrological process, climatic process, topographic process, soil process, biological process and geochemical process.

(2) **Undergroundwater**

Except for regions such as extremely high mountains and deserts, most of the Earth's surface is networked by large and small rivers systems that nourish the earth and nurture creatures. Hydrophilicity is the nature of human beings, whose ancestors had an origin from water. With the rapid increase of population, economic development and environmental pollution, water resources have become one of the serious constraints on human life and production activities. The phenomenon of water shortage is widespread all over the world, and the present situation of China's water resources is even less optimistic. China's per capita water consumption is only a quarter of the world's average, and most rivers have been polluted to varying degrees, and many urban undergroundwater sources have also been polluted to a certain extent. In the survey of hydrological data, it is important to investigate the distribution of river and lake systems and wetlands, the flow, velocity (the highest, the most withered, and the average velocity), runoff area, sediment concentration and water quality of surface water (river and stream surface), as well as mountain torrents and flood disasters. At the same time, the existing form, aquifer thickness, salinity, degree of pollution, hardness, hydrology and dynamic conditions of groundwater should be also investigated. Generally speaking, mountain areas and cities are mostly located in the upper reaches of river basins. Therefore, it is of great and urgent significance to protecting water quality and quantity, and saving water for the industrial and agricultural production and urban life in China. In urban planning and construction, we should strictly protect the water network landscape ecosystem and forest landscape ecosystem in urban areas, actively protect and conserve water sources, prevent water pollution and reasonably determine water consumption standards, use recycled water as much as possible, utilize reclaimed water and rainwater, and adopt permeable ground to replenish undergroundwater as far as possible.

(3) **Wetland**

A wetland is a unique ecosystem with high diversity on earth, which is the intersection of the atmosphere, hydrosphere, lithosphere and biosphere and zone where various stresses interact. It is estimated that there are about 8.6 million km^2 of natural wetlands in the world, accounting for 6.4% of the global land area. However, due to insufficient understanding of their important ecological service functions, wetlands are suffering from serious damage. There has been no universally accepted definition of wetland to date. In 1971, the Convention on Wetlands of International Importance Especially as Waterfowl Habitat had the following interpretation of wetlands: marsh, moisture source, peatland or water area, whether the water is natural or artificial, permanent or temporary, fresh, brackish or salt, and whether the water body is static or flowing, including the sea area with a water depth of not more than 6 meters at low tide, are called wetlands. In 1979, the wetland scientists of the United States Fish and Wildlife Service redefined wetlands in a research report entitled "Classification of Wetlands and Deepwater Habitats of the United States" after several years of investigation, survey and research: Wetlands are land in transition between terrestrial and aquatic ecosystems, where water levels are often close to the surface, or the surface is covered by shallow water.... A wetland must have one or more of the following three characteristics: ① Dominant aquatic plants grow periodically at least on the land; ② the impermeable water area soil dominates in the matrix; ③ the matrix is non-soil and is saturated with water or covered by shallow water at certain times in the growing season.

① Wetland and its types

According to the above definition, wetland is an area covered with water and aquatic plants. Some wetlands are permanent flood areas, while other wetland soils are only periodically saturated with water, but all have aquatic plants growing in a certain period of the growing season. According to the different directions of water movement, wetlands can be broadly divided into three types:

a. Basin wetlands are formed on shallow concave terrain or in the process of lake and pond filling and shallowing;
b. River wetlands are formed after flashing on both sides of rivers and streams;
c. Coast and lakeshore wetlands are formed by the sea or large lakes, of which the most famous is coastal mangrove swamp.

② Structure and function of the wetland ecosystem

a. Structure of wetland ecosystem

The structure of wetland is closely related to its moisture characteristics, which are, on the one hand, the physical properties of water and its movement (such as precipitation process, surface river and groundwater flow, direction and kinetic energy of water flow), and the chemical properties of water; on the other hand, the period of soil water accumulation, including the duration, frequency, depth and occurrence season of water accumulation. The period of soil water accumulation affects the composition of plants by affecting the germination, survival and death of each stage of plant life cycle.

A wetland is one of the most diverse and interesting ecosystems, which includes a variety of community types, such as benthic community, limnetic and marsh biomes and coastal invertebrate community. Together with small fish, these invertebrates, crustaceans and insects in particular, form the food base of birds such as waterfowls, herons and gulls, and provide various nutrients for their spawning and brooding; wetlands inhabit a variety of amphibians and reptiles, especially frogs, toads and turtles, and are also home to a rich variety of herbivores, of which micro-crustaceans filter algae from the water, snails feed on algae that grow on the leaves and other debris, geese nibble on newly emerged aquatic plants, fulica atras, mallards, and various anatidaes foraging at the surface feed on algal mats or beds.

b. Function of wetland ecosystem

A wetland is an ecosystem with high productivity. As a sediment and debris system, wetlands accumulate a lot of substances such as carbon, nitrogen and phosphorus, and conduct material exchanges with the atmosphere and landscapes. The complexity and differences of various wetlands make it difficult to give a general overview of their ecological functions. The productivity of freshwater wetlands is affected by various water factors, such as groundwater, surface runoff, precipitation, dry periods and flooding of river wetlands. Regardless of the type, wetlands are closely correlated to the overall landscape in which these are located, including size, soil properties, land use, plant species and nutrient availability (Fig. 1).

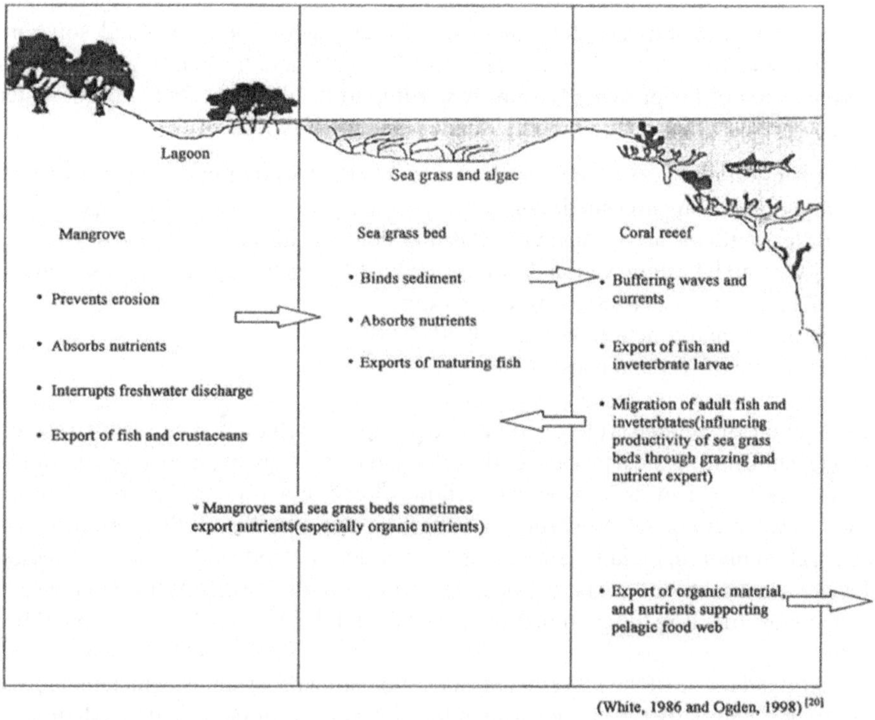

Fig. 1 Function map of coastal wetland ecosystem. *Source* White, 1986 and Ogden, 1998

③ Values of the wetland ecosystem

Wetland has important ecological and economic values, which is compared to "the kidney of the Earth" by ecologists. Wetland can retain a large amount of rainwater, snow water and surface runoff, and slowly release them into underground or sandstone aquifers. Basin wetland also has the function of regulating flood like a reservoir. Wetland also acts as a water filtration system, allowing wetland plants to absorb excessive heavy metals such as nitrogen, phosphorus, sulfur, copper and iron from surface runoff into the biological chain, many of which are deposited in the bottom layer of wetland. As a wetland has the functions of filtering heavy metals and lowering the pH value, we can properly divert urban sewage and industrial wastewater into natural or constructed wetland, in order to exert its purification function. However, it must be pointed out that if the amount of sewage and waste water exceeds the capacity of wetland purification, it will backfire and cause damage and destruction to the wetland ecosystem itself.

Wetland has many economic functions, and hardwood trees and timber from wetland have special values in horticulture. Wetland is the habitat of a large number of amphibians and reptiles, and the nesting and wintering place of a lot of important wildlife. The staggered distribution pattern of highland and lowland wetlands greatly increases the number and diversity of wildlife population. A wetland can provide a place for entertainment and resort to carry out eco-tourism, and popularize scientific knowledge, adding a lot of fun to people's lives.

④ Wetland protection

As people are driven by selfish economic interests, ignoring the important functions and values of the wetland ecosystem, the global wetland ecosystem has been severely damaged, with the area of wetlands decreased sharply. Wetlands (including forest and non-forest wetlands) once made up approximately 3% of the Earth's surface, but now a large area has been transformed into other lands for use. A large number of wetlands in the Yangtze River Basin and the Yellow River Basin in China have also been severely damaged. For example, Wuhan City in Hubei Province, which is a city of hundreds of lakes and a province of thousands of lakes, has developed water networks, many of which are wetlands. In the long-term urban development and construction, many lakes and wetlands have been filled into urban construction land, bringing great losses to the city's microclimate, biodiversity and eco-environment. In 1971, the United Nations adopted the *Convention on Wetlands of International Importance Especially as Waterfowl Habitat* in Ramsar, with the United States, Britain, Canada, and the former Soviet Union as parties to the convention. Its purpose is to stop the gradual encroachment and damage to wetlands, both now and in the future, and to identify the basic ecological role of wetlands and their economic, cultural, scientific and recreational values. The amendment to the convention was adopted at the UNESCO on December 3, 1982 in Paris, with the aim of making the convention more effective. In 1992, China became a party to the convention. The protection of wetlands is attracting the attention of governments and people all over the world.

(6) Soil

Soil is the evacuated surface of the lithosphere, which not only provides the essential nutrients and water for plants, but also serves as a habitat for soil animals.

Soil is an extremely complicated complex of living and abiotic environments, with a genesis inseparable from biological activities. Soil is the base or foundation of all terrestrial ecosystems. The biological activities in the soil affect not only the soil itself, but also the biological communities on it. Any soil characteristic is formed under the combined action of five factors of soil formation: parent material, climate, biological factors, topography and time. Parent material refers to the loose rock material that can form the soil, which is the main factor that determines the chemical composition of the soil. Climate has a great influence on the development of soil, and temperature determines the degree of weathering of rocks, and the decomposition and decay rate of organic and inorganic matters. Meanwhile, climate also affects the growth of plants and animals in a region, which are important factors affecting soil development. Topography affects the amount of water entering the soil. Compared with flat land, sloping land affects the intensity of soil erosion. More water is lost on slopes, less water penetrates into the soil, and the soil layer is thin and stunted, while there is often more water entering lowlands and flats, which increases the moisture in the deep layer of soil and enables it to fully develop. Time is an important factor of soil formation. Rock weathering and breakage, accumulation of organic matters, decay and mineralization, loss of inorganic matters in the upper layer of soil, and differentiation of soil layer all take a long time. Therefore, the formation of good soil may take 200 to 2,000 years, and develops more slowly in arid regions than in humid ones. Plant growth can reduce soil erosion and loss and affect the content of nutrients in the soil. Animals, bacteria and fungi can break down organic matters and mix with inorganic ones, which is conducive to soil aeration and water infiltration.

(1) **Soil geology**

Soil is a three-phase system composed of solid, liquid, and gas, in which solid particles are the material basis of soil, accounting for more than 85% of its total weight. Soil texture is composed of solid particles of different sizes. According to the diameter, soil particles can be divided into four categories: coarse sand, fine sand, silt and clay (Table 1).

The percentage combination of these solid particles of different sizes is called soil texture. The soil can be divided into three categories: sand, clay and loam according to its texture. Sand soil mainly includes coarse sand and fine sand, while silt and clay particles account for less than 10% of the total, which, therefore, features low viscosity, many pores, strong ventilation and permeability, and poor capacity of water

Table 1 Four types of soil particle diameters

Classification	Coarse sand	Fine sand	Silt	Clay
Numerical value (mm)	2–0.2	0.2–0.02	0.02–0.002	Below 0.002

storage and fertilizer retention. Clay soil, mainly silt and clay particles, account for approximately 60–85% of the total, with heavy texture, compact structure, strong capacity of water and fertilizer retention, but with small pores, poor ventilation and permeability, which is sticky when wet and hard when dry. Loam soil is relatively uniform in texture, neither too loose nor too sticky, in which sand clay, sand powder and clay particles account for roughly the same proportion, with good ventilation and permeability, and a certain capacity of water and fertilizer retention, making it an ideal agricultural soil.

(2) **Soil structure**

Soil structure refers to the arrangement of solid particles, the number and size of pores, and the size and number of aggregates. Soil structure can be divided into micro-aggregate structure (less than 0.25 mm in diameter), aggregate structure (0.25 to 10 mm in diameter) and various structures larger than aggregate structure. The aggregate structure is that the humus soil particles in soil are bonded into small agglomerates with a diameter of 0.25–10.00 mm, which has the characteristics of water stability that does not disperse when soaked in water. The soil with aggregate structure is well-structured soil, because it can coordinate the relationship among water, air and nutrients in the soil, and improve its nutritional properties. Aggregate structure is the foundation of soil fertility. Soil with no or poor structure has solid mass, poor ventilation and permeability, poorly developed plant roots, and the activities of soil microorganisms and animals are also restricted.

Soil texture and structure are closely correlated to the water, air and temperature conditions in soil.

(3) **Soil pH**

Soil pH is the most important chemical property of soil and a comprehensive reaction of its various chemical properties, which has important effects on soil fertility, soil microbial activity, synthesis and decomposition of soil organic matter, conversion and release of various nutrient elements, availability of trace elements, and distribution of soil animals. The value of soil pH in China is divided into five grades (Table 2).

(4) **Soil organic matters**

Soil organic matters include non-humus and humus. Soil humus is an important source of carbon and nitrogen for plant nutrition, and more than 99% of nitrogen in soil exists in the form of humus. Humus is also an important source of various mineral nutrients needed by plants. Soil organic matter can improve the physical structure and chemical properties of soil, and facilitate the formation of soil aggregate structure,

Table 2 Grading of soil pH values in China

Value grading PH	Strong acidity	Acidity	Neutrality	Alkalinity	Strong alkalinity
PH value	<5.0	5.0–6.5	6.5–7.5	7.5–8.5	>8.5

thereby promoting plant growth and nutrient absorption. In general, the more organic matter there is in soil, the more species and quantity of soil animals there are.

(5) **Soil inorganic elements**

A total of 13 kinds of inorganic elements taken by plants from soil are indispensable for the normal growth and development of any plant, including seven macro-elements, such as nitrogen, phosphorus, potassium, sulfur, calcium, magnesium and iron, and six trace elements, such as manganese, zinc, copper, molybdenum, boron and chloride. Soil must contain a certain proportion of elements necessary for plant growth and development, and its nutritional status can also be improved through rational fertilization. Inorganic elements in soil also have some effects on the distribution and quantity of animals.

With the continuous expansion of urban scale, the agricultural land on the outskirts has been constantly transformed into urban construction land. The precious soil surface layer formed through thousands of years of labor input and biological evolution will be destroyed completely. It would be a wise approach to collect the topsoil and use it for newly reclaimed land or covered on the barren agricultural land nearby. In fact, as early as 1964 when we conducted a special survey of the Neijiang City Master Planning, we proposed methods and suggestions for land reclamation that were not implemented due to the lack of incentive policy and system guarantee.

(7) Creatures

Creatures, including plants, animals and microorganisms, are very important and sensitive natural resources. With the aggravation of global environmental problems, biodiversity has been seriously threatened, and a large number of species has decreased, to which special attention needs to be given to strengthen protection.

Relatively speaking, there are abundant biological resources in mountain areas and cities. In the planning, we should investigate and understand the distribution and types of various biological resources, as well as the problems existing in the conservation and maintenance, find out problems, and propose protection measures to maintain biodiversity in mountain areas and cities.

(1) **Vegetation**

Vegetation refers to the general term for plant life that covers an area. Plant life refers to all plants in forests, grasslands, dry lands and wildernesses. As an important component of the natural environment, vegetation plays a special important role, whose social ecological value is far greater than the economic one. Through the photosynthesis of green plants, it synthesizes the inorganic substances in the natural environment into organic ones, and at the same time converts the absorbed solar energy into chemical energy for storage in organic substances, in order to combine the organic and inorganic realms into an organic whole and ensure the ecological balance and stable development of the natural environment. The ecological service function of vegetation also includes reducing or blocking the impact of external pressure such as wind, fire, plant diseases and insect pests on natural forests, providing a habitat and breeding place for wildlife, storing rainwater and intercepting runoff, filtering pollution such as noise, smoke and dust, adjusting temperature and humidity, and improving microclimate and local environment. Therefore, the basic situation of local plants should be investigated. However, some urban planners or developers often lack of understanding of the important role of vegetation, and only regard it as one of the material elements and means to beautify the environment. Therefore, the ecological service function of vegetation is often neglected in environmental planning or construction. In many places, the original vegetation is often uprooted, new plants are planted, and even ancient and famous trees are transplanted in other places, which not only destroys the original eco-environment system, and causes soil erosion and water and soil loss, but also may brings other ecological problems. Therefore, the original vegetation in this region should be used as much as possible, and the dominant species of local plants should be taken as the main objects of cultivation, in order to allow the plant community after greening to be in harmony with the local habitat, and integrated into the regional ecosystem.

China is a country lacking in forest resources. Its forest coverage rate was only 13.92% in 1996, with a very uneven distribution. Among 160 countries and regions, China ranks about 120th, and according to the average forest area owned by the population, it ranks about 121st. For a long time, the eco-environment has been seriously degraded due to excessive deforestation, which intensifies the occurrence of natural disasters such as drought, flood, hurricane and sandstorm. Therefore, it is necessary to unremittingly strengthen the protection and restoration of forest vegetation and improve the forest coverage.

Japan is also a country with more people and less land. Due to its persistence of afforestation and protection of vegetation, the forest coverage rate has reached more than 60%. The forests in Finland were severely damaged in the Second World War, but due to its intensifies efforts in forest restoration, the forest area now accounts for approximately 71% of the country's land area.

(2) **Animals**

Animals are important resources, which are of great significance to maintaining the global biodiversity and ecological balance. Animal resources can be divided into three categories: domesticated animal resources, wild animal resources and aquatic animal resources. Wildlife and aquatic animal resources are diminishing rapidly throughout the world. Therefore, it is an important and arduous task to strengthen the protection of wildlife and aquatic animal resources.

In 2002, Germany adopted an amendment to its constitution, which listed animals into constitutional protection for the first time. It became the first country in the world to give animals constitutional rights and interests, which treated animal protection in the same way as respecting and safeguarding human dignity. However, the bad conduct of hunting wild animals still exists in some parts of China despite repeated prohibitions. Japan and other countries also have the bad behavior of whaling for food. The food culture for rare delicacies is actually a loss of human morality and reason.

According to the principle of combining landscape ecology and design with nature, large areas of natural woodlands, wildernesses and wetlands are reserved in mountain areas and urban areas in mountainurban planning and construction, and animal communication and migration channels are established in the green patches and corridors of urban areas and suburban peripheral areas. In the construction of major infrastructure projects, such as hydro-junctions, truck railways and expressways, sufficient channels for animal migration must be designed and reserved. For example, along the Qinghai-Tibet Railway project to be opened to traffic, 33 animal passages have been set up, which provides convenient conditions for the free flow and migration of Tibetan antelope and other rare wild animals, making it China's first environment-friendly railway on the plateau. In the construction of expressways, migration and transportation corridors are set up for animals such as foxes and hares. In the construction of hydropower stations such as the Gezhou Dam, special "fishways" are also provided for reflux fish. In the construction of roads in urban and suburban areas, due consideration should also be given to the establishment of migration and activity channels such as culverts for animals such as hares, squirrels, snakes, and frogs.

(8) Land
(1) **Meaning of land**

Land is the most important natural resource, the basic data of human survival and the object of labor. Due to the immovability and non-renewability, land is a typical scarce resource.

Different from agricultural land, urban land is the material carrier of society, economy, culture, production and life, as well as the extremely important means of production and living elements of urban economy. Urban land, as a scarce resource, has both use and economic values. The monetary expression of land resource value is land price. Scientific understanding of land value, comprehensive assessment of land grade, and reasonable estimation of land price serve as the basis of paid land use.

Urbanization is not only a process of agglomeration of population and industries into cities, but also a process of transforming part of agricultural land into urban land. According to the present population growth rate, it is estimated that by 2025, China's population will increase to 1.6 billion. Based on the exiting cultivated land, the per capita cultivated land will be reduced to 1.22 mu. Therefore, strict policies for the protection of land resources must be implemented. The comparison of China's land resources with those of Russia, the United States and Canada is listed in Table 3, which indicates the shortage of land resources in China.

The practice of China's reform and opening up has proved that the market alone cannot solve the problem of "optimization and reasonable allocation of resources and environment". Instead, we must rely on the decision-making of the government, implement it through scientific planning and strict management, and limit spontaneous behaviors and speculation in land development.

Table 3 Comparison of Land resources between China and Russia, the United States and Canada (2004)

Ountry area (10,000 km^2)	China	Russia	The United States	Canada
10,000 km^2	960	1770	936	997
Hectare per capita	0.75	12.12	3.6	34.3
Cultivated land per capita	1.41	13.1	10.9	26.9
Comparison	1	16.3	4.86	46.35

(3) Constraints of urban development

Before deciding on land use and urban space development, various constraints in urban development should be fully considered and clearly defined, in order to propose corresponding measures, study and formulate the optimal development strategy under the restrictive conditions. The restrictive factors of urban development mainly lie in natural ecological conditions, the development of human environment, and economic, policy and decretive constraints.

① Constraints of natural ecological conditions

In any city, especially a mountainurban that is restricted by many natural conditions, its complex natural landform, geological conditions and landscape composition often impose certain constraints and restrictions on urban development. According to the concept of ecology, the use of natural resources should not damage the overall natural eco-environment system. For areas sensitive to environmental impacts and prone to geological disasters, the development should be strictly restricted.

② Constraints of the development of cultural environment

Many mountain cities in China have a very rich and centuries-old history of urban construction, historical and cultural heritage and cultural environment, in which the overall geomantic pattern, ancient city walls, river and lake water systems, street and lane systems, ancient houses, ancient wells, famous trees and various cultural traditions, religious beliefs and local cultural customs must be protected and inherited, and these are the constraints of cultural environment.

③ Economic constraints

The engineering economic construction, traffic engineering and foundation construction of mountain cities are closely correlated to the topography, geology, geomorphology, natural environment and development pattern. For example, in the construction of rugged and complicated mountain cities, the direction of road traffic will be restricted by the terrain, and the cost of basic engineering will inevitably increase. Therefore, the economic possibility must be considered for the feasibility of project construction.

④ Policy and decretive constraints

The national land improvement or regional planning puts forward policy and statutory restrictions on the overall and land use plans of each city, and the city itself must make policy and statutory restrictions. For example, there should be height limit restrictions near airports, and water conservation areas, wetlands and scenic spots in urban areas must be protected, where development activities are prohibited.

(4) **Land use intensity**

Appropriate land use intensity is of great significance to the economic efficiency of urban land supply and the feasibility of construction. For example, due to the large population and relatively little arable land in China, the land resources used for urban construction are very precious. "Treasuring and reasonably using every inch of land, and effectively protecting cultivated land" is China's basic national policy. We must carry out reasonable land use. However, if the development of land per unit area is not restricted and the plot ratio is arbitrarily increased, it may be overcrowded, which may affect and damage the use, and deteriorate the environmental quality. Therefore, it is necessary to control and define the intensity of use for the land with different use nature and requirements, in order to reasonably guide its construction and implementation.

(5) **Land use control**

In view of the grim fact that China has more people and less land, and the contradiction between people and land is prominent, a strict land governance system must be implemented. Based on the comprehensive ecological evaluation on the whole country—river basin—region—urban area—rural area from the macro, meso and micro perspectives, eco-environment maintenance and construction are carried out to restore the ecological service functions of degraded ecological land. Effective protection and management must be carried out for nature reserves, scenic spots, water source reserves, river and lake systems, wetlands and basic farmland.

Moreover, on the basis of comprehensive evaluation of construction conditions, the construction planning of land for urban and rural construction shall be carried out from regions to cities and villages, and functional organization and layout are carried out according to the different characteristics of land use, so that land for urban and rural construction can be given full play to its potential. For example, the areas with poor soil fertility and agricultural production conditions are used as production land for industrial production and storage, and the land with good sunshine and ventilation is used for residence.

3 Social Conditions

(1) Population

Population is the basic survey subject of urban planning, including permanent population, floating population, age structure, gender structure, occupational structure, family population, birth rate, mortality rate, natural population growth rate, spatial distribution of population, social stratification, distribution of nationalities, population aging, proportion of disabled people, living conditions and employment.

(2) Educational and cultural level

It includes the level of education, the educational level of residents, cultural quality, the level of various educational facilities, distribution of primary and secondary schools, enrollment rate, trend of development and changes, and proportion of university graduates.

(3) Traditional cultural value and diversity

It includes the cultural value integrity of historical and cultural buildings and their environments in urban areas, the integrity and protection of the overall environmental pattern of urban landscapes, as well as the customs and religious beliefs of different ethnic groups (Qiheng 2003).

(4) Environmental quality and pollution

It includes the quality of a city's eco-environment, the pollution of atmospheric, water and acoustic environments, and the quality and compliance of solid waste disposal.

(5) Community construction

It includes the degree of sound development of community institutions, the level, characteristics and problems of social, cultural, medical and public health services, the democratic system of communities and the degree of residents' participation in the decision-making and management of a city and their community.

4 Economic Conditions

(1) Economic development level

It includes GDP, economic growth rate, energy consumption per unit of GDP, environmental costs, per capita economic income, Gini coefficient, employment rate, unemployment rate, etc.

(2) Industry

It includes the composition and characteristics of urban industrial structure (the proportion of primary, secondary and tertiary industries), the industrial development and development potential, and the situation of leading industries.

(3) Market and real estate

It includes the degree of market development, the degree of development of commercial trade, the number of urban home buyers, the degree of housing satisfaction, the average household and per capita residential areas, etc.

(4) Infrastructure level

It includes the urban infrastructure level of road, transportation, water supply, drainage, electricity, telecommunications, heat supply, modernization, medical and public health, as well as satisfaction and existing problems.

(5) Local finance

It includes the source of local finance, fiscal revenue and expenditure, as well as the the development and changes of revenue and expenditure over the years.

5 Investigation Techniques and Approaches

(1) Traditional techniques and approaches

The traditional investigation activity is the process of transmitting the information of external things to human brain through human sensory organs. In the process of field survey, if the information observed is only based on the impression of human sense organs and brain without other means, it may be distorted or even disappear completely in the future (Nanwei 2001). Therefore, it is an important part in the process of field investigation to make records carefully. Commonly used techniques and approaches for survey recording mainly include survey record sheets, survey cards, survey icons, photographs and videos.

(1) **Survey record sheet and survey card**: Survey record sheet and survey card are traditional tools for field survey record, and also the commonly used manual record methods in survey activities. Investigators may, prior to the implementation of field investigation, design observation record sheets or observation cards in advance according to the purpose of social investigations, and then use the sheets or cards conveniently in the process of field investigation to observe urban social affairs or social phenomena in a timely manner. The basic requirements for the design of a survey record sheet or survey card are, as follows: ① The time and place of the survey should be specified in detail, which is an

important evidence to show the original survey records; ② The survey contents contained in the survey record sheet or survey card should be specific, detailed and focused, and the survey contents should be quantified as much as possible, in order to make the survey results more convincing

(2) **Survey diagram**: It refers to the activities of further analysis, thinking and research of the investigators on natural or social affairs or social phenomenon of a city surveyed in field investigation, which are timely recorded and visually analyzed in a convenient and intuitive manner by means of urban planning drawing and sketching. The diagram has an intuitive and clear expression effect, which is of great significance to enhancing the persuasiveness of the survey and ultimately forming the results of a survey report. In order to coordinate with the effective development of field survey activities, the main methods of survey diagram are, as follows: Topographic maps are used for plane positioning or measurement; topographic maps or manual drawings are cooperated to record the natural characteristics such as topography, landform and vegetation; the activity status of people in the social investigation site is recorded; manual drawings are adopted to record the relationship and characteristics of the construction and spatial environment of the survey site, etc.

(3) **Photography and videography**: In addition to the manual methods such as survey record sheets, survey cards, and survey icons, field survey techniques can also adopt modern technical means, such as photography and videography, to record and reproduce the characteristics of natural topography, landform, landscape, vegetation, architecture and spatial environment and the real state of urban social affairs and social phenomena

(2) GIS and digital techniques and approaches

With the development of modern information technology, the development and extensive application of remote sensing technology, geographic information system, global positioning system and digital technology have laid a more scientific and effective methodological basis for the scientific decision-making of urban planning. Traditional survey approaches are mostly based on qualitative and comprehensive analysis; and digital analysis techniques can better combine qualitative and quantitative analysis.

6 Comprehensive Evaluation on Construction Conditions

One of the most important part in the investigation of mountain cities is the comprehensive evaluation of Mountainurban construction conditions. In the comprehensive assessment, we should adopt the approach of combining qualitative and quantitative analysis, take the land use and environmental issues as the focus and core contents, and expand the environmental assessment to the category of complex natural, economic and social ecosystem (Liangyong 2001).

(1) Comprehensive evaluation of natural resources

(1) **Selection of ecological factors**

There are many natural ecological factors that affect the construction and development of mountain cities, but the impacts are not all equally important. It is impossible to take all relevant factors into account in the evaluation. Hence, key factors with the greatest impact can only be selected according to different regions, cities and research objects, such as geology, topography, soil, hydrology, vegetation, climate and meteorology, land use, environmental quality, biodiversity and special economic and cultural values of natural resources.

(2) **Establishment of grading standards for ecological suitability of single factors and their weights**

The original data for the evaluation of ecological factors are graded and quantified, and different weight values are assigned to the degree of influence, and larger weight values are given to the factors with greater impact.

(3) **Calculation of comprehensive suitability**

The results of single factor evaluation are superimposed and the weighted factor approach is adopted for evaluation to obtain the grading standard for the comprehensive suitability of the ecological factors of urban natural resources. For example, the comprehensive suitability is divided into five levels: very appropriate, appropriate, basically appropriate, inappropriate and very inappropriate, and a comprehensive evaluation map of land use is drawn.

The calculation formula of suitability evaluation on ecological factors is, as follows:

$$V_{ij} = \sum_{k=1}^{n} \left(\breve{B}_{kij} \cdot W_k \right)$$

where: j—Number of land use mode;

k—Number of ecological factors affecting j land use approaches;

n—The total number of ecological factors affecting j land use approaches;

B_{kj}—Suitability evaluation value of the kth ecological factor (single factor evaluation value) of the ith plot with the land use mode of j;

W_k—Weight value of the kth factor to j land use modes;

V_{ij}—Comprehensive evaluation value of the ith plot with the land use mode of j.

(2) Ecological sensitivity evaluation

In the Mountainurban land use planning, the ecological sensitivity of land needs to be analyzed to ensure the sustainable use of natural resources, and to coordinate the relationship between development and protection. Ecological sensitivity refers to the adaptability of ecological factors to external pressure or changes. There are many factors affecting the ecological sensitivity of land use, such as geological instability, gullies, landslide zones, steep slopes, diversity of vegetation and landscape, and land and landforms with special landscape and cultural value. Through the weighted superposition and clustering of single-factor evaluation, the comprehensive evaluation criteria and ecological sensitivity zoning are obtained, such as

Type I of sensitive or the most sensitive area;
Type II of sensitive area;
Type III of sensitive or low sensitive area;
Type IV of sensitive or insensitive area.

Type I of sensitive or the most sensitive area is the land prohibited for development and construction. Type II of sensitive area is the land restricted for development. Types III and IV of sensitive area are the lands suitable for development and construction.

(3) Comprehensive evaluation of construction conditions

According to the land classification in the comprehensive evaluation of natural resources and ecological sensitivity, and then comprehensively considering social, economic and human factors, the classification of land suitability for the development and construction of mountain cities can be determined. In general, these can be divided into four categories: prohibited construction land, controlled construction land, restricted construction land and suitable construction land. Hence, these can guide the mountainurban development and construction.

References

Baopu F, Climate M, Press S (1983) Mountain climate resources. Nanjing University Press, Development and Utilization, p 1996
Liangyong W (2001) Introduction to sciences of human settlements. China Architecture & Building Press, Beijing
Nanwei L (2001) Natural geography. Science Press
Qiaoli X (1994) Mountain geographic system. Central China Normal University Press
Qiheng W (2003) Research of geomantic theory. Tianjin University Press, Tianjin
Wenhua L, Jingzhu Z (2004) Review and prospect of ecology research. China Meteorological Press, Beijing
Yongchang S (2000) Urban ecology. Shanghai Normal University Press, Shanghai

Chapter 3
Site Selection

1 Principal and Experience in Ancient Cities

The planning and construction of any city is carried out in a specific natural environment, which is a combination of nature and man.

Since ancient times, China has attached great importance to the land and site selection of cities in the urban planning of cities and the construction of human settlements, and paid attention to the utilization and protection of natural mountains and rivers in the layout structure and development form of cities, leaving us precious and excellent urban cultural heritages. Ancient Chinese ancestors knew about the use of hilly land from a long time ago. On October 18, 1997, the author saw a picture of the excavation site of a Changsha kiln at the Changsha Museum in 1983, which was a dragon kiln in the Tang Dynasty. The kiln site was built on a hillside, with a length of 41 m and a width of 3 m, including a hearth and a chimney. Another excavation site was built on the east bank of the Xiang River in Datang Village, Nantuo Township, about 20 km south of downtown Changsha. In 1985, the C^{14} determination was conducted on the unearthed relics by Peking University, which was an ancient kiln site 6,900 $\pm$ 100 years ago, and 7,000 years ago after tree ring calibration. A large number of excavations indicate that these are all located on the hillsides beside rivers, which on the one hand, makes it convenient for the transportation of raw materials and products, while on the other hand, makes use of the hillside terrain to set up hearths and build chimneys on high slopes, in order to facilitate chimney ventilation, and save energy and cost.

In the early feudal society of China, the defensive function of cities must be prominent due to the annexation of vassal states and continuous wars. The cities were generally established in the places with dangerous terrain, high position, and convenient land and water transport, which were easy to defend and difficult to attack. There is an old saying in China that goes, "building cities to guard the monarch and building walls to protect the people". Shanhaiguan, Yulin, Yan'an, Tianshui, Chongqing, Langzhong, Wenzhou, Taizhou, Yantai and Penglai were all important military towns and mountain cities in ancient times. Guan Zhong, the Prime Minister

© The Author(s), under exclusive license to Springer Nature Singapore Pte Ltd. 2021
G. Huang, *Theory of Mountainurbanology*,
https://doi.org/10.1007/978-981-16-0819-3_3

of the State of Qi in the Spring and Autumn Period, made a systematic exposition of the systems, the site selection and construction of cities of the State of Ying in his famous works *Guan Zi*.

The site selection of ancient Chinese cities had an inseparable relationship with mountains and rivers. According to the principles of ancient geomancy, the differences between grand, medium and small convergence and environmental capacity determine the difference in building large, medium-sized and small cities. It is contained in the famous Chinese classics *Guanzi* • Chap. 5 *City Planning Thoughts* that "Regardless of the establishment of a country or a city, it must be built near rivers if it is not under mountains. In high terrains, it is not close to high hills to ensure sufficient water supply. In low terrains, it is not close to rivers to save the construction of ditches and dikes. Local people should rely on natural resources and favorable geographical environment. Therefore, the building of a city needs not be restricted to stereotypical norms, and the pavement of a road must not be constrained by strict standards". It is said in the *Guanzi* • *Chapter 57 Topographic Survey* that "Sages must build their capitals in places with flat terrain, stable geology, fertile soil, and abundant products, which backed mountains, with rivers or lakes passing by, so as to provide constant flows of water. The network-shaped tributaries or canals branching from the main channel are criss-crossing, with the main function of diverting and draining water. In this way, natural resources and agricultural products can be utilized to feed local people and breed livestock." This is an important guiding principle for the site selection, planning and layout of mountain cities. It was pointed out in the *Biography of Chao Cuo* in the *History of the Han Dynasty* that as early as in the Western Han Dynasty, Chao Cuo proposed that "People should investigate the harmony of Yin and Yang, taste its spring water, examine the suitability of its land, and observe the lushness of its vegetation before building a city". In the ancient Roman times, the *Ten Books on Architecture* written by Vitruvius took the health of land as the primary condition for site selection, proposed that "the first is to choose the land that is most beneficial to health, that is, it should be a highland", and determined the form and orientation of the city according to the wind direction, temperature and climate.

Throughout the history of urban development over thousands of years, many urban sites have remained unchanged once selected despite all the vicissitudes, such as Beijing, Nanjing, Chongqing, and Chengdu. There were also many cities whose locations often changed in the historical development, or because of natural disasters such as large and small mountain torrents and mudslides, wars and turmoil, or traffic barriers, especially in the mountain areas with complex natural conditions, where were more likely to occur, such as Wanxian, Fuling, Fengjie and Wushan in the Three Gorges Area of the Yangtze River. It can be observed that various factors must be taken into account in the site selection of a city, such as fine weather, favorable geographical position and culture. In particular, the site selection of mountain cities should be conducted with great caution, and multiple factors should be considered and compared with the local and surrounding environment, in terms of the situation of the mountains and rivers, geography, climate, geology, soil, hydrology, vegetation, biology and socio-economic conditions.

The site selection of a city is not only based on whether there are mountains, rivers and enough land for development in the location of the city itself, but must first focus on the overall environment of the region (Bingeng 2000). In order to explain with the concept of the geomancy theory, "to find the dragon vein" is the top priority, and importance should be attached to the form, origin and ending of mountains, which are closely related to the overall environment, geomantic pattern and the social, economic and cultural development of a region. According to the theory of geomancy, the dragon veins in China originated from the Kunlun Mountains, which is the ancestor of the dragon veins in China. From this origin, these extend eastward, and are divided into three main lines, namely the northern main line, the central main line, and the southern main line.

The northern main line refers to the mountains in the vast region north of the Yellow River, including the provinces of Qinghai, Gansu, Shanxi, Hebei, and the three provinces in Northeast China. The mountain ranges extended from the Northwest Plateau. The water west of the main ridge flew into Longmen Xihe, and the water east of the ridge flew into Youji and into the East China Sea. A branch range in the west was Hukou Fengyue, and the secondary branch range surrounded the Fenshui Basin, with Wangwu Mountain in the south and Leishou Mountain in the west. Another branch range was Heng Mountain, and the secondary branch range was Taihang Mountain which stretched for thousands of miles, with high mountains and dense forests. The longest branch was Yan Mountain, which was connected with the land of the capital and ended in Pingluan in the east.

The central main line refers to the mountains between the Yellow River and the Yangtze River, including the provinces of Sichuan, Shaanxi, Henan, Hubei, Anhui and Shandong. The mountain range originated from Shu-Han, of which a branch extended to Chang'an and ended in the central Shaanxi plain; a branch went down to the Hangu Pass, then to Song Mountain, and ended in Mount Tai. Another branch came from the north of the Han River, and went down and ended in Hangzhou.

The southern main line refers to the mountain systems in the south of the Yangtze River, including the provinces of Yunnan, Guizhou, Guangxi, Hunan, Jiangxi, Guangdong, Fujian, Zhejiang and Jiangsu. Its main range originated from Min Mountain, which was sandwiched by two rivers, with one branch on the left, forming many branches in the north of the rivers. The branch on the right was scattered to the provinces of Hunan, Fujian and Guangdong, and ended in Zhejiang Province and Nanjing City in the east. One of the branches was Guiling, surrounded Xiang River, descended northward, and ended in Mount Lu. Another branch surrounded Poyang Lake from south to east, passed by Mount Huang in Anhui Province, and ended in Nanjing City. One branch was divided into another branch in the Tianmu Mountain, and ended in Zhejiang Province, whose north ended in Shaoxing City, and south in Fujian and Guangdong provinces, including the Lingnan Mountain System (Fig. 1). The three main lines are divided into large, medium and small main lines, which are further divided into large, medium and small branches. This reciprocating downward fractal formed mountain and water systems covering the land of China. This is the overall systemic view of the ancient theory of geomantic omen on China's geographical features and landscape patterns. Therefore, the geomancy theory tells

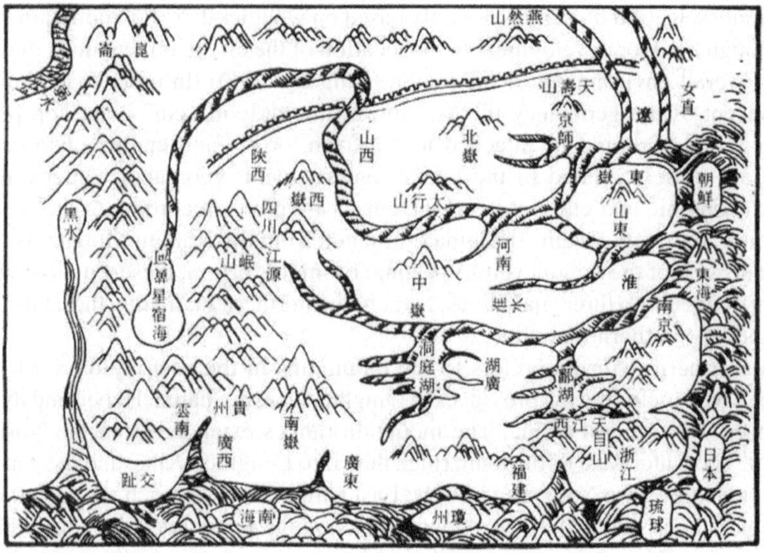

Fig. 1 The Diagram of the Three Major Dragon Veins of China Contained in the *San Cai Tu Hui*. *Source* Cheng Jianjun, Kong Shangpu. *Geomancy and Architecture*. Jiangxi Science and Technology Press. October 1992, p. 20

us, "If you want to know the situation of a city, you must first investigate the veins of geomantic omen" ("What is the Evidence of Homestead Selection" in the *Keynotes for the Q&A of Site Selection* (Volume 1) written by Qingjiangzi, the block-printed edition of Huaide Hall in the 38th Year of the reign of Emperor Qianlong). The grand convergence of mountains and rivers must form a city, the medium convergence of mountains and rivers must form a town, and the small convergence of mountains and rivers must form a village. In his doctoral thesis "Research on Geomancy and Construction of Landscape Cities in ancient times", Yang Liu compares the dragon vein of mountains and rivers in China to a big tree, under which the cities and towns across the country are the fruits, and illustrated the relationship between fractal and isomorphism with a diagram (Fig. 2).

Since ancient times, China has attached great importance to the site selection of capitals to the planning of cities, counties, towns and villages, with a whole set of theories and approaches as guidance, such as the *Book of Songs*, the *Rites of Zhou*, the *Guan Zi*, the *Book of Changes*, the *Book of History* and the *Theory of Geomancy*. According to historical records, many famous historical Figures and literary giants in ancient times made outstanding contributions to urban planning and site selection, who were good at making use of mountains and rivers to create immortal achievements for generations to come. The examples are too numerous to mention one by one: in the legend of King Yu's Flood Taming, there was a moving story that he was too busy to go home, even passing by his own house; Li Bing, Governor of Shu in the Qin Dynasty, led the people of Sichuan Province in building

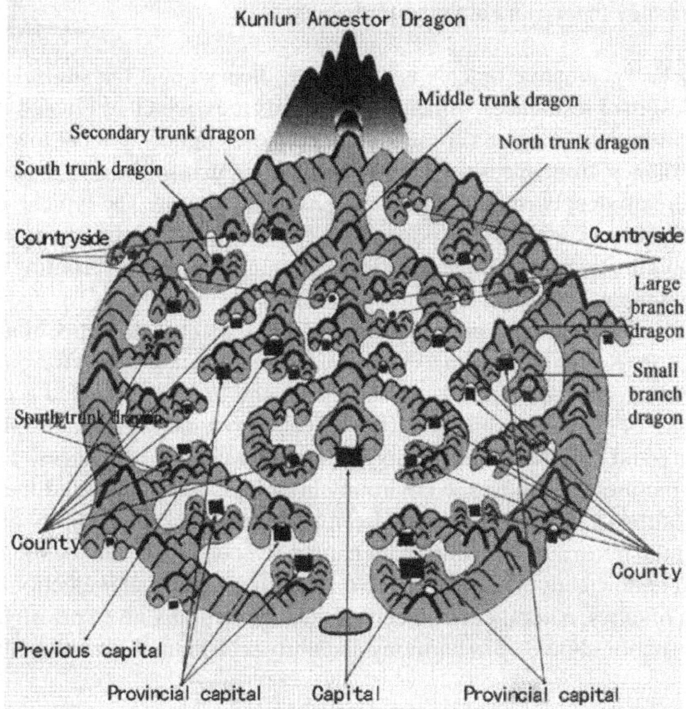

Fig. 2 Network Diagram of Urban System. *Source* Yang Liu. *Study on Geomantic Thought and Construction of Ancient Landscape Cities*. Doctoral Dissertation of Chongqing University. 2005, p. 208

the Dujiangyan irrigation project, bringing benefits to all descendants; Guan Zhong, Prime Minister of the State of Qi in the spring and autumn period, decided on the site selection of the City of Lu; General Zhang Liang of the Han Dynasty suggested to choose Chang'an as the capital; Xie Lingyun ran Yongjia Prefecture in the Jin Dynasty; Guo Pu in the Jin Dynasty weighed soil to determine the location of "Big Dipper City" (Yongjia Prefecture); Wu Zixu's ran Pingjiang Mansion in the late Spring and Autumn Period, Yu Wenkai to Chang'an and Luoyang in the Sui and Tang dynasties, Bai Juyi and Su Shi to Hangzhou in the Tang and Song dynasties respectively, Fan Zhongyan to Yan'an and Huai'an in the Song Dynasty, Yuan Zhen to Yuezhou in the Tang Dynasty, Ouyang Xiu to Yangzhou in the Song Dynasty, Liu Yuxi to Suzhou in the Tang Dynasty, and Liu Bowen to Nanjing in the Ming Dynasty; Li Chunfeng selected the locations of Linhai and Taizhou in the Tang Dynasty; Guo Shoujing planned and built the capital of the Yuan Dynasty, etc.

(1) Geomancy theory and architectural science

Geomancy, or "geomantic omen", is an ancient theory about the site selection and planning design of residences, villages, towns and cities, which originated in China's early agricultural civilization. From ancient Chinese capitals to local towns, mountains and villages, from ancestral temples and garden mausoleums to residences, the designs were all deeply influenced by the geomancy theory. The typical geomancy patterns of ancient towns and residences summarized by Mr. Cheng Jianjun et al. (Fig. 3) fully illustrates the emphasis on the geomantic pattern of "fronting rivers and backing mountains" and "facing south with the back to the north" in ancient times (Jianjun 1992). From the overall layout of many existing cities in China, such as Beijing, Xi'an, Kaifeng, Luoyang, Nanjing, Chengdu, Yangzhou, Fuzhou, Suzhou, and Langzhong, we can find the capital pattern and order established according to the principles of geomancy. Furthermore, we can also find the geomantic venation in the planning and construction of many villages, towns, and residences. The reason is that the picturesque landscape environment in China has cultivated the aesthetic sentiment of the Chinese people, and also fostered the concept of "unity of man and nature", which integrates humanity with nature. Respecting and advocating nature is an inherent endowment of Chinese people. Therefore, in the site selection, planning and design of cities, natural environment should be naturally given priority, and artificial construction should be in harmony and unity with natural environment (Qiheng 2003).

The concept of "unity of man and nature" and "living in harmony with nature" in the geomancy theory cannot be understood as only passively adapting to nature, but also includes the renovation and positive transformation of poor natural terrain and environment, which must be based on meticulous investigation and research, and mastering the laws of nature and scientific principles. For example, in ancient China, there were a lot of practices in renovating undesirable terrain, artificially piling up soil, cultivating the land vein, opening channels for water diversion, digging lakes for water storage, and avoiding poor flow of water, etc. Furthermore, we have accumulated extremely rich experience, which is worthy of our inheritance and development.

The geomancy theory is in the same line with the philosophical thoughts of Chinese Confucianism, Taoism and Buddhism, such as the *Book of Changes*, *Tao Teh King* and the *Theory of Yin and Yang and Five Elements*, which is not only an aesthetic concept, but also a philosophy of life. The *Book of Changes* has an important influence on the development and integration of science and culture in the world, with Yi-ology and geomancy research institutions all over the world. When the author visited Germany in 1987, he had academic exchanges with Professor Clemens Geißier, Director of the German Institute of Urban Planning, and Professor W. Landzettel, Director of the Institute of Regional Architecture and Residential Planning at the University of Hanover, who all mentioned that they had benefited from the I-ching and geomancy theories in China. In Germany, it was also very popular for scholars to study the ecological living environment with reference to the Chinese geomancy theory. On the shelves in their offices, there are many books

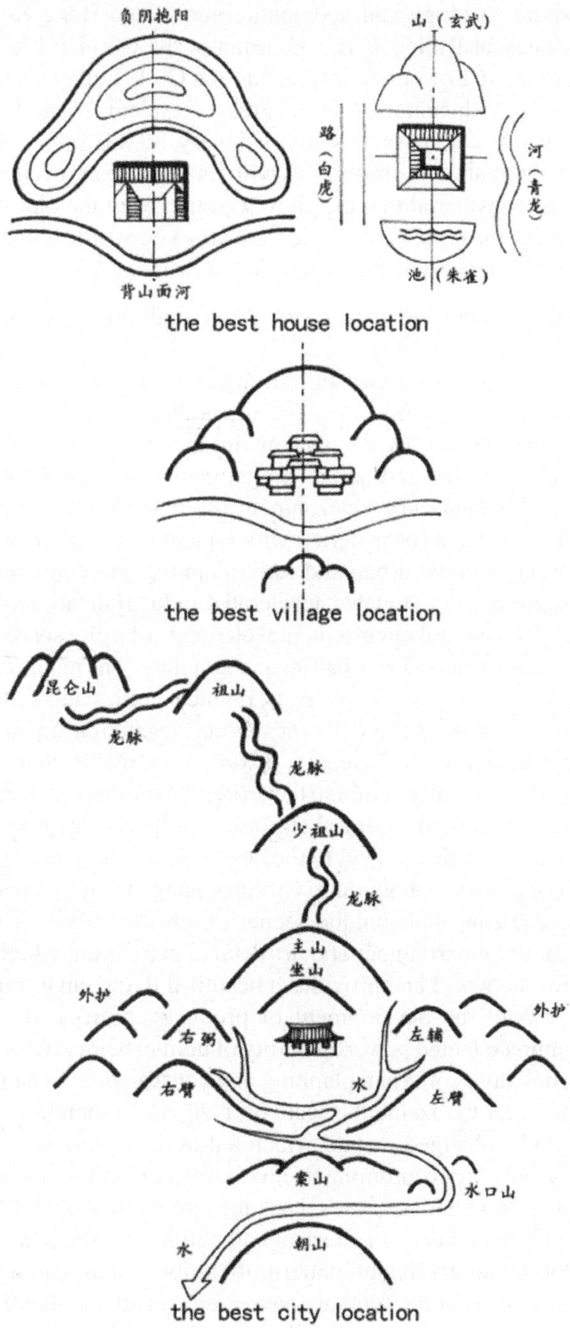

Fig. 3 The Best Location of Residences, Villages and Cities in the Geomancy Concept. *Source* Cheng Jianjun, Kong Shangpu. *Geomancy and Architecture*. Jiangxi Science and Technology Press. October 1992, p. 24

about Yi-ology and geomantic omen from Hong Kong, Taiwan and South Korea. Professor R. Schubert, Chairman of the German-Chinese Cultural Exchange Association in Lower Saxony, told us that G. W. Leibniz, a famous German mathematician in the 17th century, had the greatest esteem for the I-ching theory. In the process of creating calculus and studying binary, he was surprised to find that the binary system had remarkable similarities with the eight trigrams in the *Book of Changes* in China, and considered that the eight trigrams were the oldest binary notation in the world. In contrast, the Chinese descendants know little about the traditional culture of their motherland, which really makes us ashamed.

(2) Geomancy theory (Fengshui) and modern environmental ecology

Geomancy advocates the harmonious concept of "unity of man and nature". From the perspective of modern environmental ecology, the geomancy theory, with its unshakable reasonability, deeply contains the philosophy of mutual tolerance and mutualism between man and nature. In geomancy, man and nature, artificial architecture and natural landscape environment are considered as an integral organic system, and the schemata (or patterns) with typical significance is extracted and standardized, in order to guide urban and rural planning and construction, which is still of positive significance to today's accelerated industrialization of China.

The natural environmental elements of a city serve as an important foundation for the construction of urban spatial structure. The natural topography such as mountains, rivers, lakes, coasts, plateaus, ravines, gullies and valleys, and the landforms such as forests, grasslands, wildernesses and vegetation are important natural elements for the formation of characteristic urban landscape. If we respect the natural environment, and organically combine the artificial buildings with natural environment conditions, just as the old saying that goes, " properly integrate the natural resources and the characteristics of land to the best". Hence, we can create a good living environment. The cities of Chongqing, Guilin, Changshu and Wenzhou, with many ancient poems eulogizing their unique sceneries, are the result of the ingenious combination of natural environment and artificial construction, which creates a "paradise on earth" for the world and provides a beautiful living environment.

With the development of productive forces, the progress of science, and the unprecedented powerful ability of human beings to transform nature, it seems that the constraints of urban planning and construction on natural conditions can be ignored. In order to expand the scale of a city, its mountains, rivers, lakes and water systems can be changed at will, which led to the destruction of its natural eco-environment system, more prominent environmental problems, and the gradual disappearance of urban characteristics, causing serious destruction to many excellent urban space environments and historical and cultural heritage in China and the complete loss of the urban geomantic pattern of "unity of man and nature" created by predecessors in history. Some scholars have even considered that the geomancy theory represents feudal, conservative and backward thoughts, and advocate to repudiate it totally (Guangyu 2002).

In the development and evolution of the geomancy theory for thousands of years, it is undeniable that there are some feudal superstitions, conservative, pedantic and outdated things, which certainly need to be criticized and rejected. However, this does not affect us to actively absorb the essence of the geomancy theory. We should take advantage of its scientific and reasonable elements in a historical and dialectical materialist manner, carry forward them, criticize and abandon the elements of feudalism and superstition, in order to improve our understanding of Chinese traditional architectural culture, cherish the precious historical and cultural heritage of urban planning and architectural science that has been passed down to date, and better serve the cause of China's urban modernization drive.

(3) Case study of site selection of ancient cities
(1) **Langzhong**

Langzhong City, located at the northern margin of Sichuan Basin and in the middle reaches of Jialing River, is the most important town in northern Sichuan and a national historical and cultural city, just like a poet that goes "The best ancient city in the painting of the 300-li scenery of Jialing River" (li, kia Chinese unit of length, 1 li = 0.31 m). Langzhong City was built in the Qin Dynasty with a history of more than 2,300 years, which was a major town in the past dynasties in Southwest China under the rule of feudalism. This served as a seat of provincial, municipal and administrative governments successively. In the middle of the Warring States Period (330 B.C.), the capital of Bazi was moved from Chongqing to Langzhong. During the Three Kingdoms Period, Liu Bei sent Zhang Fei to guard Langzhong for seven years (214 to 221 A.D.). This was the capital of Sichuan Province for 17 years in the early Qing Dynasty, which was moved to Chengdu in the 8th year of the reign of Emperor Shunzhi (1661).

Langzhong City was intersected and converged by Daba Mountain, Jianmen Mountain and Jialing River system, with Panlong Mountain (the main mountain) in the north and Jinping Mountain (Chao'an Mountain) in the front, surrounded by water on three sides, and the other side backed by mountains. With the geographic features of "fronting rivers and backing mountains" and "facing south with the back to the north", the site selection and layout of the city were organically and harmoniously combined with the geomantic pattern of the natural landscape (Fig. 4).

Langzhong City was the center of Chinese astronomical research in the Han Dynasty, with a gathering of talents and prosperous cultural development. According to the *Records of the Grand Historian of China*, the *History of the Han Dynasty* and the *History of the Later Han Dynasty*, during the reign of Emperor Wu of the Han Dynasty, the famous astronomer Luo Xiahong was sent to Chang'an to work with Sima Qian on the compilation of the *Taichu Calendar* and the *Armillary Sphere.* Langzhong was also the birthplace of two famous astronomers, that is, Ren Wensun and his son Ren Wengong, and the three generations of astronomers Zhou Shu, his son Zhou Qun and his grandson Zhou Ju. During the Han and Tang dynasties, geomantic studies were popular, and many astronomers and mathematicians were also masters of geomancy. For example, Yuan Tiangang and Li Chunfeng once moved

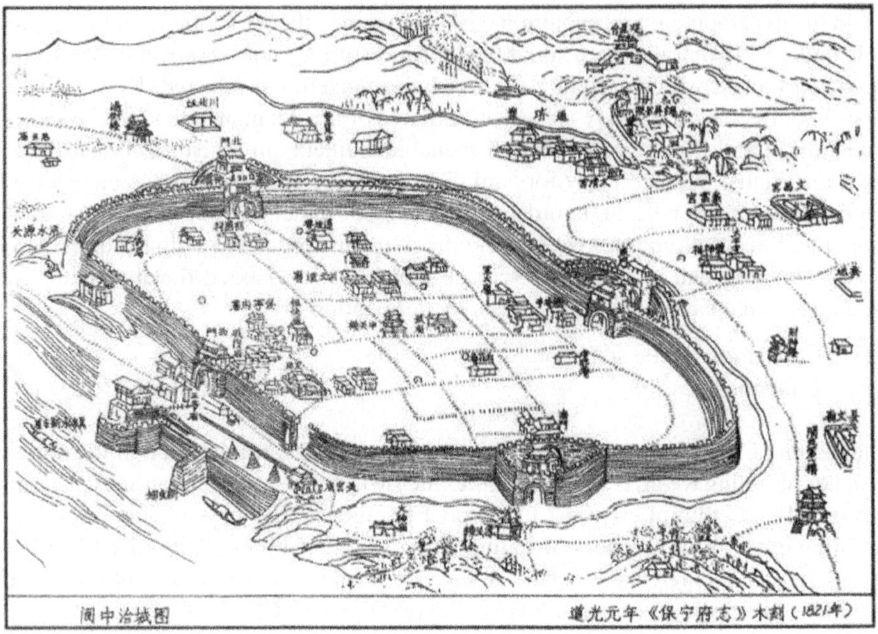

Fig. 4 Map of Langzhong City in the Qing Dynasty. *Source* Ying Jinhua, Fan Bingeng. *Famous Historic and Cultural Cities in Sichuan Province*. Chengdu: Sichuan People's Publishing House. October 2000, p. 174

to Langzhong. There were countless literati and moguls who traveled to Langzhong. There were numerous scenic spots and historic sites in and around the city, which was a holy place of religion and tourism through the ages (Fig. 5).

(3) **Wenzhou**

According to the text of the Wenzhou master plan: "The city was not established in the first year of the reign of Emperor Ming of the Eastern Jin Dynasty, which was first planned to build in Jiangbei. Guo Putook the soil and weighed it, which was light, then crossed the river and climbed onto a mountain in the northwest (now Guogong Mountain), when he saw that many peaks were staggered and shaped like the Big Dipper, with Huagai Mountain as its mouth. He told the villagers that if the city was built outside the mountain, it might gather wealth and prosperity, but there would be wars and floods. If the city was beside the mountain, enemies would dare not to invade, and the city could maintain long-term peace and stability. The city was nicknamed "Big Dipper City". In addition, an auspicious omen appeared in the city that a white deer crossed the city with flowers in its mouth. Wherever it went was full of birds' twitter and fragrance of flowers. Hence, the city was renamed "Deer City". When the city was built, 28 wells were dug because of 28 constellations in the sky according to their corresponding positions. The city had water symbolizing five elements, and

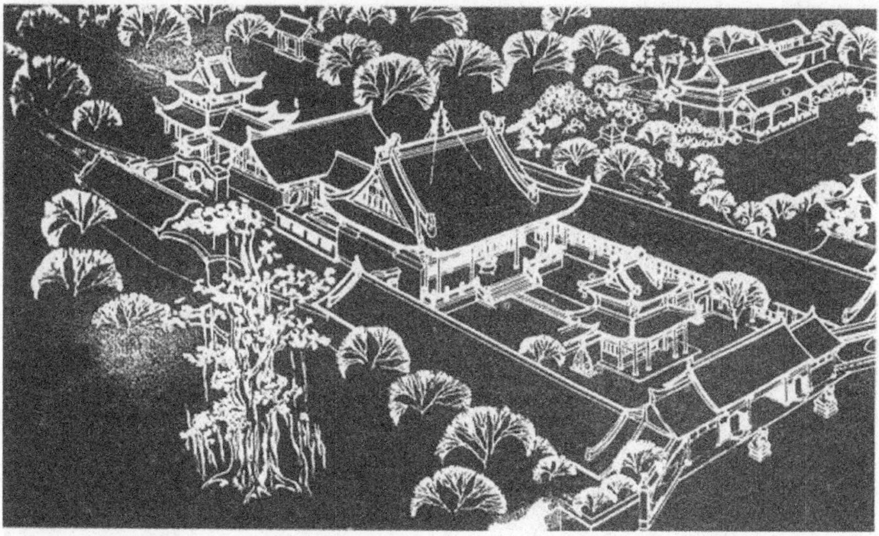

Fig. 5 Panoramic Aerial View of Zhangfei Temple. *Source* Ying Jinhua, Fan Bingeng. *Famous Historic and Cultural Cities in Sichuan Province.* Chengdu: Sichuan People's Publishing House. October 2000, p. 179

was connected with five Big Dipper mountains, namely Songtai Mountain, Guogong Mountain, Haitan Mountain, Huagai Mountain and Jigu Mountain" (Fig. 6).

The city was built across mountains, with brick and stone walls, 7 li (a Chinese unit of length, 1 li = 0.31 m) long from east to west, 5 li wide from north to south, and 18 li in circumference. The city wall was 11.67 m high, 4 m wide and 9923 m long, covering an area of about 3.8 km^2. It had seven gates: Zhenhai Gate (east gate) in the east, Rui'an Gate (grand south gate) and Yongning Gate (small south gate) in the south, Laifu Gate (triangle gate) in the southwest and Wangjiang Gate (Shuo Gate) in the northwest. A semicircular wall was additionally built outside the gates respectively, so the city was also known as "Moon City". The city was surrounded by water on all sides, with Huichang Lake and Ou River as the natural wall in the north and the south, trenches excavated manually in the east and the west, and a moat in the west. The city had sufficient water supply with Fugui Lake in the east and the west, which was connected with excavated rivers and canals, forming a criss-cross layout of water systems and entering the Ou River at the Fengen Water Gate. Lane, canal and bridge in the city formed a water and land transportation network, just like an ancient poet that goes "Looking down from a platform, you can see many boats, and there are also many small bridges in the waterways". The mountains and sea were integrated, showing picturesque scenery of a coastal city with forests like landscape paintings (Fig. 7).

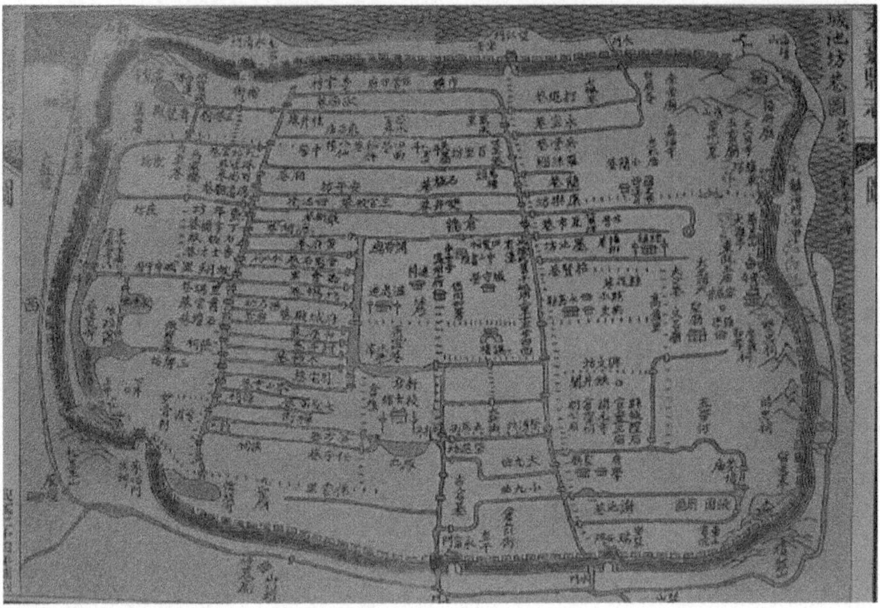

Fig. 6 Map of Streets and Lanes in Wenzhou in the 8th Year of the reign of Emperor Guangxu in the Qing Dynasty. *Source* Wenzhou Municipal People's Government, *Atlas of Master Plan of Wenzhou City*. May 1998, p. 10

(4) **Xuyong**

Xuyong County, located in the mountain area on the southern edge of Sichuan Basin and at the junction of Sichuan, Yunnan and Guizhou provinces, is the first batch of historical and cultural cities announced in Sichuan Province, where there is an old saying of "a cock awakening three provinces".

Xu Yong County, formerly known as Yongning, belonged to Yelang State from the Warring States Period to the early Han Dynasty. In the 6th year of the reign of Emperor Wu in the Han Dynasty, Qianwei County was established, which Xuyong belonged to. After the Ming Dynasty, Xuyong became an important gateway to the border among the three provinces and a military hub in southern Sichuan. The ancient city was located on the upper reaches of Yongning River, and the river flew through the city from the south to the north, forming the east and west cities. The east city, commonly known as "Soil City", belonged to the administrative region of Xuanfusi in Yongning; the west city, commonly known as "Stone City", was under the jurisdiction of Sichuan Province. Both the east and the west cities were built in the Hongwu 5th year of the reign of Emperor Zhu Yuanzhang in the Ming Dynasty (1372). According to the records of *Xuyong County Chronicle*, the two cities had seven towers, nine gates and a water gate, with a perimeter of 1,470 and 2,200 m (later expanded to 3,520 m), respectively, between which there were two bridges

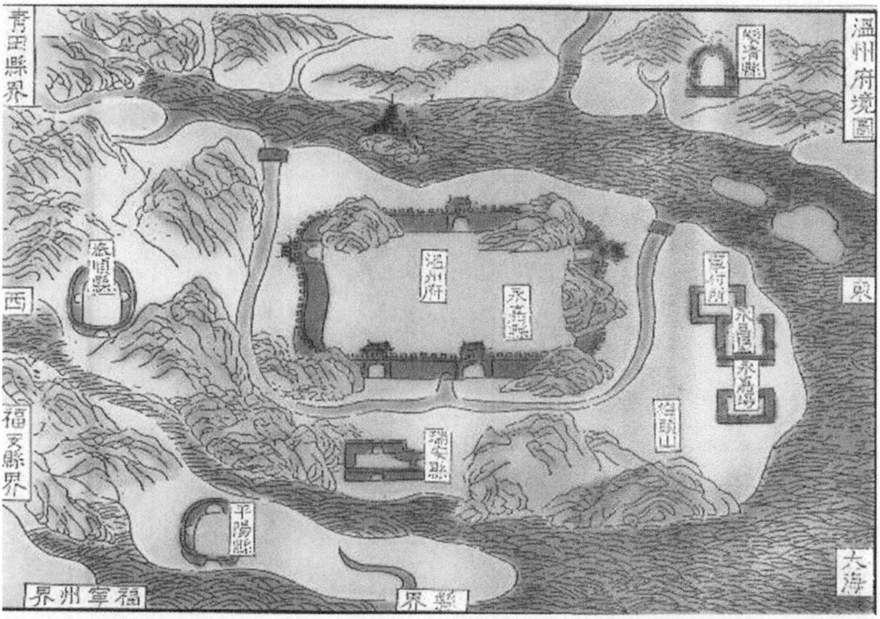

Fig. 7 Map of Wenzhou Prefecture in the 33rd year of the reign of Emperor Wanli in the Ming Dynasty. *Source* Wenzhou Municipal People's Government, *Atlas of Master Plan of Wenzhou City*. May 1998, p. 11

on the Yongning River (Fig. 8). In the ancient city, there were Shizi Mountain in the southeast, Qing Mountain in the northwest, Yongning River in the south, Chishui River in the southeast and Yufu Pass in the east, with the river flowing through Luzhou into the Yangtze River in the northeast. There are three ancient roads from Xuyong to Yunnan and Guizhou provinces: one to Qujing, Yunnan, one to Weixin, Yunnan, and the other through Moni and Chishui to Bijie, a major city in Guizhou Province. "Yufu Pass" is 10-li away from Yuyong, which were written by Yang Shen, a famous poet in the Ming Dynasty and inscribed with the couplet that "In the ruling of Han and ethnic minorities for thousands of miles, Yufu Pass is the most important one in Guizhou and Sichuan provinces".

Xueshan Pass, 75 km away from Yufu Pass, is an impregnable ancient pass near the Chishui River, with the main peak of 1,800 m above sea level, overlooking Qian Mountain in the south and the Yangtze River in the north. Xueshan Pass is endowed with precipitous terrain and magnificent weather. As a poet by Zhu Wannian, a successful candidate in the highest imperial examination, goes, "Yufu Pass is grand, the mountain road is like a snake circling up to the sky, the green mountains both sides of the bank are magnificent, and the blue water is like a long rainbow". In addition, there is the Spring and Autumn Temple, a national cultural relic protection unit in the city, which is a wood-structured palace-style building (i.e. the Shaanxi Guild Hall)

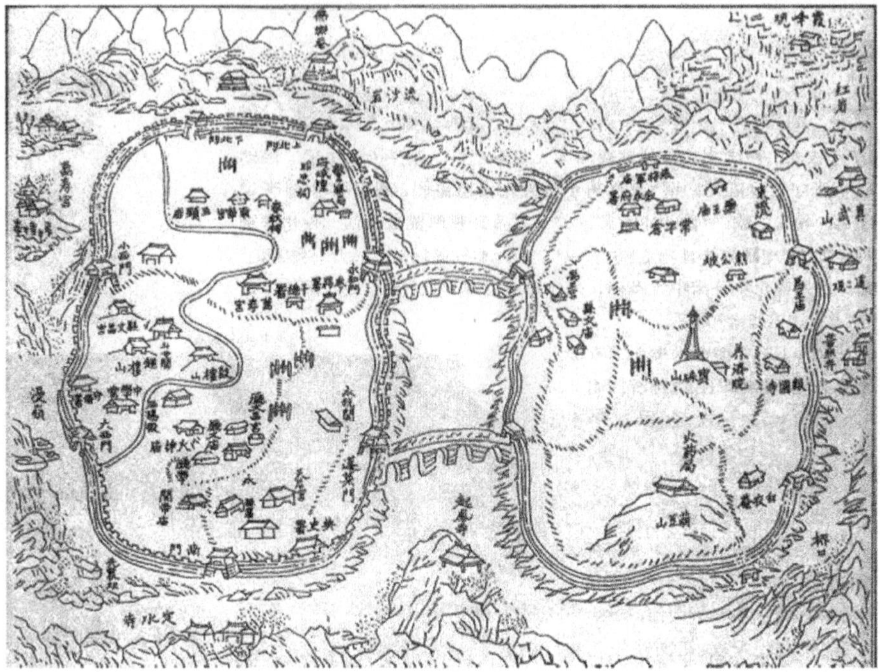

Fig. 8 Map of "Two Cities" of Xuyong in the Qing Dynasty. *Source* Ying Jinhua, Fan Bingeng. *Famous Historic and Cultural Cities in Sichuan Province*. Chengdu: Sichuan People's Publishing House. October 2000, p. 424

built during the Ming and Qing dynasties. This was named the Spring and Autumn Temple because Guan Yu liked to read the *Spring and Autumn Annals* (Fig. 9). The temple, which covers an area of 4,000 m^2, with a construction area of 2,600 m^2, has extremely fine, dainty and exquisite architectural carvings. According to the *China Attractions Dictionary*, "The Spring and Autumn Temple has completely preserved the essence of sculpture, painting and architectural art in the late Qing Dynasty". The corridor railings were carved with eight famous scenes of Xuyong: morning glow in the Temple of Longevity, moon reflected in the river between Penglai Bridge and Yonghe Bridge, ferry at dusk on Tielu Benchland, sound of morning bell in the Dingshui Temple, snow glistening on red rocks in the Red Rock Dam, connected clouds and mist on the Manling Slope, spring scenery on Baozhu Mountain, and picturesque scenery of roads and river along the Liusha Rock, which truly reflect the essence of scenic spots in Xuyong and the style and features of an ancient city in the Ming and Qing dynasties in the most exquisite way.

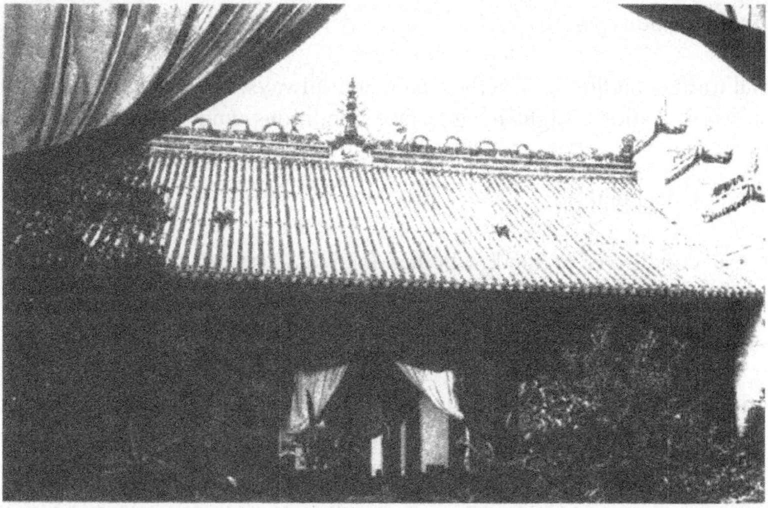

Fig. 9 Spring and Autumn Temple in Xuyong

2 General Consideration for Site Selection

(1) Regional conditions

The geographical location, longitude and latitude of a city site and its relationship with the surrounding regional environment;

(2) Climatic conditions

Climatic conditions such as sunlight, temperature, humidity and wind direction;

(3) Location

The specific location and altitude of a city site;

(4) Land

The size of land area, the situation of the east, south, west and north, gradient, aspect, slope direction, slope length, river and lake surface, landscape changes and development space, etc.

(5) Hydropower conditions

Conditions of water resources, surface water, groundwater, water quality, water reserves, lowest and highest flood levels, mountain torrents and flood inundation, conditions of electricity consumption and degree of satisfaction, etc.

(6) Traffic conditions

Regional traffic conditions, whether there are railways, highways and expressways, distance from stations, highway interface conditions, and water and air transport conditions, etc.

(7) Geological conditions

Geological conditions of land, whether there is adverse geological distribution, bearing capacity of foundation, etc.

(8) Animal and plant conditions

Vegetation conditions, growth, biodiversity, etc.

(9) Building materials and labor

The characteristics, grade and market supply of local building materials; labor market, technical conditions, etc.

(10) Cultural conditions

Cultural diversity, historical and cultural relics, education of local residents, cultural level, ethnic composition, customs, etc.

(11) Sanitary conditions

Local health conditions, whether there has been a local epidemic disease, etc.

3 Case Study of Modern Mountainurban Site Selection

The new Yunyang County is a fully flooded city of the Three Gorges Project. As the original county was located on an ancient landslide, with complex geology and difficult migration to un-flooded area, the new site was selected in Shuangjiang Town, 30 km away from the old county in the upper reaches of the Yangtze River. The new county, 38 km away from Wanzhou City, is located at the confluence of the Yangtze River and and the Pengxi River, with a land area of more than 20 km^2. The county is mainly located on the apricus south and east slopes, with stable geological structure and changing topographic relief. There is a "dragon vein" on the middle ridge, which extends to the confluence of the two rivers. On the top of the ridge, there is a provincial cultural relic protection unit named Mopan Village, which is a historical site of resisting the invasion of the Yuan Dynasty in the Southern Song Dynasty. Surrounded by mountains and luxuriant trees, the selected site of the new county is full of artistic expression, making it an ideal land for new county construction (Fig. 10).

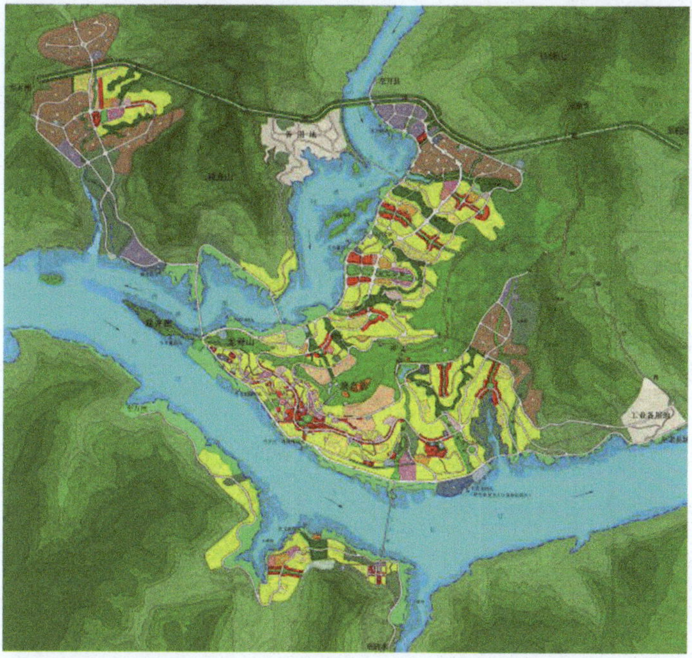

Fig. 10 Landscape Pattern of New Yunyang County. *Source* Text of Master Plan of Yunyang County (Moderator: Huang Guangyu). 2005

References

Guangyu H (2002) Mountainurbanology. China Architecture & Building Press, Beijing

Jianjun C (1992) Geomancy and architecture. Jiangxi Science and Technology Press

Jinhua Y, Bingeng F (2000) Famous historic and cultural cities in Sichuan Province. Sichuan People's Publishing House, Chengdu

Qiheng W (2003) Research of geomantictheory. Tianjin University Press, Tianjin

Chapter 4
Typology

China's urban planning and construction over the past 5,000 years has a long historical and cultural tradition. Since its beginning in the Xia Dynasty and the development in the Shang and Zhou dynasties, a relatively complete urban planning system had been formed, and experienced 2,000 years of development in the feudal society, which accumulated extremely rich experience in urban planning and construction, formed a unique and complete urban structure system, with a variety of types reflected in mountain cities (Guangyu 2002).

1 Mountain Cities with Different Landforms

(1) Hilly mountain cities

Such mountain cities are the most widely distributed in the vast areas of China at the first-, second-, and third geomorphic steps from above 50 m to below 3,000 m above sea level.

(1) **Shallow hilly mountain cities in the plain region**

Such mountain cities are generally distributed in the hilly and coastal areas in the plain region at the middle and lower reaches of the Yangtze River and the Yellow River at the third step in China, such as Hong Kong Special Administrative Region, Macau Special Administrative Region, Zhuhai City, Guangzhou City, Nanjing City, Hangzhou City, Wenzhou City, Zhoushan City, Qingdao City, Dalian City, Kaohsiung City, Tainan City, Yantai City, Penglai City, etc. These mountain cities have gentle topographic relief, superior natural conditions, few adverse geological disasters, relatively low difficulty in development and construction, a long history of construction, and advanced economic and cultural development (Fig. 1).

© The Author(s), under exclusive license to Springer Nature Singapore Pte Ltd. 2021

G. Huang, *Theory of Mountainurbanology*,
https://doi.org/10.1007/978-981-16-0819-3_4

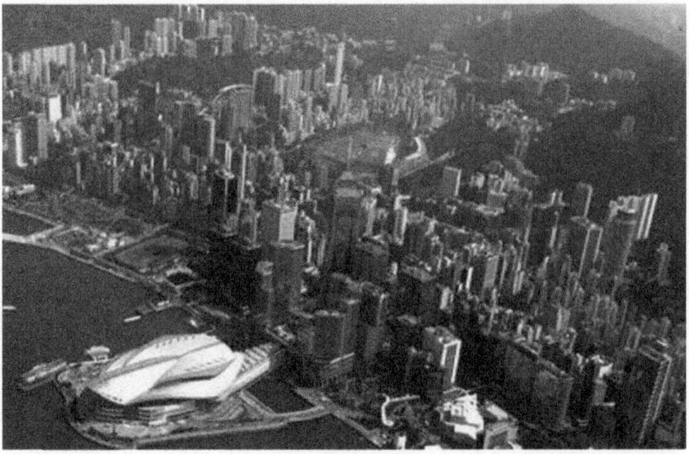

Fig. 1 Hong Kong. *Source* Website of Hong Kong Tourism Board http://www.discoverhongkong.com

(2) **Shallow hilly mountain cities in the basin region**

Such mountain cities are generally distributed in the Sichuan Basin at the second step in China below 2,000 m above sea level, such as Chongqing Municipality, Wanzhou City, Fuling City, Mianyang City, Yibin City, Luzhou City, Zigong City, Ya'an City, Xichang City, Leshan City, etc. These mountain cities have superior water and land transportation conditions, a long history of construction, large topographic relief, and many adverse geological disasters, making development and construction relatively difficult (Guangyu and Yong 2002) (Fig. 2).

(3) **Plateau hilly mountain citiesPlateau hilly mountain cities**

Such mountain cities are generally distributed in the mountain areas at the first step in China with an altitude of 2,000 to 4,000 m (Figs. 3 and 4).

(2) River valley mountain cities

Such mountain cities are mainly distributed in the valley areas of the Yangtze River and the middle and upper reaches of the Yellow River, such as Lhasa, Panzhihua, Lanzhou, Tianshui, etc. With the convenience of water conservancy, these cities have convenient transportation. The urban land is generally distributed along one or both sides of the river (Figs. 5 and 6).

(3) Ravine mountain cities

Such mountain cities are generally distributed in the Loess Plateau region, such as Yulin City, Yan'an City and Baoji City in the northwest Loess Plateau region (Figs. 7 and 8).

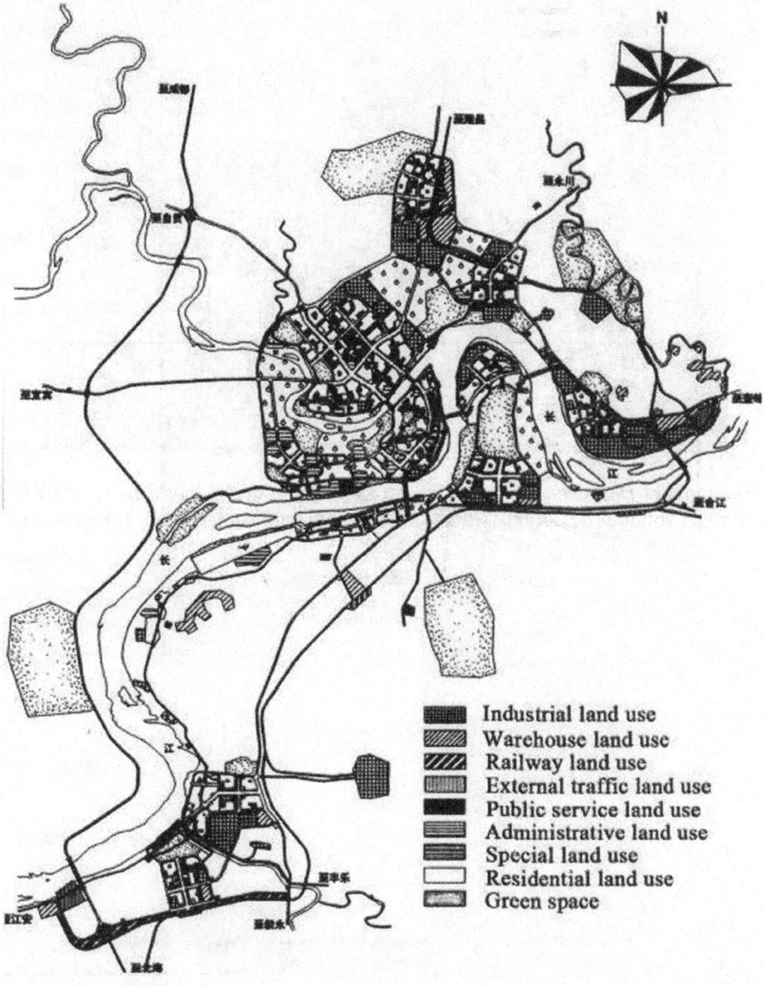

Fig. 2 Schematic Diagram of Urban Layout of Luzhou City. *Source* Ying Jinhua, Fan Bingeng. *Famous Historic and Cultural Cities in Sichuan Province*. Chengdu: Sichuan People's Publishing House. October 2000, p. 105

2 Mountain Cities with Different Functions

(1) Resources development-oriented mountain cities

There are a variety of mineral resources in mountain areas, in which mountain cities that arise from the exploitation of mineral resources occupies a large proportion, such as Zigong, Panzhihua, Gejiu, Jingdezhen, Tongling, Dongchuan, Benxi, Fushun, Baise, Gujiao, etc.

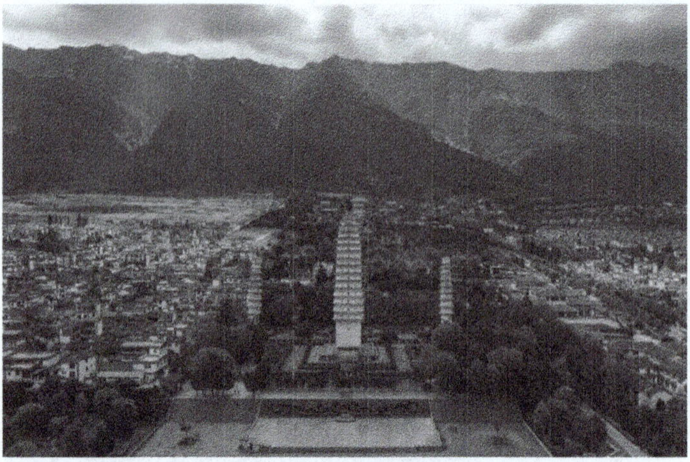

Fig. 3 Panorama of Dali City. *Source* Editorial Department of *Atlas of Cities of China. Atlas of Cities of China* (Volume II), Beijing: China Cartographic Publishing House, June 1994, p. 298

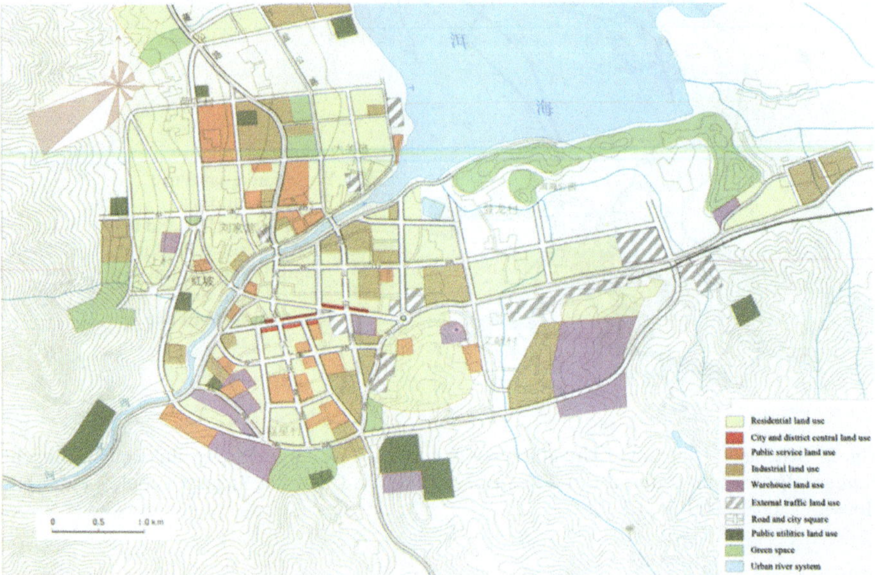

Fig. 4 Schematic Diagram of Urban Layout of Dali City in the 1990s. *Source* Editorial Department of *Atlas of Cities of China. Atlas of Cities of China* (Volume II), Beijing: China Cartographic Publishing House, June 1994, p. 299

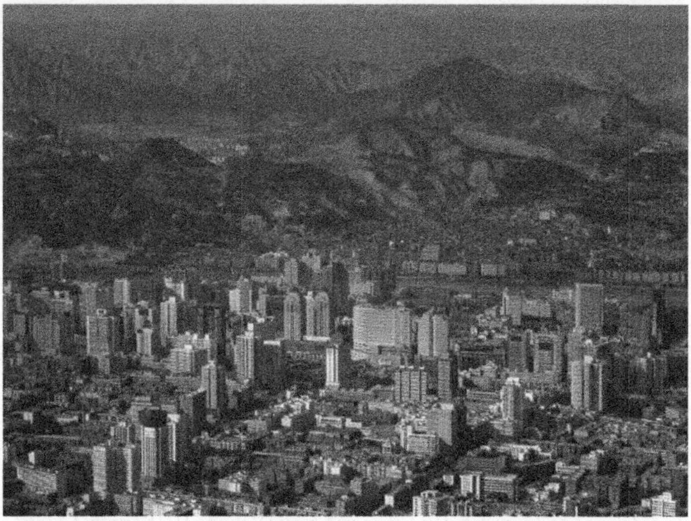

Fig. 5 Panorama of Lanzhou City. *Source* Lanzhou Tourism Website http://www.huanghetour.com/d

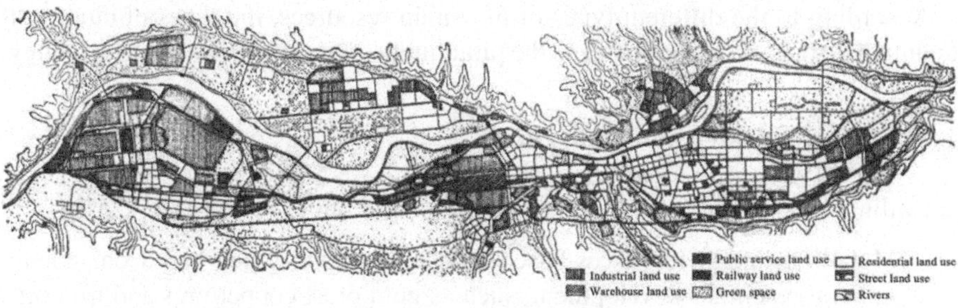

Fig. 6 Schematic Diagram of Urban Layout of Lanzhou City. *Source* Duan Jin. *Urban Spatial Development Theory*. Nanjing: Jiangsu Science & Technology Press, August 1999, p. 6

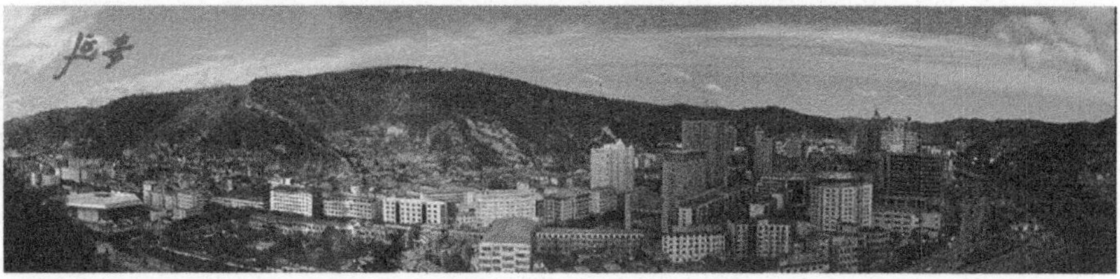

Fig. 7 Panorama of Yan'an City. *Source* Editorial Department of *Atlas of Cities of China*. *Atlas of Cities of China* (Volume II), Beijing: China Cartographic Publishing House, June 1994, p. 322

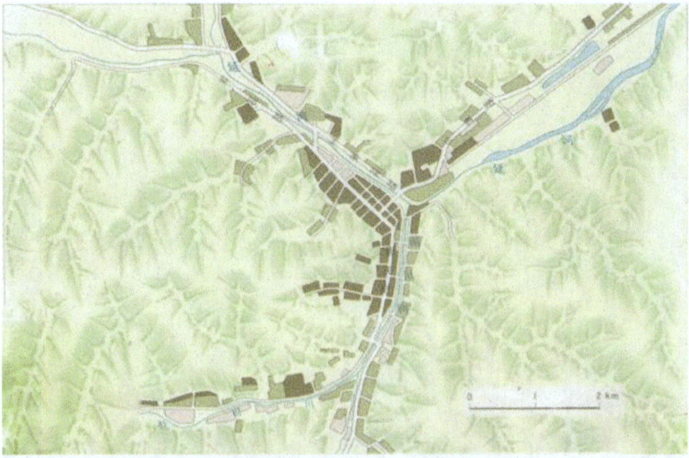

Fig. 8 Schematic Diagram of Urban Layout of Yan'an City in the 1990s. *Source* Editorial Department of *Atlas of Cities of China. Atlas of Cities of China* (Volume II), Beijing: China Cartographic Publishing House, June 1994, p. 323

According to the different types of mountain resources, the types of cities with mountain resources development as the pillar industry can be divided into (Guangyu 2003):

(1) **Mining-industry mountain cities**

According to mineral type, the cities can be divided into:

① Metallic mineral resources-based cities: Mountain cities based on metallic mineral resources development, such as gold ores, copper ores and iron ores, including mountain cities built for the development of copper mine resources such as Dongchuan City and Tongling City, and those built for the development of vanadium and titanium resources such as Panzhihua City (Figs. 9 and 10).

② Non-metallic mineral resources-based cities: Mountain cities built for the development of coal mines, salt mines, clay mines, and limestone mines, such as Huainan City for coal, Zigong City for salt, Jingdezhen City for porcelain, and Zibo City for pottery, etc. The layout and structure of these cities have different characteristics due to their different characteristics of distribution of mineral resources. However, their common feature is that the settlements or cities are distributed in a decentralized manner with the distribution of minerals, such as Zigong City, the salt capital, and Jingdezhen City, the capital of porcelain (Fig. 11).

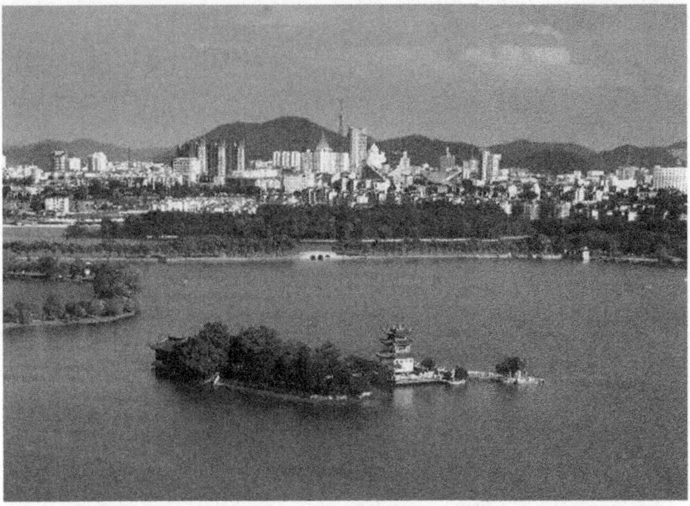

Fig. 9 Panorama of Tongling City. *Source* Editorial Department of *Atlas of Cities of China. Atlas of Cities of China* (Volume II), Beijing: China Cartographic Publishing House, June 1994, p. 344

(2) **Energy-based cities**

With the development of hydropower resources in China, the hydropower hub cities with the development of hydropower resources as the leading industry have developed rapidly, such as Sanmenxia City, Yichang City, Yibin City, Danjiangkou City, Pengshui County, etc., (Fig. 12).

(3) **Forest industrial cities**

Industrial cities mainly engaged in wood processing, such as Yichun City of Jilin Province (Figs. 13 and 14).

The common characteristics of mountain cities based on resources development are as follows:

① A city built with the development of non-renewable mineral resources as the dominant industry has different stages of urban development, including formation, development, maturity, aging and transformation, along with the development of mineral resources throughout the city's whole development process. The scale and construction speed of urban development and the layout structure of urban planning have a great dependence on the reserve enrichment and distribution of mineral resources.

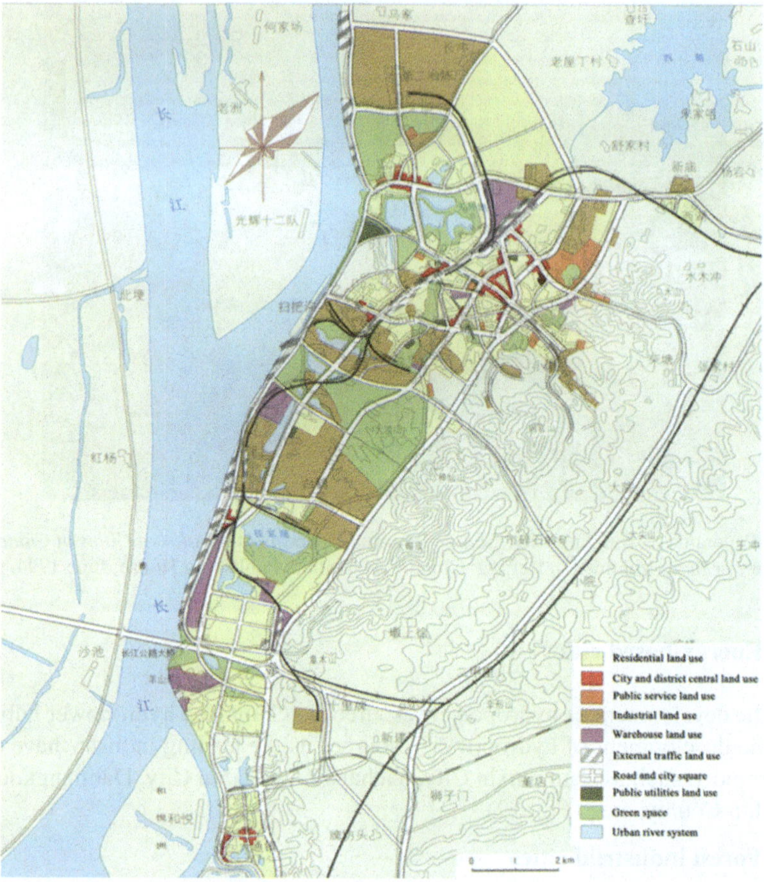

Fig. 10 Schematic Diagram of Urban Layout of Tongling City in the 1990s. *Source* Editorial Department of *Atlas of Cities of China*. *Atlas of Cities of China* (Volume II), Beijing: China Cartographic Publishing House, June 1994, p. 345

② Due to the high demand for water, electricity and transportation in the mining and processing of minerals, there are also different requirements for site selection, land selection, planning layout, and transportation mode selection and organization. Since the mining and processing of minerals generally cause great pollution to the environment, it is necessary to strengthen the recycling and comprehensive utilization of "wastes" in the production process, including waste residues, tailings, waste gas and wastewater, and develop the recycling economy to make the "wastes" emitted in the ecological process important raw material resources. At the same time, attention should be given to the reasonable proportion of the primary, secondary and tertiary industrial structures, and the occupational and

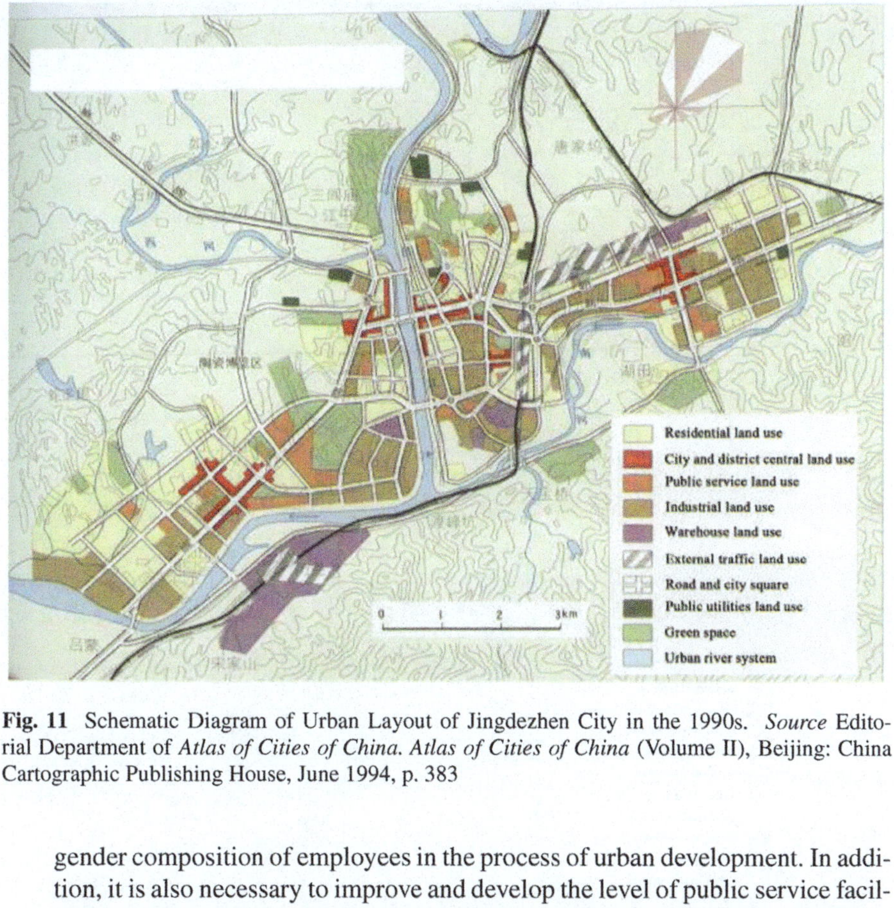

Fig. 11 Schematic Diagram of Urban Layout of Jingdezhen City in the 1990s. *Source* Editorial Department of *Atlas of Cities of China. Atlas of Cities of China* (Volume II), Beijing: China Cartographic Publishing House, June 1994, p. 383

gender composition of employees in the process of urban development. In addition, it is also necessary to improve and develop the level of public service facilities in the city and increase its comprehensive service functions; plan ahead and consider industrial transformation in a timely manner before urban development enters the aging stage, in order to ensure the sustainable urban development.

It is worth noting that with the development of economic globalization and modern high-tech and information technology, some resource-based cities has greatly weakened their dependence on the raw materials of mineral resources in terms of site selection and layout, turned to the sites close to the cities of product consumption, and arranged their industries and sites in the large-scale port areas with dense technical information, good cooperation conditions and superior transportation conditions. For example, the Shanghai-Baoshan Iron and Steel Complex in China utilizes the superior technical information and port and railway transportation conditions in Shanghai, and connects large joint enterprises to solve raw material supply by long-distance transportation of ores from Australia and other countries, so as to improve scale effect and economic benefits.

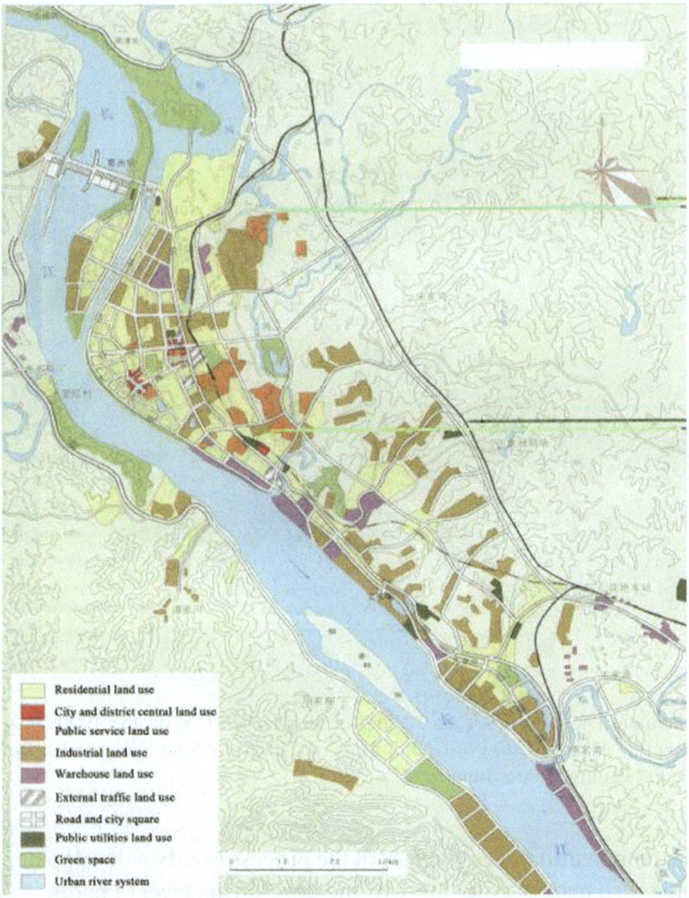

Fig. 12 Schematic Diagram of Urban Layout of Yichang City in the 1990s. *Source* Editorial Department of *Atlas of Cities of China. Atlas of Cities of China* (Volume II), Beijing: China Cartographic Publishing House, June 1994, p. 108

(2) Regional transportation hub mountain cities

Such mountain cities form the mountain transportation hub in the development process of regional transportation, such as Zhuzhou City, Huaihua City, Yingtan City, etc., all of which have risen with the development of railway transportation (Fig. 15).

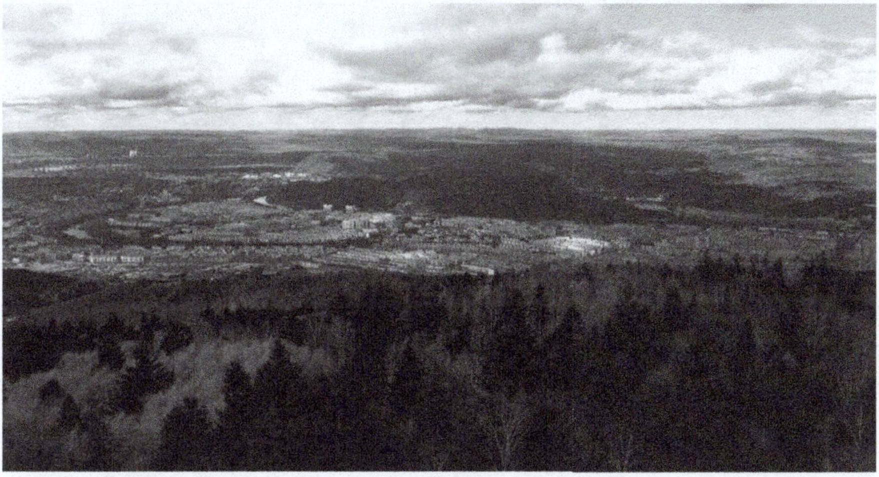

Fig. 13 Landscape Map of Yichun City. *Source* Website of Department of Housing and Urban-Rural Development of Heilongjiang Province. Institute of Urban Planning & Design Survey of Heilongjiang Province. 2002

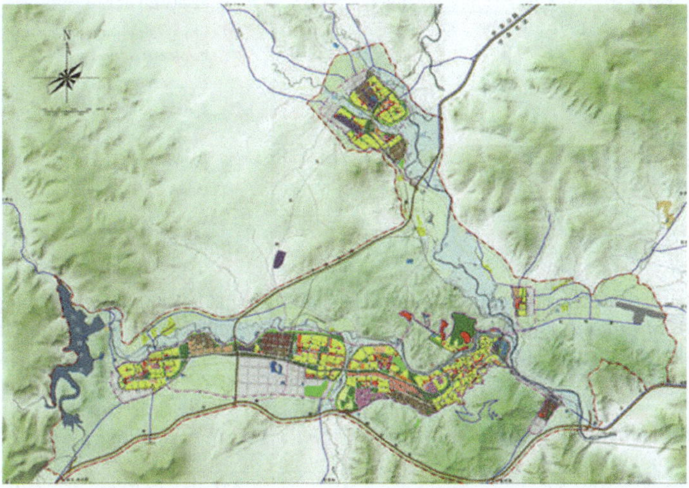

Fig. 14 Schematic Diagram of Urban Layout of Yichun City. *Source* Website of Department of Housing and Urban-Rural Development of Heilongjiang Province. Institute of Urban Planning & Design Survey of Heilongjiang Province. 2002

(3) Port mountain cities

Such mountain cities were built to meet the needs of frontier defense construction and port trade development, such as Pingxiang City, Fangchenggang City, Suifenhe City, Manzhouli City, Dandong City, etc., (Figs. 16 and 17).

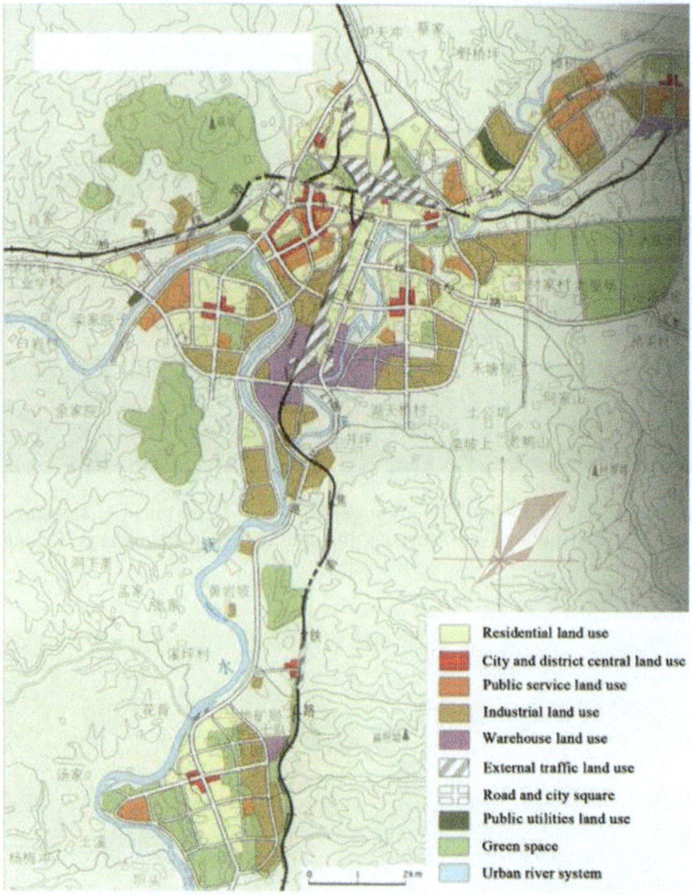

Fig. 15 Schematic Diagram of Urban Layout of Huaihua City in the 1990s. *Source* Editorial Department of *Atlas of Cities of China*. *Atlas of Cities of China* (Volume II), Beijing: China Cartographic Publishing House, June 1994, p. 156

(4) Memorial mountain cities

Such mountain cities have emerged from the memorial sites of important history and culture, such as Zunyi City, Yan'an City, Jinggangshan City, Xibaipo City, Hancheng City, etc., (Figs. 18 and 19).

(5) Mountain cities of scenic tourism or religious worship

The cities and tourism service bases have emerged due to the development of scenic tourism resources or religious worship, such as Lhasa City, Leshan City, Emei City, Jizu Mountain, Beidaihe City, Shanhaiguan City, Chengde City, Sanya City, Dujiangyan City, Huangshan City, Zhangjiajie City, Tai'an City, etc., (Figs. 20 and 21).

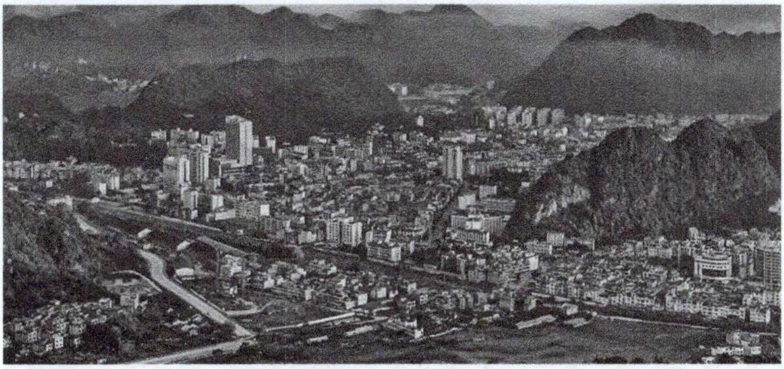

Fig. 16 Landscape Map of Pingxiang City. *Source* Editorial Department of *Atlas of Cities of China*. *Atlas of Cities of China* (Volume II), Beijing: China Cartographic Publishing House, June 1994, p. 226

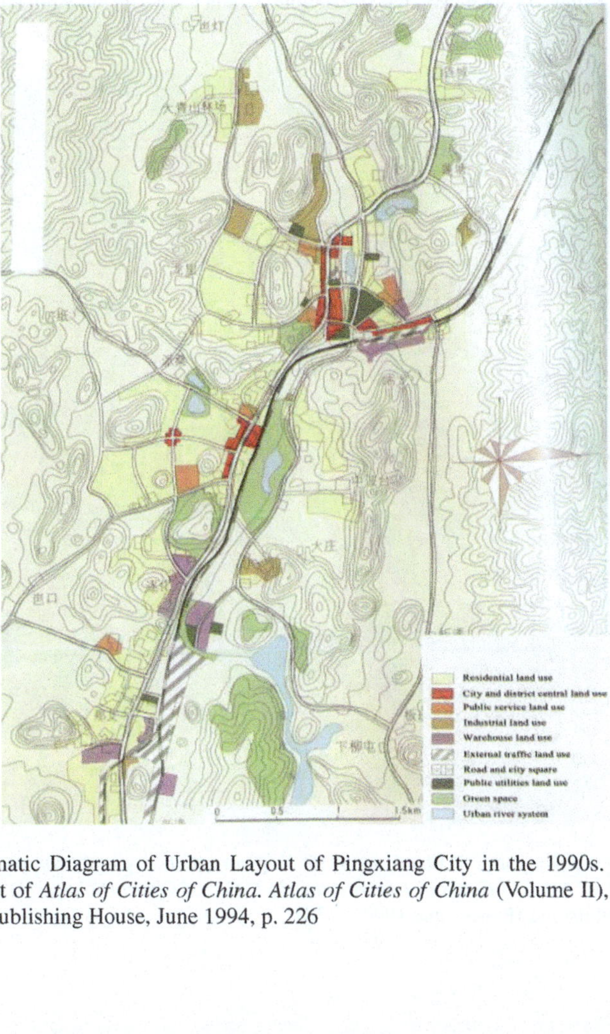

Fig. 17 Schematic Diagram of Urban Layout of Pingxiang City in the 1990s. *Source* Editorial Department of *Atlas of Cities of China*. *Atlas of Cities of China* (Volume II), Beijing: China Cartographic Publishing House, June 1994, p. 226

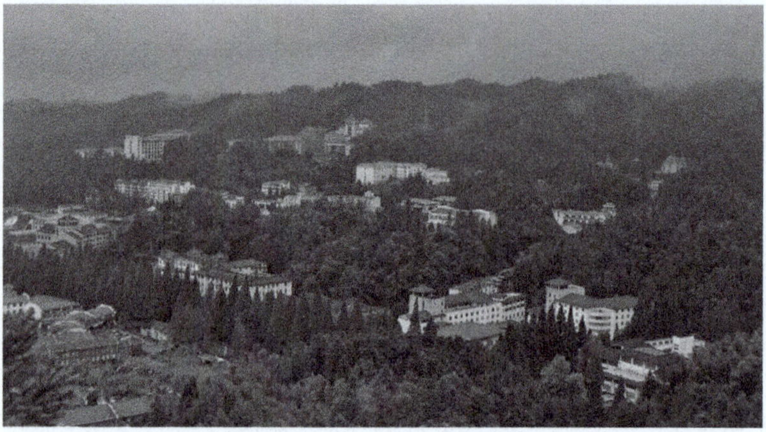

Fig. 18 Landscape Map of Jinggangshan City. *Source* Editorial Department of *Atlas of Cities of China. Atlas of Cities of China* (Volume II), Beijing: China Cartographic Publishing House, June 1994, p. 381

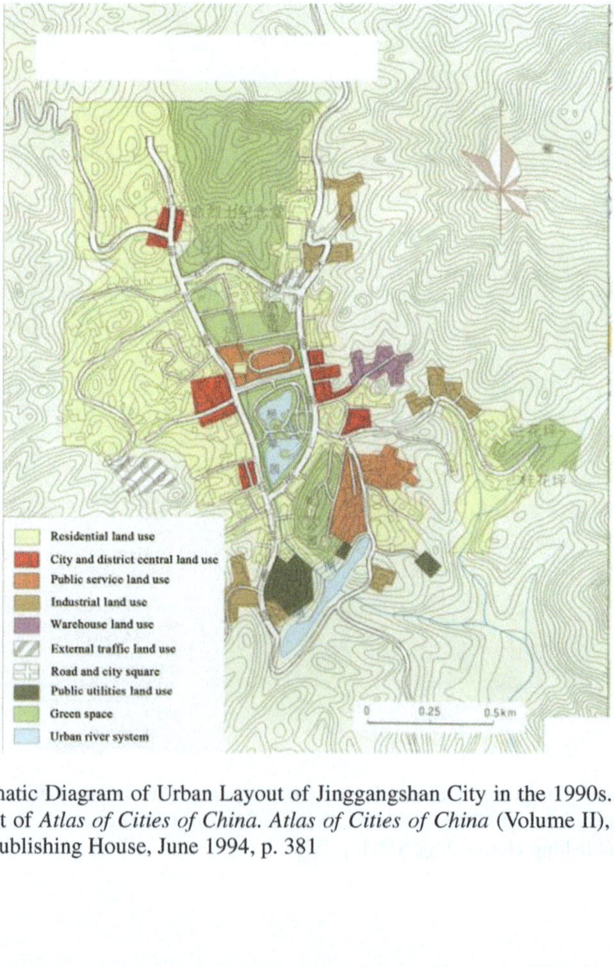

Fig. 19 Schematic Diagram of Urban Layout of Jinggangshan City in the 1990s. *Source* Editorial Department of *Atlas of Cities of China. Atlas of Cities of China* (Volume II), Beijing: China Cartographic Publishing House, June 1994, p. 381

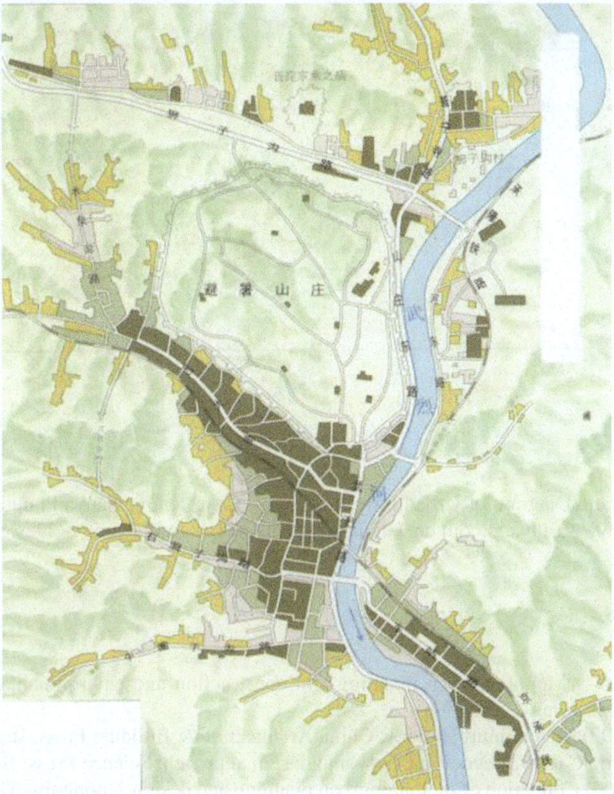

Fig. 20 Schematic Diagram of Urban Layout of Chengde City in the 1990s. *Source* Editorial Department of *Atlas of Cities of China. Atlas of Cities of China* (Volume II), Beijing: China Cartographic Publishing House, June 1994, p. 40

(6) Comprehensive mountain cities

Such mountain cities are provincial capital centers focusing on administrative functions, such as Chongqing Municipality, Kunming City, Guiyang City, Nanjing City and Hangzhou City, which often have the functions of economy, culture, transportation and information centers. Furthermore, the above six different types of mountain cities are mostly based on one function, with some other functions as well, such as religious and memorial cities, which generally have the functions of scenic tourism or recuperation (Guangyu 1994).

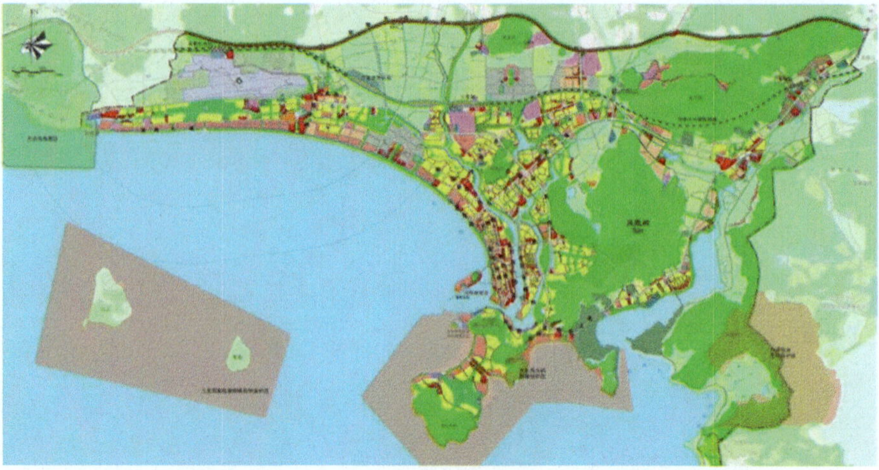

Fig. 21 Schematic Diagram of Urban Layout of Sanya City. *Source* Sanya Planning Information Network http://www.sanyaup.com.cn, China Academy of Urban Planning & Design 1999

References

Guangyu H (1994) Mountainurban planning and construction and environmental development. Science Press, Beijing
Guangyu H (2002) Mountainurbanology. China Architecture & Building Press, Beijing
Guangyu H, Yong C (2002) Ecocity: theory and design approach. Science Press, Beijing
Guangyu H (2003) Collection of mountainurban planning and design. Chongqing University Press, Chongqing

Chapter 5
Planning

1 General Principle of Mountainurban Planning

(1) Principle of safety

The principle of safety is the most foundmental requirement for the planning and construction of mountain cities. Safety is reflected in all aspects of mountainurban planning, such as land selection, land use, overall layout, industrial and residential organizations and municipal engineering facilities like road traffic system, as well as urban disaster prevention and control. Since the 1970s, with the deepening of the global environmental crisis, more and more attention has been given to the environmental security problems caused by the eco-environment of human society, which has drawn great concerns.

(2) Principle of nature protection

Learning from nature, as the essence of traditional Chinese culture, is not only an important concept of traditional Chinese Confucianism, Taoism and Buddhist, but also a major principle of traditional Chinese aesthetics, architecture, landscape architecture, calligraphy, painting, medicine, and political and agro-economic activities. Nature is the mother and mentor of human beings who are born and live in nature. Everything about human beings, including human beings themselves, is endowed by nature. The traditional Chinese philosophy is the integration of celestial bodies and human bodies, just like an old saying that goes "Heaven and earth is a great universe, and man is a small one". Lao Zi, a famous philosopher in the Spring and Autumn Period, considered that the origin of the universe is "nature" which gives birth to the universe, and the universe gives birth to all things. Therefore, there is the concept of "Man regards the earth as the law, the earth regards the heaven as the law, the heaven respects the operation law of all things on earth, and the operation law respects the law

© The Author(s), under exclusive license to Springer Nature Singapore Pte Ltd. 2021
G. Huang, *Theory of Mountainurbanology*,
https://doi.org/10.1007/978-981-16-0819-3_5

of nature" in the *Tao Te Ching*. In the *Book of Changes*, it is recorded that "Everything has the two sides of Yin and Yang, both of which interact and form a new harmony". Xuncius, a famous philosopher in the Warring States Period, said, "The law of nature is permanent, which does not exist because of Yao (a benevolent monarch), nor does it disappear because of Jie (a brutal monarch). It will be auspicious to adapt to it with measures that lead to stability, and dangerous to adapt to it with measures that result in chaos". All this emphasizes the respect for nature and the harmonious coexistence of man and nature. The planning and construction of mountain cities should be based on the basic principle of respecting nature, coordinating the relationship between man and nature, stepping out of the traditional values of "anthropocentrism", and promoting the transformation of traditional behaviors and values towards the ecological culture of environmental friendliness, resource conservation, and social harmony (Huang 2002).

(3) Principle of system

The weather (climate), earth and human (society) in mountain cities form a complex large system that is mutually conditional and reciprocally determined, which is far more systematic and interdependent than that in plain cities. From the perspective of ecology, the complex ecosystem composed of society, economy and nature in mountain cities has formed a complex relationship between the complex structures and functions and the flow and transformation of substances, energy and information. Therefore, in urban planning and construction, it is necessary to consider all kinds of complex social, economic and natural factors and ecological and environmental security factors in a more comprehensive manner (Richard 2003).

(4) Principle of economy

The economic principle is not only the economy of input and output, but also the economy of environment, space and time.

The land resources that mountain cities can provide for urban construction are particularly valuable, which also requires a lot of investment in people, money and materials for construction. In order to achieve greater benefits with less investment and resource consumption, it is necessary to implement comprehensive measures such as compact layout, intensive use of land, energy conservation, investment saving, reduction of engineering and environmental costs to meet economic requirements. For example, most of the traditional mountain settlements in China are developed in the form of "high density", which is not only due to the needs of economic, social and cultural exchanges, but also an inevitable result of people's coordination with the mountain natural environment in the long-term production and life. As the land resources in mountain areas are limited, only through concentrated and compact settlements can the limited land resources be used reasonably, and can cooperation and exchanges be conducted effectively to save time and improve efficiency. From the investigation of the mountainurban distribution of settlements, it can be concluded that people try to make use of relatively barren hillside land to build villages and towns, and fertile soils for farming and planting, so as to achieve the purpose of "It is necessary to adapt to the changes of the weather in the four seasons, and to exert the potential of all things as much as possible on the earth, and use human power in the country, so that everything can grow smoothly and the food crops can be propagated" (Jia Sixie, Northern Wei Dynasty). Just as Guo Xi, a famous painter in the Song Dynasty, said in his works the *Elegance of the Bamboo and Spring·Theory of Traditional Chinese Landscape Painting*, "A village is backed by flat ground, not by mountains. Backing on flat ground is to facilitate farming, not relying on mountains because it is too far for farming. However, some villages are backed by mountains, because there must be a piece of land for cultivation between the mountains" (Guo Xi, Song Dynasty).

(5) Principle of cultural inheritance and diversity

China's diverse natural landscapes nourish the big family of the Chinese nation and foster profound regional and multi-ethnic cultures, such as Wuyue culture, Jingchu culture, Bayu culture in the Yangtze River Basin, and Qilu culture, Central Plains culture, plateau culture and grassland nomadic culture in the Yellow River Basin. These colorful cultures present distinct regional and national characteristics in urban and rural human settlements.

Mountain areas and cities are distributed in the vast land of the country, which inevitably have diverse cultural and biological gene banks. In the long-term interaction between man and nature, all ethnic groups in mountain settlements have created a vibrant local culture and ecological living environment that meet the requirements of sustainable development, which require us to conscientiously summarize, inherit and carry forward.

The principle of diversity contains a variety of meanings, including not only biodiversity, but also the diversity of urban society and culture. Biodiversity includes species diversity, genetic diversity, and ecosystem diversity; the diversity of urban society and culture includes the diversity of urban population composition and industrial structure, the diversity of architectural culture and the diversity of landscape culture. Mountain areas are often multi-ethnic clusters, such as Lhasa City, Xining City, Kunming City, Guiyang City, Dali City, Lijiang City, etc., where multi-ethnic cluster cultures, different religious beliefs, customs and habits constitute a colorful mountainurban cultural landscape.

(6) Principle of development

In urban planning and construction, we should not only consider the needs of present urban construction, but also plan for future development. Therefore, in mountainurban planning, spatial and temporal planning should be integrated from the perspective of development and dynamic changes, so as to leave room for development in the planning.

2 Mountainurban Spatial Structure

(1) Basic development patterns of mountainurban spatial structure

The basic development patterns of mountainurban spatial structure include cluster development pattern, tandem development pattern, green-core ring development pattern and network development pattern.

The urban spatial structure is the material carrier of urban social and economic development. A good urban spatial layout structure is bound to create favorable conditions for the city's social and economic development and the work, life and physical and mental health of its citizens; otherwise, it will act as a hindrance. Scientific and reasonable planning and organization of urban spatial structure can not only create a safe and harmonious living environment, but also save land and save energy, in order to obtain the maximum economic benefits. Therefore, it is the most strategic work in urban planning and construction practice.

In the current development process of mountain areas and cities in China, the contradiction between man and land is particularly prominent. Therefore, the core issue of the mountainurban spatial layout structure is how to correctly handle the relationship between concentration and decentralization. Concentration is a manifestation of urban civilization and efficiency, and decentralization is the basic feature of mountain areas and natural environment, which are a pair of contradictory unities (Huang 2002). The organic combination of the two is an important guarantee for coordinating the contradiction between man and land in mountain areas and cities and building a sustainable city in the process of urbanization development.

(2) Types of spatial structure of mountain cities

Due to the differences in natural conditions of mountain areas and cities such as geographical location, altitude, terrain slope, climate, precipitation and sunshine, the mountainurban spatial structure also has diversified development patterns.

(1) **Centralized compact structure**

Under the condition of small urban scale and relatively complete construction land, it is an ideal choice to adopt a concentrated and compact spatial structure pattern, which can effectively organize urban production and life, save urban construction land, and reduce urban construction investment and operation costs. Especially in the early stage of urban development, cities tend to develop in a compact and concentrated manner due to the relatively small urban scale and less prominent contradiction between man and land.

(2) **Cluster structure**

It is commonly found in hilly areas or mountain valleys with undulating terrain, mountains and rivers. The undulating urban land is divided by mountains, rivers, gullies, and valleys, such as Yibin City, Luzhou City and Mianyang City in Sichuan

Province, Guiyang City and Zunyi City in Guizhou Province, Nanping City in Fujian Province, Lijiang City and Dali City in Yunnan Province, Yan'an City, Ankang City and Lueyang City in Shaanxi Province, Jiangjin City and Fuling City in Chongqing Municipality, Zhuhai City and Huizhou City in Guangdong Province, and Zhangjiajie City and Zixing City in Hunan Province.

① **Single-center cluster structure**

It is suitable for small and medium-sized cities in hilly and mountain areas with a small urban scale, such as Mianyang City, Yibin City, Zigong City and Neijiang City in Sichuan Province in the middle of the last century.

Yibin City is located in the southern margin of Sichuan Basin, the hilly area of southern Sichuan and the confluence of Jinsha River, Minjiang River and Yangtze River. It is the beginning of the Yangtze River, which is known as the "First City of the Yangtze River", and is also a famous historical and cultural city in China, which was called "ancient Bo state" in the Spring and Autumn Period and the Warring States Period. Based on the natural topography of Yibin and the present development foundation, the master plan of Yibin City undertaken by the research team from 1981 to 1982 determined that the Cuiping District of Yibin City should be the urban center, and four clusters, including Baishaba, Jiuzhouba, Nananba and Tianchiba, should form a district (urban center) along the Jinsha River, the Minjiang River, and the Yangtze River. The overall layout and structure development pattern of the four clusters has laid the foundation for the overall spatial development pattern of the cluster landscape city (Fig. 1). With the size of urban population of 200,000 in 1990 and 300,000 in 2000, the population of each cluster is 30,000 to 80,000. As Yibin is a regional administrative center and a water and land transportation hub in southern Sichuan, it has the potential for resource development, leaving much room for the development of the urban spatial layout structure.

② **Multicenter cluster structure**

In hilly and mountain cities with rich and varied terrain combinations, the single-center cluster structure has developed to the multicenter cluster structure due to the further development of population scale and urban economy and culture, such as Yibin City, Neijiang City, Mianyang City, Ankang City, Yan'an City, Nanping City and Benxi City at the end of last century and in the beginning of this century.

Since the 1990s, with the further improvement of the social, economic and cultural level and the further expansion of the scale, the single-center cluster structure of Yibin City has gradually developed towards the direction of the multi-center cluster structure, forming a large-scale composite urban cluster. In 1997, the master plan layout of Yibin City developed into a multi-center cluster spatial structure. The city consists of one main center, five sub-centers and nine clusters, with a population of 300,000 in the near future and 400,000 in the long term (Fig. 2).

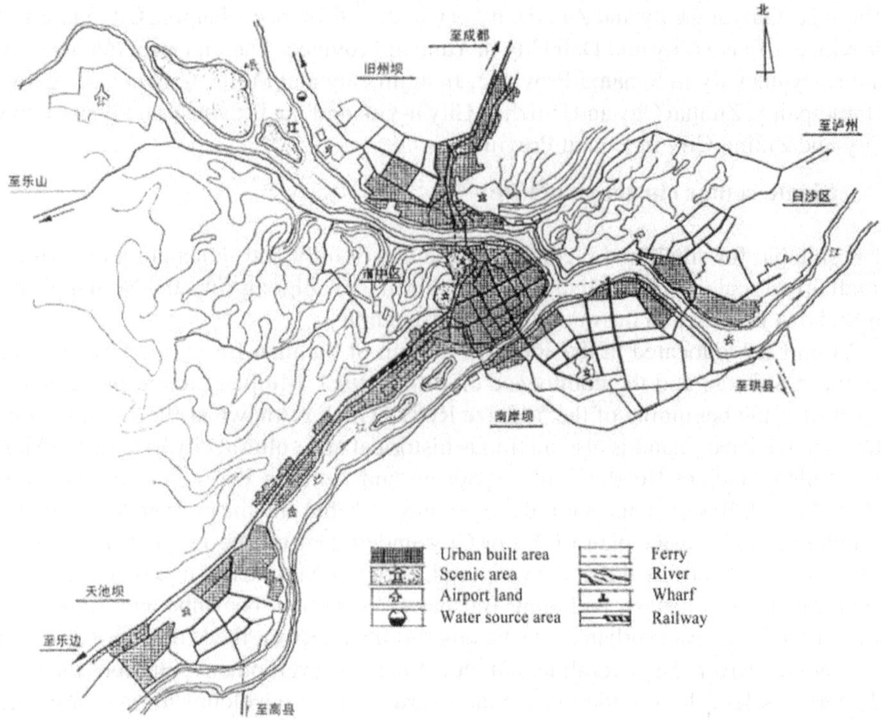

Fig. 1 Master Plan of Yibin City in 1982. *Source* Ying Jinhua, Fan Bingeng. *Famous Historic and Cultural Cities in Sichuan Province*. Chengdu: Sichuan People's Publishing House. October 2000, p. 84

(3) Band structure

① Single-center linear structure

Due to the limitation of topography or natural landform, urban land extends along hills, valleys or rivers in a zonal development. Generally, this type of cities is not large in scale and relatively simple in urban structure, such as Shizhu County and Youyang County in Chongqing Municipality, Jishou City in Hunan Province, Duyun City in Guizhou Province, Cenxi City and Pingxiang City in Guangxi Zhuang Autonomous Region.

a. Master plan and spatial structure of Shizhu Tujia Autonomous County

Shizhu Tujia Autonomous County is located on the eastern edge of Chongqing Municipality, with Lichuan County of Hubei Province in the east, Fengdu County in the west, Pengshui County in the south and Zhong County in the north. Nanbin Town is located on the second terrace of the upper reaches of Long River, a tributary of the

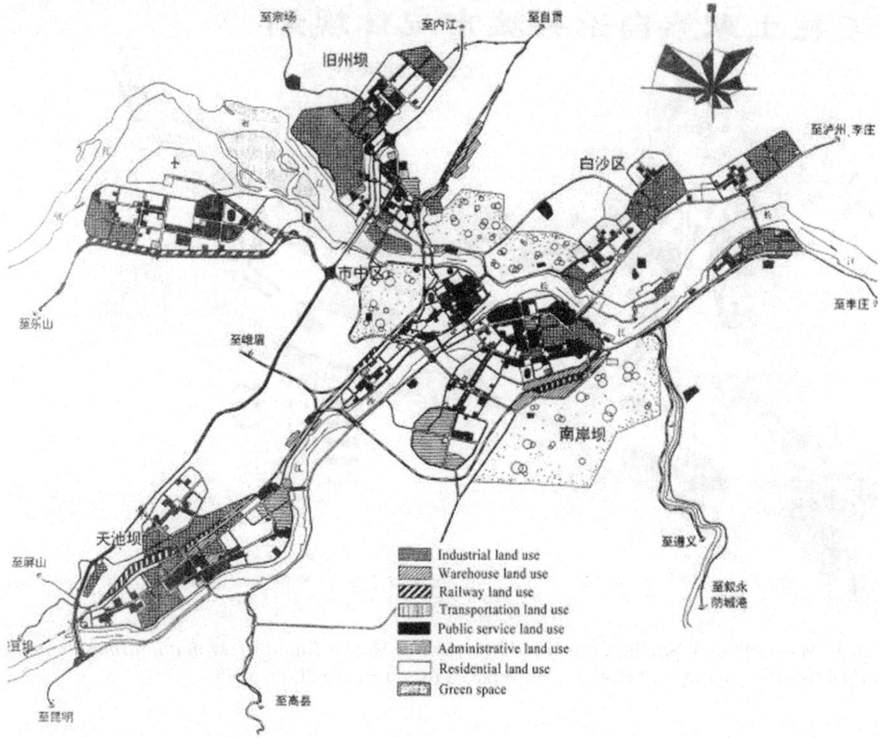

Fig. 2 Master Plan of Yibin City in 1997. *Source* Ying Jinhua, Fan Bingeng. *Famous Historic and Cultural Cities in Sichuan Province.* Chengdu: Sichuan People's Publishing House. October 2000, p. 86

Yangtze River. Shizhu County, surrounded by water on three sides and mountains on all sides and with Yudai River passing through, is a linear landscape county with excellent eco-environment and typical mountain characteristics. In 1981 and 1996, the research team successively undertook the compilation of the master plan of Shizhu County. Guided by the thought of sustainable development and ecological construction, the overall spatial pattern of a small landscape garden county suitable for the local economic development level and reflecting regional ethnic cultural characteristics was established (Fig. 3). Along the Long River, the county is in a single-center linear layout, with obvious traffic direction of urban development (Figs. 4, and 5).

b. Master plan and single-center development spatial structure of Rong County

Rong County is located in the south of Sichuan Province, with rich natural environment resources and a shallow hill landform. According to the characteristics of environmental resources, the construction and development theme of Rong County is determined in the plan as follows: Rong County will be built into a landscape city in southern Sichuan with ecological, cultural and tourist features. Based on the natural

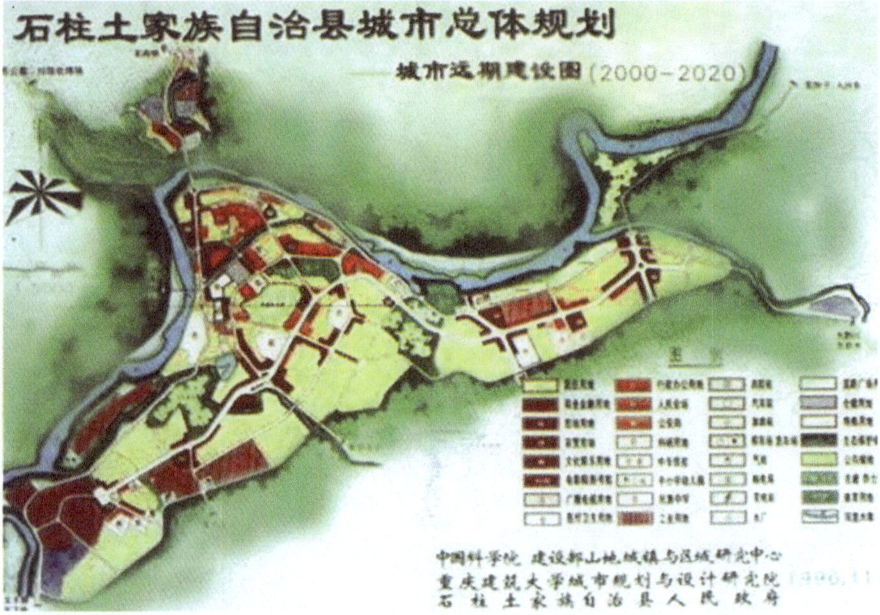

Fig. 3 Master Plan of Shizhu County in 1996. *Source* Huang Guangyu. *Mountainurban Planning and Design*. Chongqing: Chongqing University Press. January 2003, p. 95

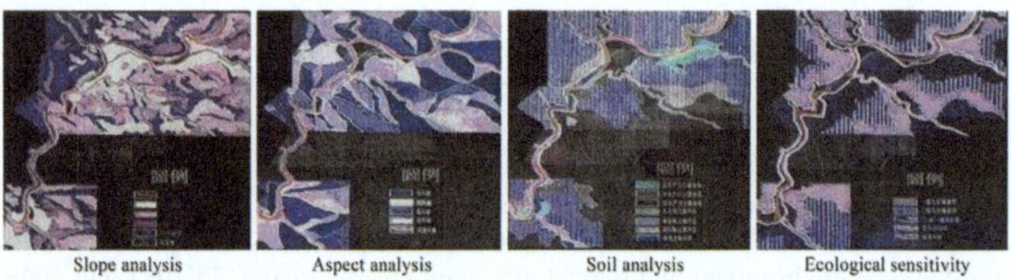

Fig. 4 Computer-generated Reverse Grayscale Map of Ecological Factors of Natural Resources in Shizhu County. *Source* Huang Guangyu. Mountainurban Planning and Design. Chongqing: Chongqing University Press. January 2003, p. 96

landscape pattern, construction status, land use conditions, economic development direction and regional environment characteristics of Rong County, a single-center urban layout structure is formed in the master plan, in which three rivers basins cross over the county that is surrounded by a green mountain, and three major areas (old county, Hexi, Hedong) are organically integrated, forming a Y-shaped green ecological corridor. The planned population of the county will reach 180,000 in 2020 (Fig. 6).

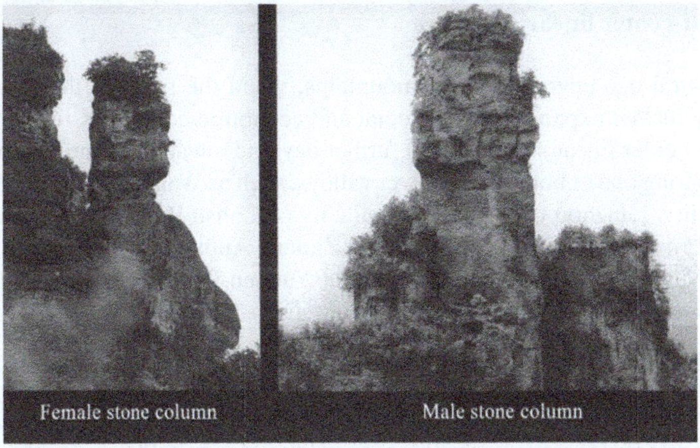

Fig. 5 Geographical Indication of Shizhu County: Man and Woman Stone Pillars. *Source* Shizhu County People's Government. *Millennium Stone Pillars Landscape Garden*. October 2004, cover page

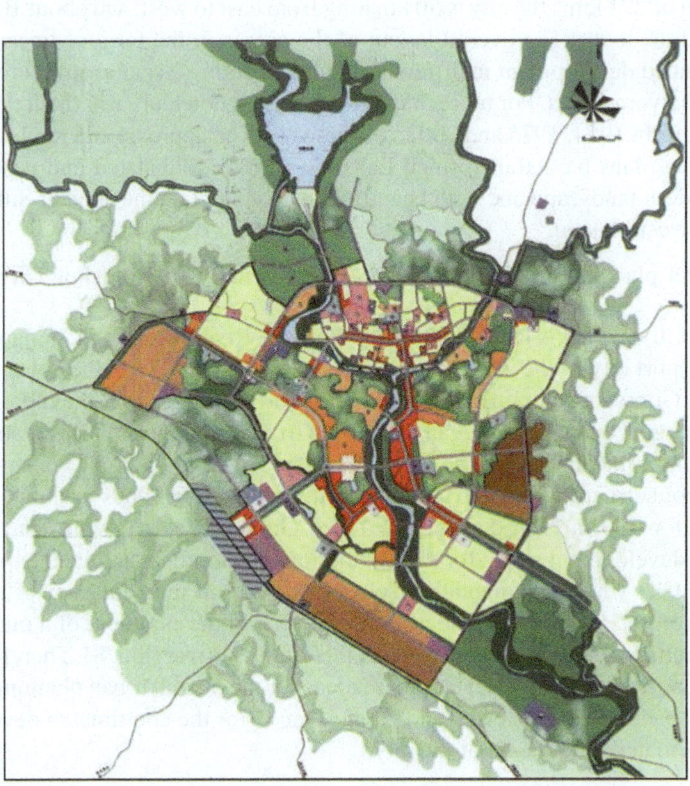

Fig. 6 Master Plan of Rong County in 2000. *Source* Text of Master Plan of Rong County (Moderator: Huang Guangyu). 2000

② **Multi-center linear structure**

In the natural eco-environment of mountains, when the scale of the single-center linear city further expands and the social and economic conditions further develop, the single-center linear structure will further develop towards the multi-center linear structure along one or both sides of river valleys, such as Wanzhou City of Chongqing Municipality, Lanzhou City and Tianshui City of Gansu Province, Guiyang City of Guizhou Province, Guilin City of Guangxi Zhuang Autonomous Region, Lhasa City of Tibet Autonomous Region, Gejiu City of Yunnan Province, Fushun City, Benxi City and Dandong City of Liaoning Province.

a. Lanzhou

Lanzhou City, the capital city of Gansu Province, is located on the border area of the Qinghai-Tibet Plateau, the Loess Plateau, and the Mengxin Plateau, with an altitude of 1,520 m and a relative height of 600 m. The Yellow River passes through the city from west to east, with the mountains from north to south facing each other. With an urban area of 221 km^2, the city is 50 km long from east to west, and about 10 km wide from north to south. The overall layout of the city is in the form of linear clusters, with balanced development in different regions, forming a spatial structure of linear clusters in river valleys. For more than 50 years, Lanzhou had carried out three urban master plans in 1954, 1975 and 2002, respectively. The approval and implementation of the three plans have transformed Lanzhou from a dilapidated and desolate city into a modern landscape one with beautiful environment on the ancient Silk Road in the Northwest Plateau.

b. Master plan and multi-center linear spatial structure of Wanzhou City

Wanzhou City, which is located along the Yangtze River in eastern Chongqing, is an important port city in the Three Gorges Reservoir Region. With the construction of the Three Gorges water conservancy project, 80% of the land in the old urban area was submerged, and the city began to develop from a single-center linear structure to a multi-center linear structure along the Yangtze River. The master plan and single-center layout structure of Wanxian City made in the early 1960s is shown in Fig. 7 (the plan for 200 m high dam of the Three Gorges Project), which has reserved sufficient development space for the population development and land expansion of Wanzhou City (the implementation scheme for 175 m medium dam of the Three Gorges Project), and laid a foundation for the spatial development of a multi-center linear structure layout along both sides of the Yangtze River (Fig. 8). Therefore, there must be long-term strategic development ideas for the overall urban planning of a city to "leave space, organize space, and create space" for the coordinated development of various material elements of the city.

(4) **Sugar gourd structure**

The sugar gourd structure is also known as tandem or beaded development structure. Due to the limitations of the mountain natural topography, the cities have multiple

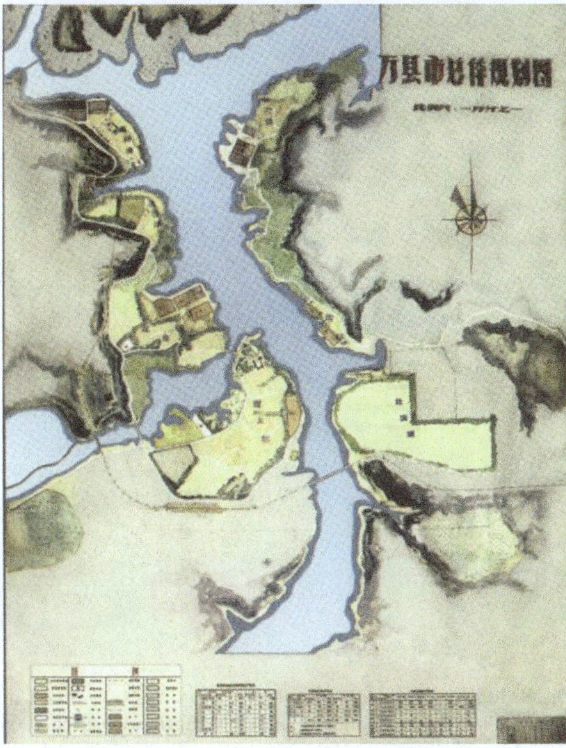

Fig. 7 Master Plan of Wan County in 1996. *Source* Huang Guangyu. *Mountainurban Planning and Design*. Chongqing: Chongqing University Press. January 2003, p. 3

clusters (or areas) along the main traffic lines and narrow and long valleys, extending longitudinally and developing in tandem, such as Neijiang City and Zigong City (Fig. 9).

Zigong City is located in the south of Sichuan Basin, in the shallow hill area along the Fuxi River, a tributary of the Tuojiang River. Thanks to the history of well salt mining, Zigong City has become a famous "millennium salt capital".

The salt brine and natural gas resources of Zigong City are scattered under the terrain of shallow hills, and the distribution structure of urban settlements and the distribution of salt wells are inextricably correlated to the topographic characteristics, just like a saying going that "The city is built for salt, boats are sailed in downstream, and people choose land and water for living". Therefore, the city forms a tandem type urban spatial structure with two centralized well clusters of the Ziliujing District and the Gongjing District as the main body.

The city is built in a gentle hilly area, with the relative height difference of generally between 30 and 50 m. The gentle hills are undulating, the narrow river surface twists and turns, and hundreds of softly contoured and lush mountains are intertwined with the city, making it in the embrace of green mountains and waters and forming a unique scenery line of a "millennium salt capital" with the towering "crown block" (brine mining derrick), which brings people a moving scene of the urban and rural

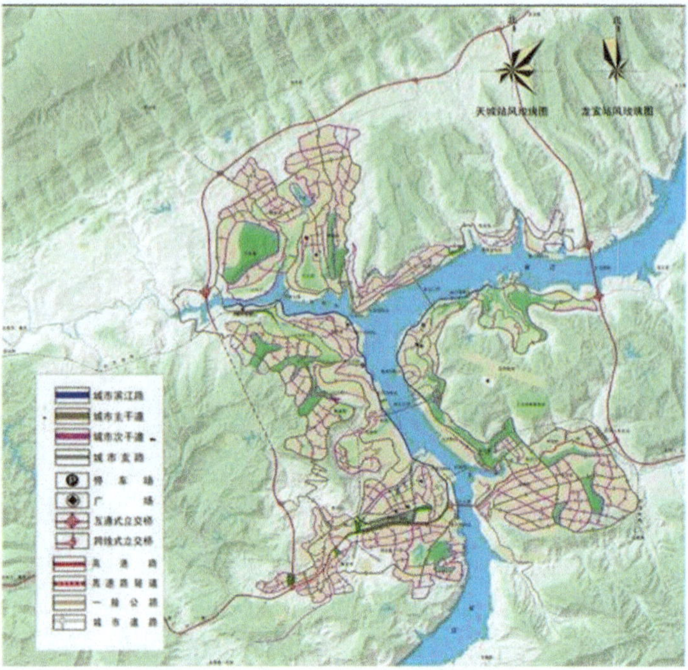

Fig. 8 Master Plan of Wanzhou District in 2003. *Source* China Academy of Urban Planning & Design, Planning & Design Institute of Wanzhou District. 2005

integration of the beautiful mountains and rivers in southern Sichuan with the production and life of the salt capital (Fig. 10). The derrick of "crown block" of Ziliu Well is shown in Fig. 11 (due to the modernization of production technology, all the "crown blocks" have been dismantled except some are reserved as historical cultural relics). The bustling scene of salt boats gathering at the riverside of Fuxi River in history is shown in Fig. 12.

(5) **Vine-with-melon structure**

Due to the combined effect of natural conditions and socio-economic conditions, the pattern of cluster urban structure or linear urban structure is likely to be further developed and mutated. For example, a city extends along the river direction or the main traffic axis, residential areas extend along mineral deposits due to the development of mining industry, or there are topographic and geomorphic constraints in complex mountains, gullies and forest vegetation between clusters, which further develops the cluster or linear structure into the "vine-with-melon" spatial structure pattern. In mountain cities, due to the complexity of natural conditions, there are many mountains, gullies, woodlands, and agricultural land interspersed in the composition of urban land. Therefore, in the labor structure of urban residents, there are often many residents engaged in agricultural and forestry production, or holiday tourism services

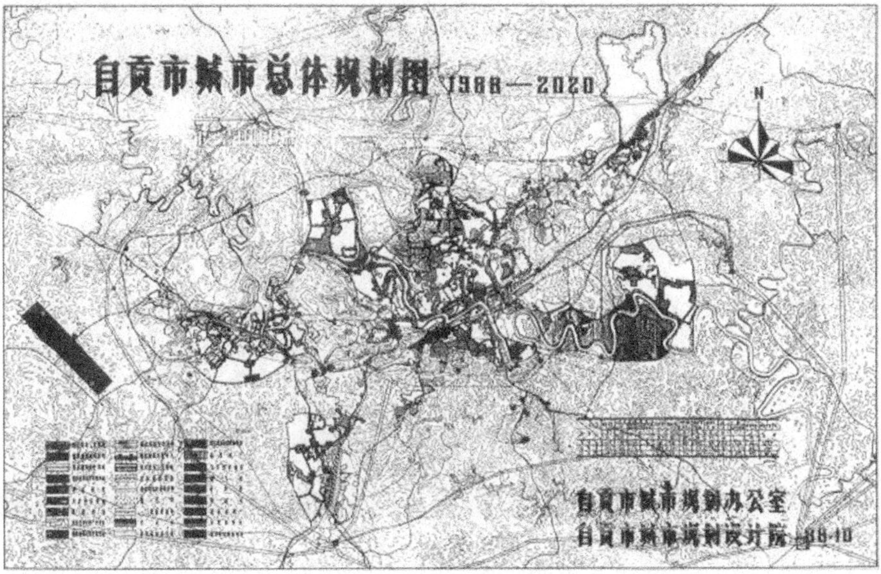

Fig. 9 Master Plan of Zigong City in 1988. *Source* Ying Jinhua, Fan Bingeng. *Famous Historic and Cultural Cities in Sichuan Province*. Chengdu: Sichuan People's Publishing House. October 2000, p. 76

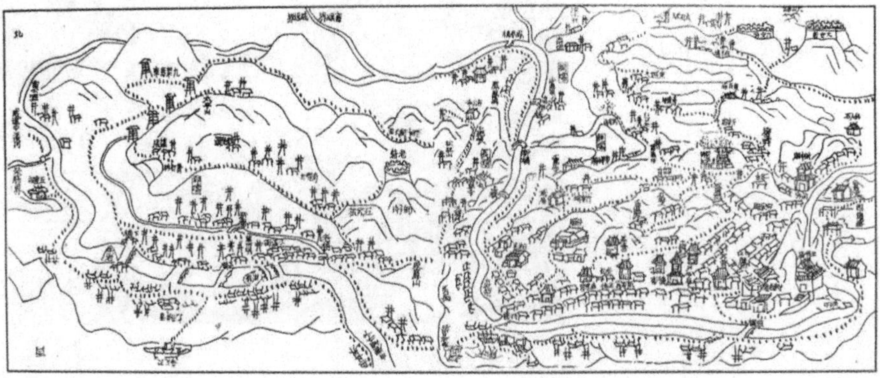

Fig. 10 Map of Ziliu Well and Gong Well in the Qing Dynasty. *Source* Ying Jinhua, Fan Bingeng. *Famous Historic and Cultural Cities in Sichuan Province*. Chengdu: Sichuan People's Publishing House. October 2000, p. 59

in addition to those doing urban work. Furthermore, there are various ways of labor and employment for residents, such as Neijiang City and Panzhihua City of Sichuan Province (Fig. 13).

Panzhihua City is located in the southwest corner of Sichuan Province, in the mountain area at the junction of Sichuan and Yunnan provinces, bordering Huili

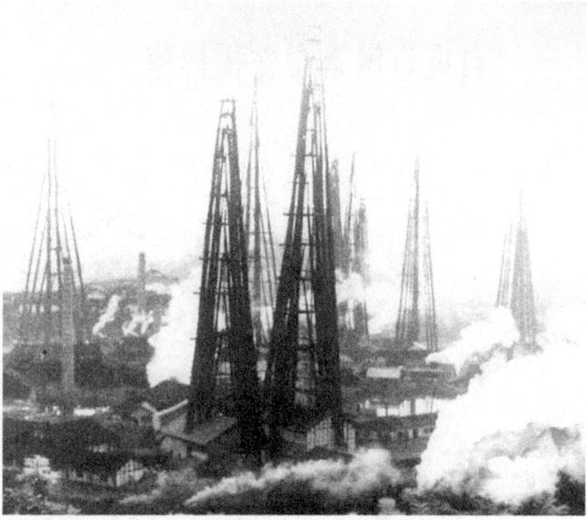

Fig. 11 "Crown Block" of Ziliu Well. *Source* Zigong Municipal People's Government. *Famous Historic and Cultural City in China-Zigong*. 2002, p. 13

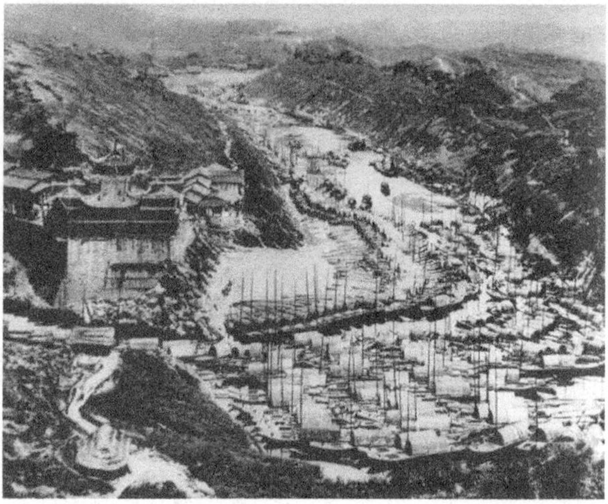

Fig. 12 Salt Boats Gathering at the Riverside of Fuxi River in History. *Source* Zigong Municipal People's Government. *Famous Historic and Cultural City in China-Zigong*. 2002, p. 17

County of Liangshan Prefecture in the east, Yongren County of Yunnan Province in the south, Huaping County of Yunnan Province in the west, and Dechang and Yanyuan counties of Liangshan Prefecture in the north. Panzhihua is the only city in China named after a flower. As the Jinsha River and Yalong River are deep and glaciated, an alpine canyon is formed, which belongs to the landform of middle mountain valley. The terrain is high in the northwest and low in the southeast, which is rugged with a great height difference. There are multi-level tablelands, piedmonts, valley peaks, stacked skirts and alluvial fans on both sides, which constitute a complex terrain and increase many difficulties in urban planning and construction. Urban

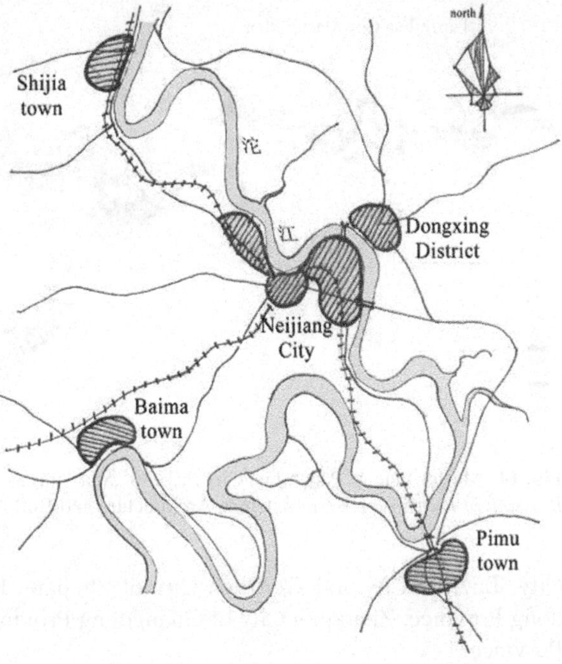

Fig. 13 Schematic Diagram of Urban Layout of Neijiang City. *Source* Drawing by the author

construction land is mainly distributed at an altitude of 1,000–1,300 m, with a slope of 10–20%, and in some areas it reaches 30%, with an average slope of about 15%.

Under the influence of many factors such as solar radiation, atmospheric circulation and geographical environment, Panzhihua City has formed a unique island-like three-dimensional climate type with the southern subtropics as the base zone, featuring abundant heat, distinct dry and rainy season, more sunshine, obvious vertical difference, and diverse microclimate. The precipitation increases and the temperature decreases with altitude, forming a typical three-dimensional mountain climate. In the revision of the urban master plan in 1997, the topography, geomorphology, natural eco-environment and the present situation of Panzhihua City were first fully investigated and analyzed, forming a distinctive "organic, loose, concentrated, zonal and balanced" subtropical river valley zone with the vine-with-melon layout structure, allowing the urban development to have strong adaptability to the regional eco-environment. After 40 years of hard work, Panzhihua City has built a landscape garden city with rich characteristics of south subtropical climate, beautiful environment, economic prosperity, and vitality (Fig. 14).

(6) **Green-core structure**

The cities form a pattern of "green-core" ecotype spatial structure around mountains, forests, wetlands, and lakes, such as Leshan City of Sichuan Province, Shaoxing

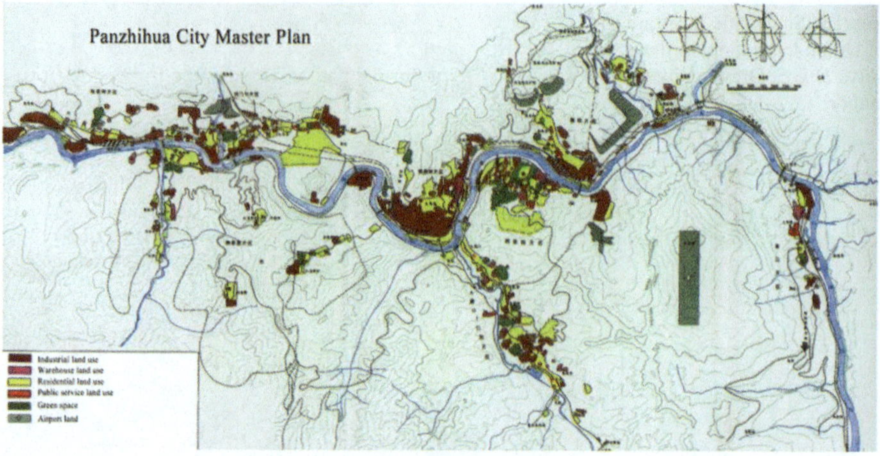

Fig. 14 Master Plan of Panzhihua City in 1986. *Source* Bao Shixing, Chen Jiayun. *Forty Years of Blossom of Panzhihua*. Beijing: China Architecture and Building Press. December 2005, p. 74

City, Taizhou City and Wenzhou City of Zhejiang Province, Yantai City of Shandong Province, Zhaoqing City of Guangdong Province, and Zunyi City of Guizhou Province.

① **Single-center green-core ring ecotype spatial structure**

It is applicable to hilly and mountain cities with rich landform and undulating terrain.

Master plan and spatial development structure of Leshan City in Sichuan Province: Leshan City, located in the southwest of Sichuan Basin, is surrounded by mountains and waters with the convergence of three rivers and unique natural environment. It has had a good reputation since ancient times that "The best scenery in the world is in Sichuan Province, and the best scenery in Sichuan Province is in Jiazhou (now Leshan)" (Figs. 15, 16 and 17). In 1987, the compilation of the master plan of Leshan City lasted more than 10 months. Through the comparative study of multiple schemes, combined with Leshan's unique natural environment and historical and cultural foundation, the city adopted the spatial structure of the multi-center complex urban cluster to form the concept of a large regional environmental circle, in which the city, landscape and nature are integrated (Fig. 18). The pattern of single-center compact green-core ring ecotype structure was adopted for the main urban area, and the city is composed of four circles, namely, green core, urban ring, water ring and mountain ring, from the inside to the outside (Fig. 19). Both the inner core and the outer periphery of the city are endowed with superior natural environmental conditions, realizing the ecological concept of "unity of man and nature".

② **Multi-center green-core ring ecotype structure**

In the new century, China's urbanization process is further accelerated, the pressure on population, resources and environment is further increased, and the competition

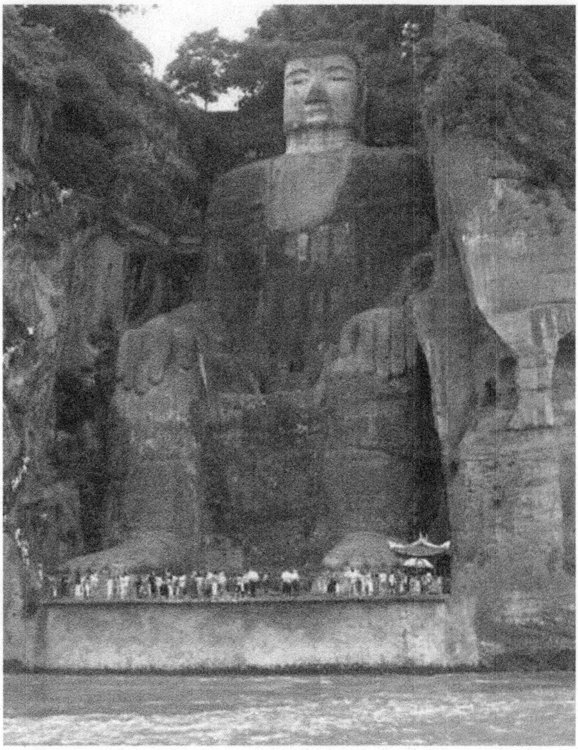

Fig. 15 Leshan Giant Buddha. *Source* Photograph by the author in 1987

between cities is further intensified, and the connection is also further strengthened. We should pay attention to the conservation of urban and regional eco-environment and sustainable future development, carry out the Scientific Outlook on Development, strive to achieve the goal of the human society to build a harmonious society, and the the pattern of green-core ring ecotype urban spatial development structure will be further emphasized.

The expanded pattern of green-core ring ecotype spatial structure is suitable for mountain and plain cities.

With the gradual expansion of urban scale, the further urban social, economic and cultural development, or the adjustment of administrative divisions, the green-core ring ecotype structure will further develop in the direction of a multi-center green-core ecotype structure, forming a relatively complex development pattern of compound metropolitan cluster. This basic pattern was adopted by Yantai City and Jining City of Shandong Province, Shaoxing City, Wenzhou City and Taizhou City of Zhejiang Province in their new rounds of urban planning. The green core can be composed of green mountains, such as Yantai City, wetlands and lakes, such as new Kaixian County under construction in Chongqing Municipality, or a large complex ecotype green core composed of several hills, wetlands and water surfaces, such as Wenzhou City, Taizhou City and Shaoxing City.

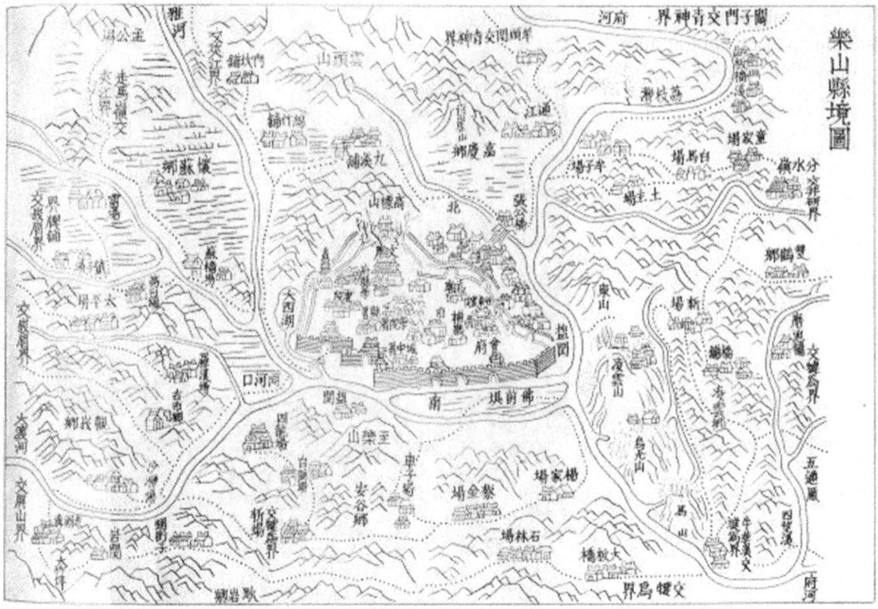

Fig. 16 Map of Leshan County in the Qing Dynasty. *Source* Huang Guangyu. Proceedings of the '97 International Symposium on Sustainable Development of Human Settlements in Mountain Areas, Beijing: Science Press. December 1997, pp. 115, 118

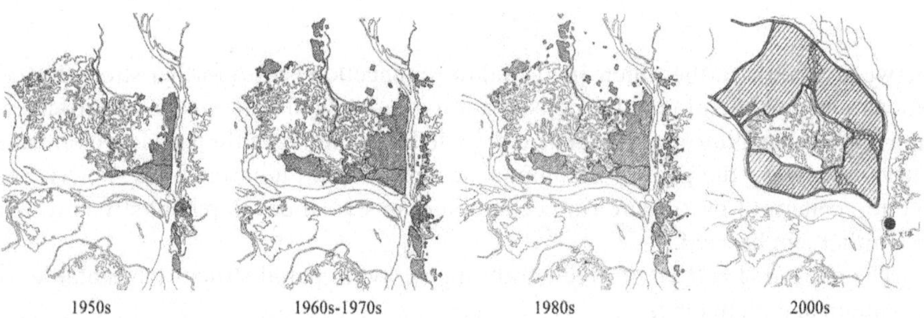

Fig. 17 Urban Space Expansion Map of Leshan City in Different Periods. *Source* Ying Jinhua, Fan Bingeng. *Famous Historic and Cultural Cities in Sichuan Province*. Chengdu: Sichuan People's Publishing House. October 2000, p. 148

The expanded Shaoxing City incorporated Keqiao District and Paojiang District into its unified planning. The overall urban spatial development strategy refers to the theory of green-core ring ecotype urban structure pattern of Leshan City, and adopts the urban structure pattern of green core plus cluster (Fig. 20), forming four clusters around natural mountains, lakes and wetland reserves in Jing Lake with an area of about 60 km^2, including Shaoxing Center cluster, Keqiao Textile Trade Center cluster,

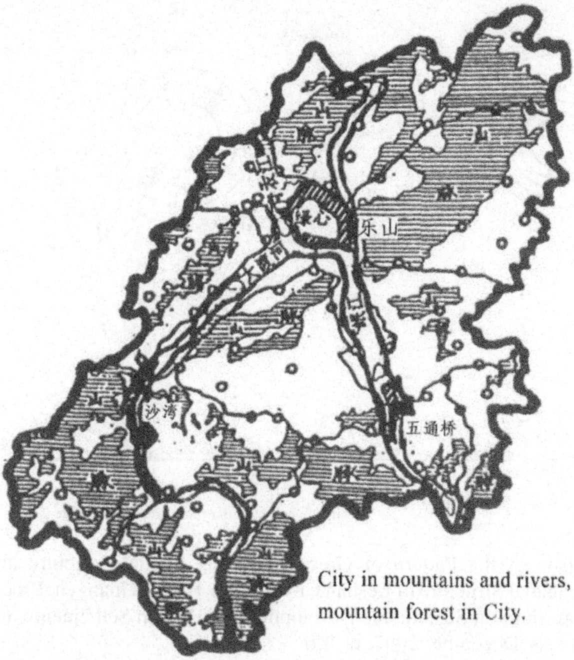

Fig. 18 Spatial Structure Map of Leshan City. *Source* Huang Guangyu. Proceedings of the '97 International Symposium on Sustainable Development of Human Settlements in Mountain Areas, Beijing: Science Press. December 1997, p. 119

Paojiang high-tech industrial cluster and Jiangbin industrial cluster. The spatial development structure of this complex urban agglomeration with a green-core ring cluster has been fully affirmed through multiple rounds of expert demonstration, which will play an important supporting role of a material spatial structure in the construction of Shaoxing's urban area, the harmonious development between man and nature, and the coordinated development among city, economy and natural evolution.

(7) **Palm structure**

This pattern was proposed mainly for the problem of high threshold and marginal cost of urban construction caused by the restriction of mountainous and geomorphic conditions in urban development: urban development can make good use of natural conditions such as water systems, mountains and gullies to divide the city into different clusters, with obvious direction of the main spatial development in urban functional areas.

① **Planning and design of New Hexi Region in Zhuzhou City**

The New Hexi Region is located in the west of Zhuzhou City, facing the old urban area across the Xiang River. With the further development of the city, the Zhuzhou

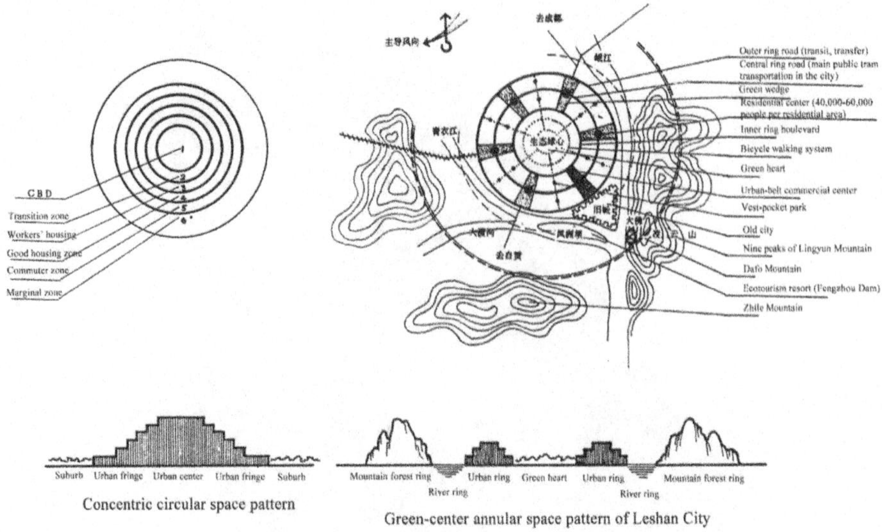

Fig. 19 Comparison of the Pattern of Green-core Ring Spatial Structure and the Pattern of Concentric-circle Spatial Structure of Leshan City. *Source* Huang Guangyu. Proceedings of the '97 International Symposium on Sustainable Development of Human Settlements in Mountain Areas, Beijing: Science Press. December 1997, p. 120

Municipal Government decided to develop across the Xiang River in the early 1990s and build the new Hexi region. In combination with the natural topography of Hexi and its relationship with Zhuzhou's main city, the plan adopts the fan-shaped radial layout structure to integrate the old and new cities into an organic whole and coordinate their development. With Tiantaishan Park as the palm, the road system extends from five radial trunk roads to the Xiang River, and from the Xiang River to the city center in combination with the radial trunk roads to form a trumpet mouth with a wedge-shaped green space, forming a ventilation corridor and a landscape corridor, in order to facilitate the introduction of river wind, adjust the climate of the new city and view the new city from the river or the river and the old city from the center of the new city (Fig. 21).

② Master Plan of Renshou County in Sichuan Province

The master plan is proposed to avoid the random expansion of the county to the surrounding area that may occur in the process of urbanization and the adverse impact on the eco-environment and urban transportation. With the old city as the main body and combined with the natural conditions, five clusters of 20,000–60,000 people of different sizes are distributed along the main roads of the city in a fan-shaped way. There are river and lake water systems, gullies, farmland and garden green space between the clusters to form a permanent isolation zone and an open space with rich natural landscape and superior ecological conditions, showing a palm-like spatial structure (Fig. 22). The balance is achieved between work and residence

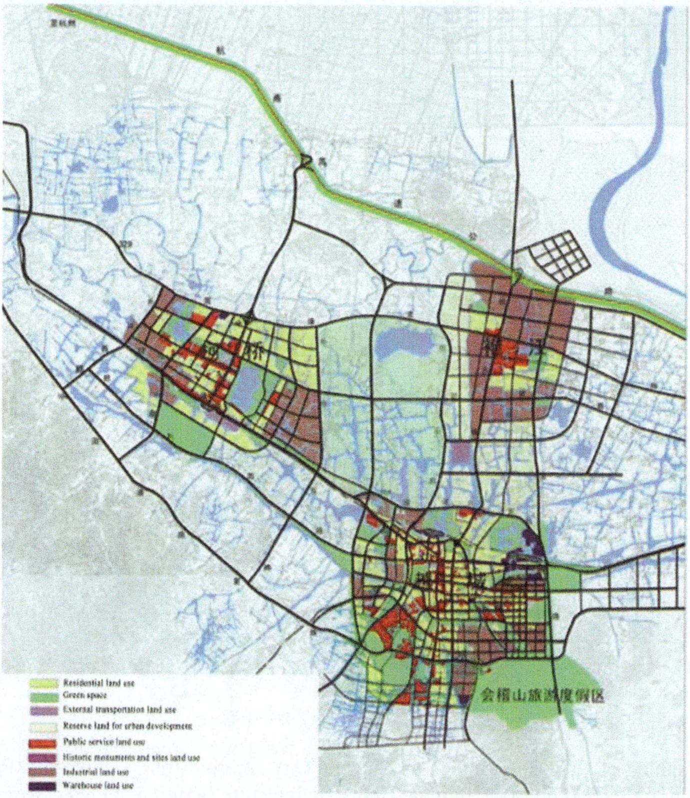

Fig. 20 Master Plan of Shaoxing City in 2001. *Source* Website of Planning Bureau of Shaoxing City http://www.sxgh.gov.cn/news/. Urban Planning and Design Institute of Shaoxing City, 2001

and between production and life, and convenient public transport links between the palm areas. The internal traffic of the palm area is mainly walking, supplemented by bicycle traffic, which minimizes the demand of residents for using cars in the cluster, thereby greatly purifying the internal traffic environment of the county and creating a safe, sanitary, comfortable and quiet living and working conditions.

(8) **Dendritic structure**

To build a city in hilly and mountain areas with complex terrain and developed gullies, the natural conditions such as mountains, gullies, and water systems are utilized in the urban layout structure to select the location of urban or industrial production base in groove or highland between gullies or valleys, forming a pattern of dendritic spatial development structure, and the municipal infrastructure such as roads and transportation are arranged along ravines, such as Shiyan City in Hubei Province and Wuzhou City in Guangxi Province.

Master plan and dendritic spatial structure of Shiyan City in Hubei Province:

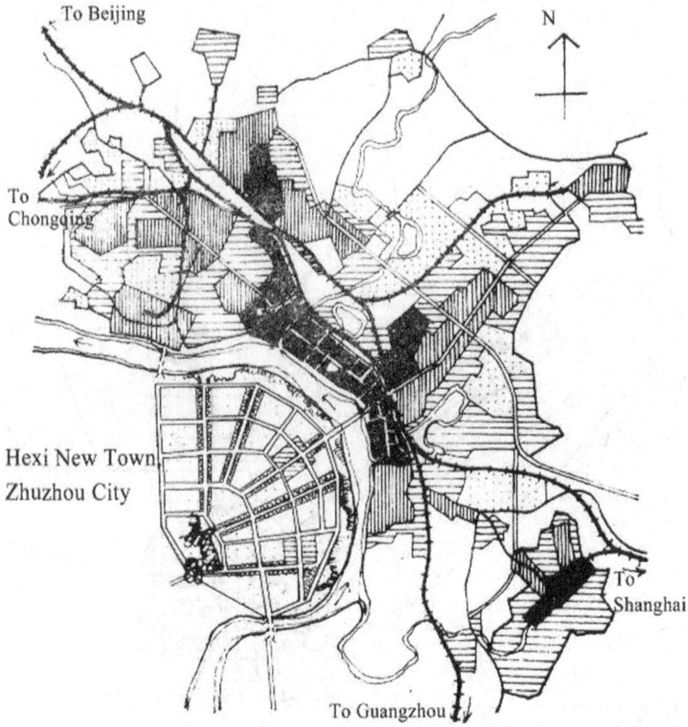

Fig. 21 Planning of New Hexi Region in Zhuzhou City. *Source* Tongji University. Principles of Urban Planning (Second Edition). Beijing: China Architecture & Building Press. November 1991, p. 192

Shiyan City is an important economic, political, technological and cultural center in northwest Hubei Province, which is located in the northwest of Hubei Province, on the south bank of the Han River, in the middle and low hilly areas at the north foot of Wudang Mountain, adjacent to Xiangfan City in the east and Baihe County in Shaanxi Province in the west. It is an emerging industrial city based on the automobile industry with undulating mountains and developed gullies, which gradually forms a development pattern of multi-center dendritic spatial structure with characteristics of mountain gardens, and the combination of concentration and dispersion (Fig. 23).

(9) **Constellation structure**

It is the further development and expansion of the multi-center cluster spatial structure. This kind of mountain cities may form a multi-center or constellation layout structure combining large dispersion and small concentration due to the huge urban scale and the complex natural topography and landforms in the area where the city is located, or the development of mining resources and the relatively scattered distribution of mining sites, such as Chongqing Municipality and Liupanshui City.

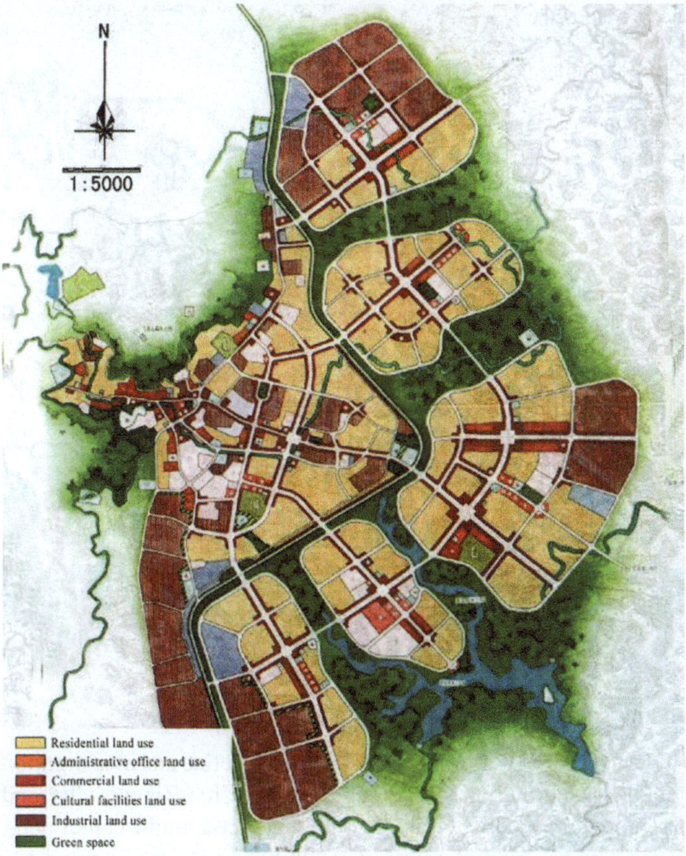

Fig. 22 Master Plan of Renshou County in 1995. *Source* Huang Guangyu. *Mountainurban Planning and Design*. Chongqing: Chongqing University Press. January 2003, p. 87

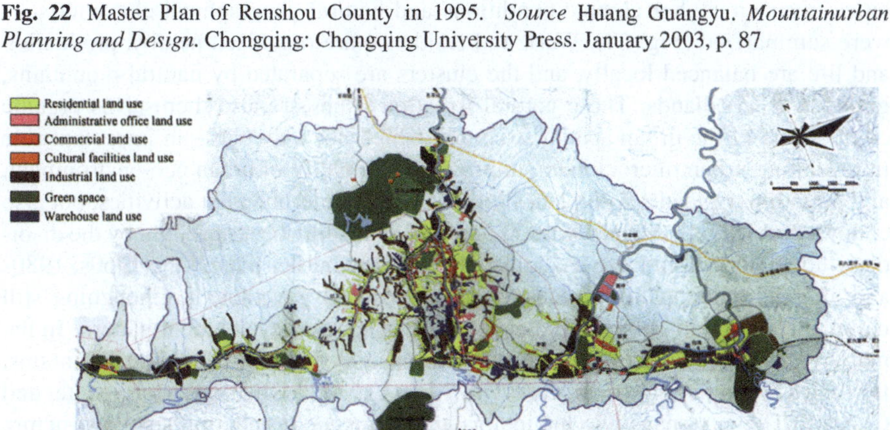

Fig. 23 Mater Plan of Shiyan City in 1997. *Source* Ju Haibin, Yuan An. Reflections on Shaping the Personality of Shiyan City. *Planners*. Periodical Office of Planners. February 2006, p. 55

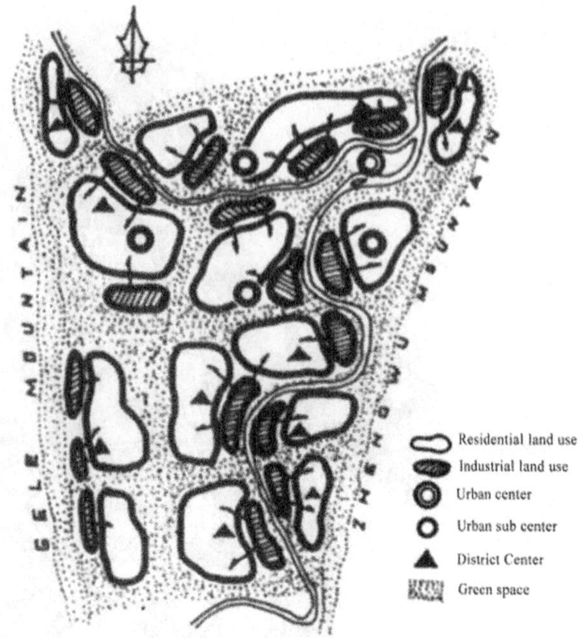

Fig. 24 Concept Map of the Layout Structure of the Central Area of Chongqing Municipality. *Source* Drawing by the author

Master plan and spatial structure of Chongqing Municipality:

From the 1970s to the 1980s, through the planning practice and investigation of mountain cities such as Chongqing Municipality, Yibin City, Zigong City and Neijiang City, the "organic, loose, concentrated, zonal, balanced, multi-center and clustered" mountainurban layout principles and development patterns adapted to mountain cities, river cities, hot climate conditions and natural geographical characteristics were summarized (Fig. 24). Work and residence are organized nearby, production and life are balanced locally, and the clusters are separated by natural mountains, gullies and woodlands. These natural isolation zones are also themselves organic components of the urban spatial structure, which not only plays an important role in regulating urban microclimate, improving the quality of urban eco-environment, and providing residents with places for leisure and entertainment activities, but also endows the city with distinct features to avoid many drawbacks brought by the disorderly urban sprawl. In its twice adjustments on the master plan in the 1960s, 1980s and after its municipality directly under the central government, Chongqing still maintained its multi-center and clustered urban spatial development pattern. In the master plan of 2004, the city expanded to the troughs of the east and west mountains, forming a constellation structure of "One Main City, Five Districts, Multi-center and Cluster". The city spans two mountains and forms a constellation spatial structure of "One Main City, Five Districts, Multi-center and Cluster" (Fig. 25).

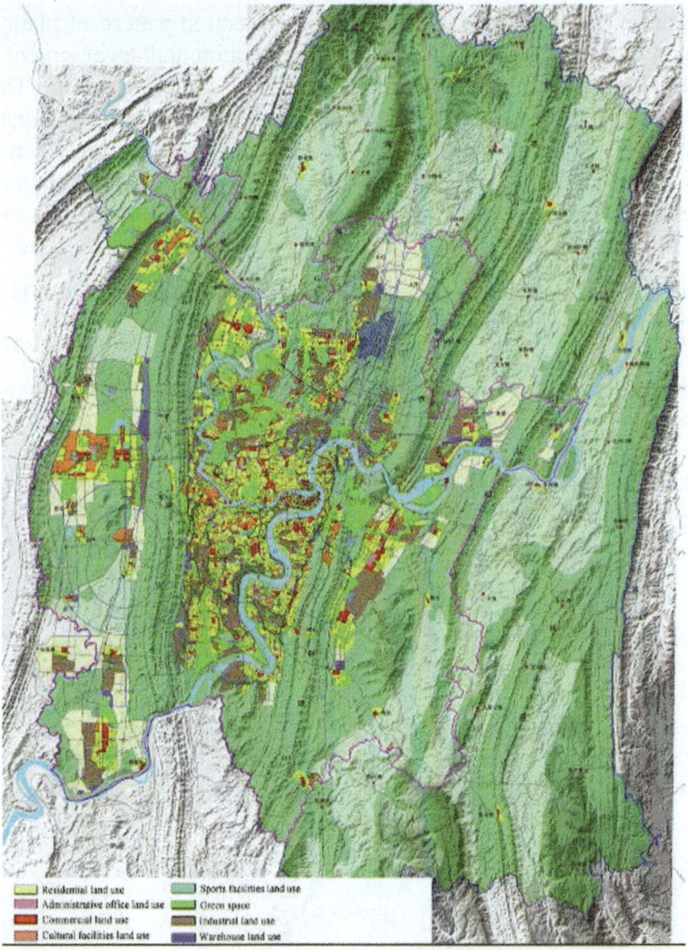

Fig. 25 Spatial Structure Expansion Map of Urban Planning of Chongqing Municipality in 2005. *Source* Text of Master Plan of Metropolitan Area in Chongqing

(10) Separated structure of old city and new district

In order to effectively protect the old city with rich historical and cultural values and coordinate the contradiction between the rapid social and economic development of the city and the protection of historical heritage, the protection of the old city was separated from the development of the new district to form a pattern of dualistic urban spatial development structure of "new district-old city". Hence, these two are not only relatively independent, but also interconnected into an organic whole, ensuring the coordinated and sustainable development of the city's society, economy and culture. This structural pattern is the result of long-term in-depth investigation and study on the mountain natural environment and historical and cultural environment, and comparative study with the protection pattern of plain historical cities, which has a strong adaptability to the protection and development of mountain historical cities with fragile natural and cultural eco-environment. This type of cities includes Lijiang City and Dali City in Yunnan Province, Wutongqiao District, Shawan District, Guang'an District, Xiaoxi Ancient Town, Taiping Town in Leshan City, Sichuan Province, Ciqikou Ancient Town in Chongqing Municipality, Tucheng Ancient Town, Zhenyuan Ancient Town and Qingyan Ancient Town in Guizhou Province, Fuchuan County in Guangxi Zhuang Autonomous Region, etc.

① **Pattern of spatial development structure of separating old city and new district in Lijiang City**

In 1983, the master plan of Lijiang City determined the overall development strategy of "conscientiously protecting the old city, actively developing the new area, and gradually developing the rich scenic resources in the surrounding areas" (Figs. 26 and 27).

The plan clearly put forward that the principle of taking tourism as the main direction of industrial development and separating the protection of old city from the development of new district was a dualistic structural pattern for the protection of

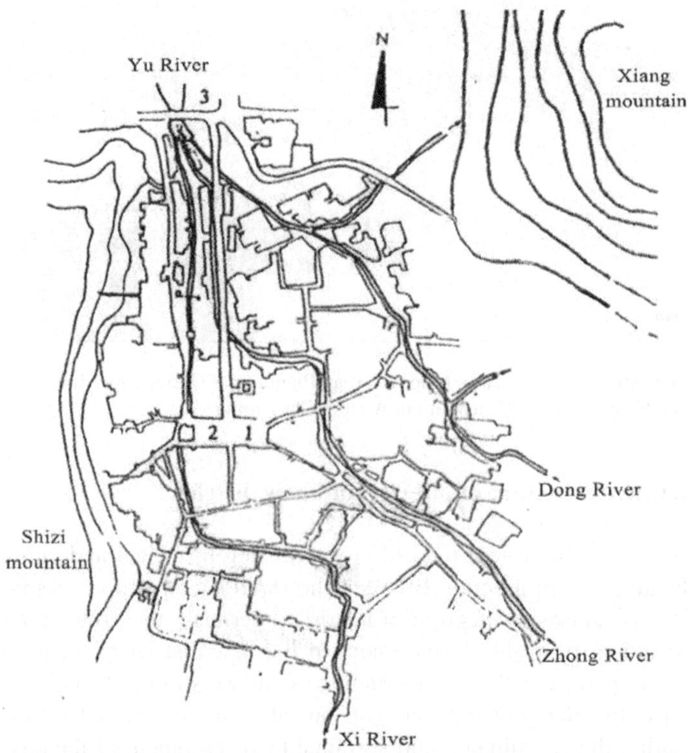

Fig. 26 Master Plan of Dayan Town in Lijiang City. *Source* Huang Guangyu. *Mountainurban Planning and Design*. Chongqing: Chongqing University Press. January 2003, p. 21

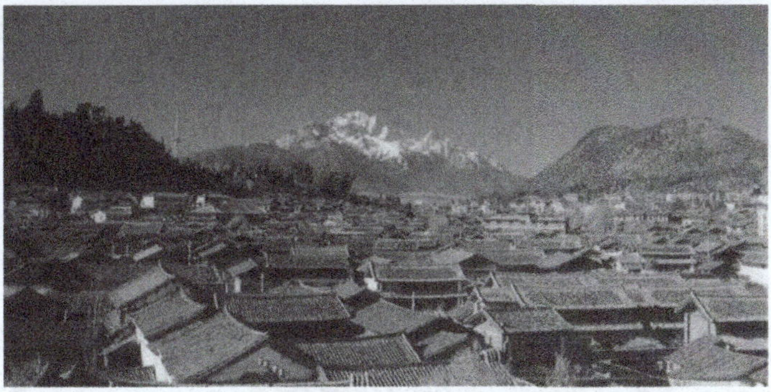

Fig. 27 Panorama of Dayan Town in Lijiang City. *Source* Self drawing

historical cities proposed earlier in China. The overall layout structure of "One Town (Dayan Town), Two Blocks (old city and new district) and Two Points (Nankou and Changshui industrial points)" (Fig. 28) was formulated. It embodies the coordination

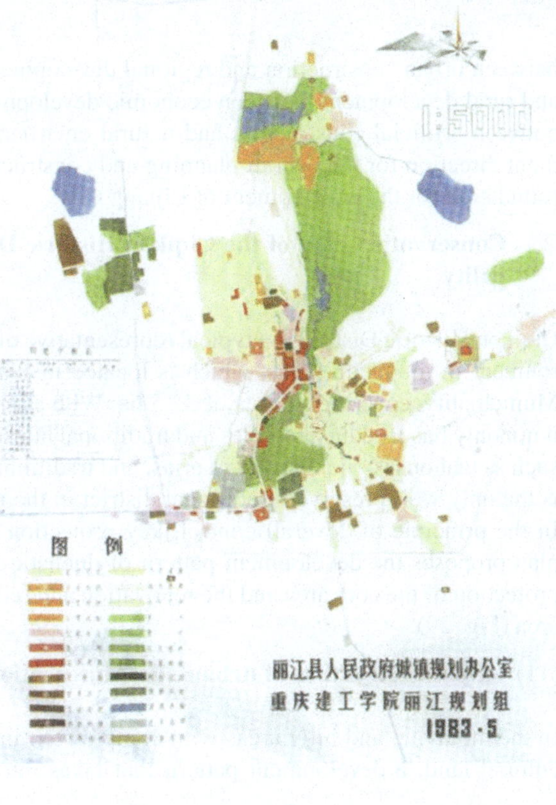

Fig. 28 Master Plan of Lijiang City. *Source* Photograph by the author in 1984

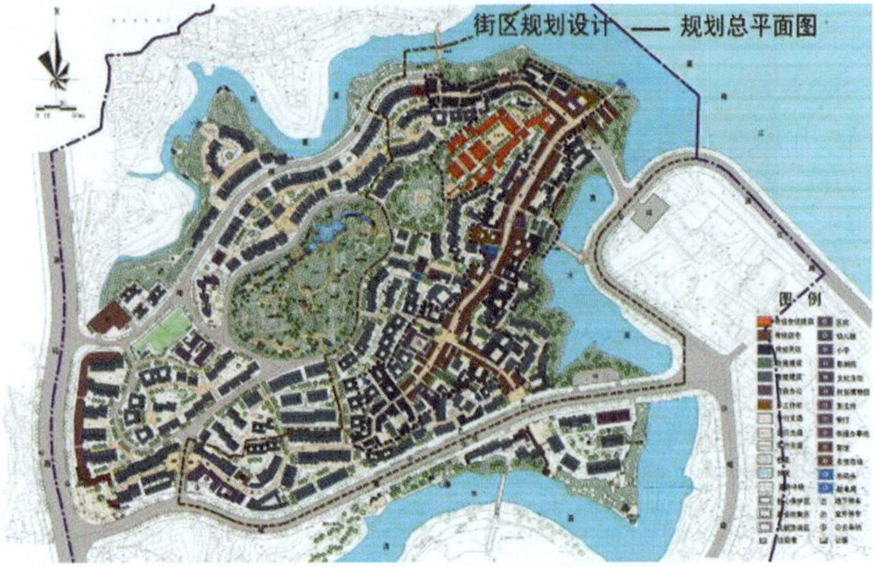

Fig. 29 Protection Plan of the Ciqikou Historic District in Chongqing. *Source* Collection of Institute of Urban Planning & Design, Chongqing University, May 2003, p. 15

between urban construction and regional development, between urban construction and rural development, between economic development and social civilization, and between artificial construction and natural environment, and sets a clear development direction for the overall planning and construction of the city and lays a solid foundation for the development of Lijiang City.

② Conservation plan of the Ciqikou Historic District in Chongqing Municipality

Ciqikou Historic District is a typical representative of the traditional settlement environment in mountain areas, which is located in Shapingba District of Chongqing Municipality, covering an area of 32.5 ha. With a history of more than 1,800 years, it not only has Buddhist temples and traditional houses, but also retains folk cultures such as teahouses, Sichuan opera arias, and traditional handicraft workshops, which is the only well-preserved traditional district in the main urban area of Chongqing. In the principle of "overall control, key protection and unified coordination", the plan proposes the development pattern of dualistic spatial structure with the strict protection of the core area and the renovation and reconstruction of the coordination area (Fig. 29).

(11) Spatial structure of urban-rural integration

In the mountain and hilly areas with complex terrain and scarce and precious agricultural land, a development pattern that takes into account the co-existence and

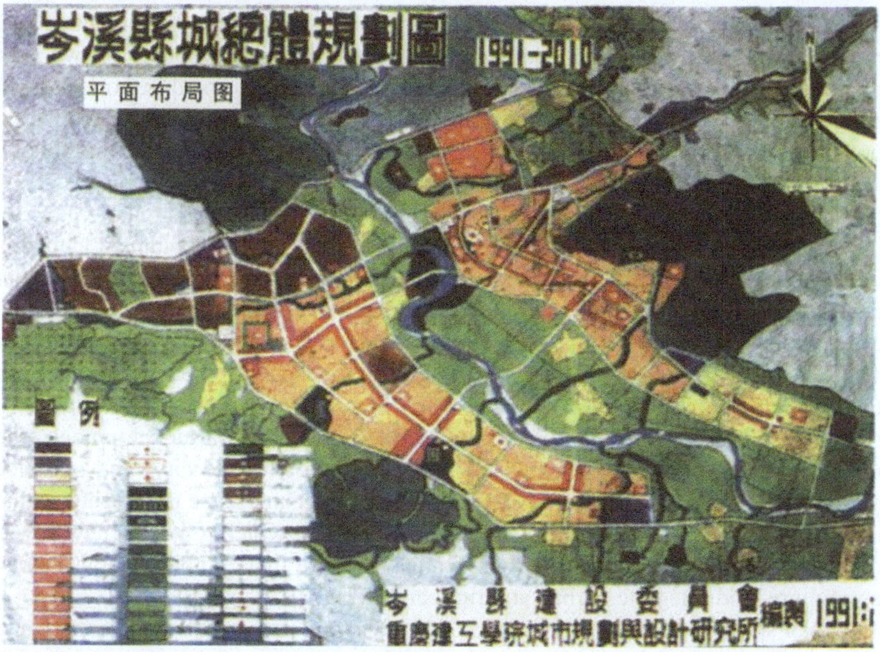

Fig. 30 Spatial Structure of Urban-rural Integration of Cenxi County. *Source* Huang Guangyu *Mountainurban Planning and Design*. Chongqing: Chongqing University Press. January 2003, p. 60

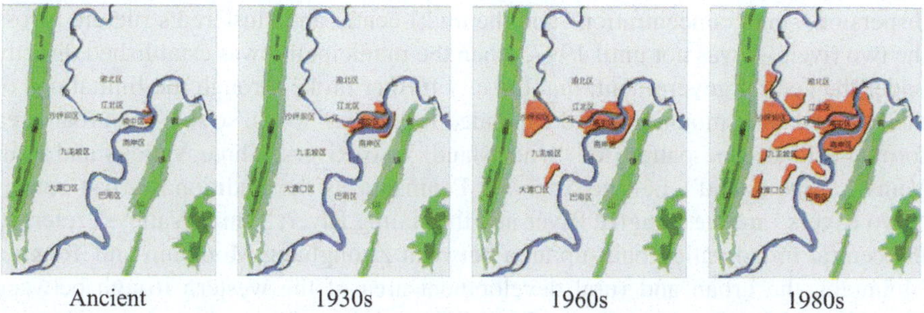

Fig. 31 Urban Spatial Expansion Map of Chongqing Municipality in Different Periods. *Source* Repainting based on the text of Master Plan of Metropolitan Area in Chongqing in 2005

common prosperity of urban economy and agricultural economy is adopted. Cenxi County is located on a triangular hilly valley terrace surrounded by mountains on three sides in the southwest of Guangxi Zhuang Autonomous Region. The Yichang River flows through the center of the trough from west to southeast, dividing the triangular platform into north and south parts. The valley terrace is divided into two levels, of which the elevation is 89–94 m, 94–100 m for the urban built-up area,

93.4 m for 10-year flood, and 90–92.5 m for the farmland along both banks of the river valley. As the exit of the Yichang River lies in the narrow valley of two high mountains in the west, a large flood detention area is often formed in flood season. A large flood occurs every 3–5 years, and a flooded area is formed in the trough on both sides of the river in annual flood season, which is a high-yield agricultural land. The built-up area is about 3 km^2, with an urban population of 40,000, which is 75 m^2 per capita. The planned population is 100,000, with a planned land area of 9 km^2. In the long term, the population can be expanded to 160,000, with a planned land area of 20 km^2. Combined with the characteristics of natural ecology, the planning layout retains the agricultural farmland, natural mountain forests and water systems of the catchments on both sides of the river. The city develops parallel to the terraces on both sides, avoiding the danger of flooding. Agricultural residential areas are arranged in the northeast of the old city in a concentrated way to facilitate the agricultural work of farmers. The leeward side of the northwest is for industrial development. The urban public activity center is divided into one main city (old city) and one vice district (new district) in east and west wings, forming a landscape urban pattern that integrates mountains, water, fields, forests and the city, which reflects a new development pattern of the mountainurban spatial structure that combines industry and agriculture, and integrates urban and rural areas (Fig. 30).

Urban spatial structure is not static, but a dynamic process of development and change. The main factors affecting its development and change are the social, economic and technological conditions. For example, the mountainurban Chongqing was built in the early stage at the confluence of two rivers where water transport is convenient. With the increase of population, economic development and technological progress, Chongqing gradually developed into a clustered layout of "large dispersion, small concentration" and the multi-center and clustered structure across the two rivers. It was not until 1997, when the municipality was established directly under the central government, that the city further broke through the limitations of the east and west mountains and expanded to a larger space, so that the urban area formed a landscape pattern of "One Island, Two Rivers, Three Valleys and Four Veins". "One Island", i.e. the Yuzhong Peninsula, is the traditional urban center; "Two Rivers" are the Yangtze River and the Jialing River; "Three Valleys" refers to the central broad valley built-up area between Zhongliang Mountain and Tongluo Mountain, the urban and rural development area of the western trough between Zhongliang Mountain and Jinyun Mountain and the urban and rural development area of the eastern trough between Tongluo Mountain and Mingyue Mountain; "Four Veins" refer to the Jinyun Mountain Range, Zhongliang Mountain Range, Tongluo Mountain Range and Mingyue Mountain Range from west to east. The overall mountainurban landscape pattern of "One Main City and Five Districts", covering an area of 5,473 km^2, is formed with mountains and valleys parallel and interlaced and two rivers running through (Figs. 31 and 32).

(3) Characteristics and problems of the spatial structure of mountainous cities

Combined with the above mountainurban spatial structure development patterns, there are the following common characteristics and problems.

Fig. 32 Analysis Chart of the Urban Landscape Pattern of Chongqing Municipality. *Source* Text of Master Plan of Metropolitan Area in Chongqing in 2005

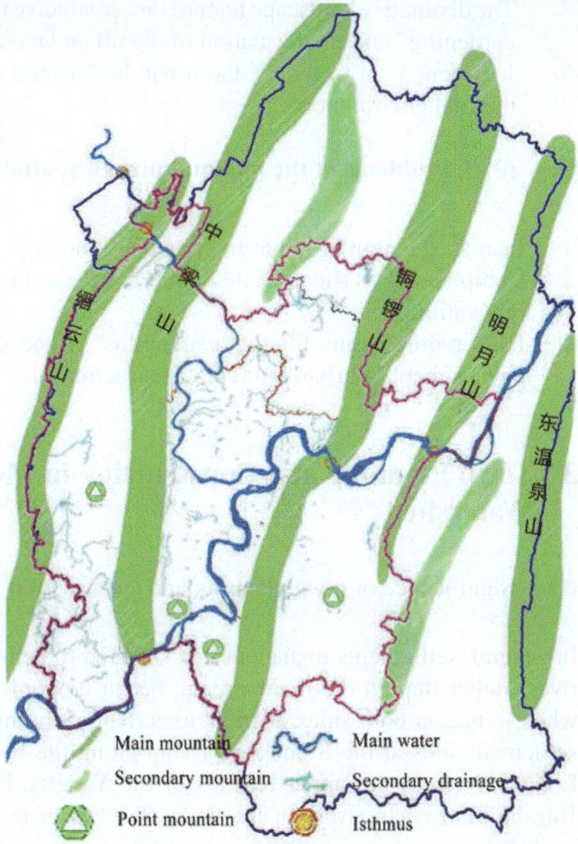

(1) **Characteristics of the mountainurban spatial structure**

① The combination of centralization and decentralization makes the urban spatial development flexible, leaving more room for urban development.

② The urban and rural areas are interlaced with superior environment, which is conducive to the ecological balance between urban development and natural evolution.

③ Work and accommodation are combined, and production, work, life and residence are balanced in the vicinity, which is convenient for production and convenient life.

④ The balanced distribution conducive to the passenger and freight traffic flow and flow direction can decrease the area of static and dynamic traffic land, reduce the excessive concentration of traffic pressure and environmental pollution in the downtown area, and purify the urban traffic environment.

⑤ It is conducive to disaster reduction, prevention and emergency evacuation.

⑥ The distinctive landscape features are conducive to the realization of "landscape gardening" and the formation of the urban landscape garden pattern.
⑦ It is beneficial to reduce the urban heat island effect and improve the urban thermal environment.

(2) **Problems of the mountainurban spatial structure**

① Increased length of roads and pipelines and large investment in infrastructure
② Complicated vertical design and large workload of outdoor environment renovation
③ High requirements for the adaptability to and coordination with the natural environment in urban artificial construction

3 Town Planning and Construction in Mountainous Watershed

(1) Significance of town planning and construction in mountainous watershed

In general, settlements in ancient times tended to be located in the tributaries of big rivers rather than in the main stream, because people felt more secure and livable when living on both sides of tributaries than in the main stream. For example, the settlement sites at the Stone Age found along the banks of the Daning, Daxi and Longxi rivers in the upper reaches of the Yangtze River basin, and the Weishui, Jingshui and Chan rivers in the upper reaches of the Yellow River Basin are the evidence.

Mountain towns are large in number and widely distributed. In the social and economic development of mountain areas, the planning and construction of small towns is of particular significance. The mountain area is vast, and the differences in natural conditions and economic and cultural development are very obvious. Therefore, it is necessary to make a good plan for small towns according to their local conditions and the characteristics of the geographical and ecological units of small watersheds. For example, the Yangtze River, the Yellow River, the Pearl River, the Nen River and the Nu River are different in terms of their natural landforms and climatic conditions in the upper, middle and lower reaches. These are the same as the material conditions, national habits, and living styles resulting therefrom. All this will affect the spatial layout structure, planning and construction of small towns, as well as local building materials and construction approaches. As the natural geographical and climatic conditions and eco-environment conditions are different between the mountain watersheds in the south and in the north, among hills, basins and peak forest karst areas, and among the mountain valleys of the Qinghai-Tibet Plateau, the arctic-alpine and arid area, the loess hills and gully area, the Loess Plateau and the the Yunnan-Guizhou Plateau, we must adopt different ways of planning, designing, construction technology and building approaches for ecological towns, such as energy conservation, land saving and basic farmland protection, according to local conditions.

(2) Planning and construction examples of small towns in mountainous watershed

Planning and design of small towns in mountain areas under complex terrain conditions in the Jialing River Basin—Planning, design and construction practice of comprehensive demonstration sites for the construction of Guandu Mountain Town in Wanyuan County

Wanyuan County is a poor county in the mountain area of Sichuan Province, and also one of the revolutionary bases of Daba Mountain. Guandu Town is located on the northeast edge of Wanyuan County, at the junction of Sichuan, Shaanxi and Hubei provinces, and on the steep slope along the Houhe River, a tributary of the Jialing River. Guandu Old Town, with a population of 1,850, is located in a landslide area and needs to be completely relocated. The "Design of Comprehensive Demonstration Site for the Construction of Guandu Mountain Town" conducted from 1985 to 1986 was the first comprehensive pilot project for the construction of mountain towns designated by the China Spark Program, and also a pilot project of China-German cooperative research. The systematic research and beneficial exploration were carried out in various aspects such as land selection, planning, construction and design, wall reform, building materials development, building structure system, construction management, spreading technological knowledge to farmers, mass participation and construction fund raising for towns in mountainous watershed. A systematic study was conducted on energy conservation, land saving and eco-environmental protection of agricultural cultivated land in mountain areas. A series of ecological technologies and approaches from planning and design to construction management has been created, featuring "adjusting measures according to local conditions, unified planning, user participation, self-built public assistance and self-built mutual assistance". It is of practical significance to determine how to plan and build small towns with mountain characteristics, and promote the construction and development of mountain towns in the vast mountain areas in China.

The total land area of the new town planning area is 174 ha. The planned population (up to 1990) was 5,000, with a planned land area of 35 ha. By 2000, the planned population reached 8,000, with a planned land area of 61 ha. The planning area is characterized by complex terrain and large gradient, which is 20–30% (9–13.5°) in most sections and 100% (45°) in local sections, with a relative height difference of nearly 100 m. There are also three large gullies in the central area, and the terrain is broken, which has the typical topographic and geomorphic features of mountain areas.

(1) Comparison of the schemes for new town site selection

There are two options for the construction land of a new town, one is the hillside wasteland and the other is the fertile flat land. Through the comparison and demonstration of the site investigation scheme, it is decided to adopt the hillside scheme. The comparison of the schemes for new town site selection is shown in Table 1.

Table 1 Comparison of the schemes for New Town Site selection

No.	Comparison items	Hillside scheme	Flat scheme
1	Distance from the old town	2 km	1.8 km
2	Distance from the county seat	10 km	10.5 km
3	Traffic conditions	Connected with the transit traffic Hanyu Road, and close to the railway station	Cross-border traffic with tens of thousands of highways and railways passing through
4	Land conditions	Broken-up land with a large gradient at 10–50%	Flat land with a small gradient, below 10%
5	Geological conditions	Mud mixed with conglomerate, bearing pressure 3 kg/cm^2	Sand mixed with pebble, with strong resistance to pressure 5 kg/cm^2
6	Electricity conditions	Easy access to power supply	Easy access to power supply
7	Water conditions	Close to Hou River, convenient for water use	Close to Hou River, convenient for water use
8	Flooding conditions	Not inundated by flood	Below the 50-year flood level
9	Occupation of cultivated land	No need for land requisition fee for the use of barren slopes and poor lands	Occupying 405 μ of fertile farmland with a land acquisition fee of 4.05 million yuan
10	Utilization of existing buildings	Public buildings such as hospitals, hotels, district offices and post offices available	No existing building
11	Utilization of existing infrastructure	470 m-long zigzag road available to connect to the train station	No existing foundation construction
12	Whether there is room for development	Yes	The land is divided into two parts by the highway, which hinders the development
13	Other conditions	Located on the leeward side of the cement plant, with 110 kV high-voltage lines passing through	Located in the downwind side of the cement plant and Qingyu Iron-making Plant

(2) Overall layout planning concept and its characteristics

① Combination of old and new to form a town center with supporting facilities

Most of the farmers, after selling agricultural and sideline products in the market or acquisition point, buy the means of production or daily necessities nearby, some of whom also have the need for cultural entertainment, diet, savings, mailing, etc. In addition to meeting the spiritual and material environmental requirements of town residents, the planning should also meet the basic requirements of farmers for entering the city. Considering the flow of farmers to the city from all directions and the characteristics of taking different modes of transportation into the city, the approach of combining with the existing public construction service facilities is adopted. The hinterland on both sides of the zigzag-shaped road is used for the centralized planning of the public construction activity center with supporting facilities, and the mountain terrain is used for the diversion of people and vehicles. According to the actual mountain conditions, various public construction projects are arranged intensively and compactly. In this way, clear service functions and convenient purchase and sale of farmers are achieved, which avoids the problem of usually taking streets as markets in the town and the contradiction between mixed market and road functions (Fig. 33).

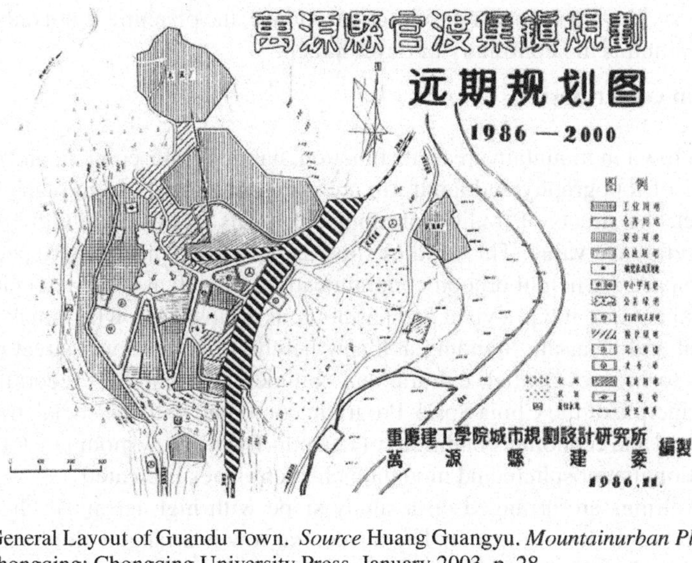

Fig. 33 General Layout of Guandu Town. *Source* Huang Guangyu. *Mountainurban Planning and Design*. Chongqing: Chongqing University Press. January 2003, p. 28

② **Saving land and reasonable layout is one of the important principles of town and village planning and construction.**

Through comparative analysis, the new town site is selected scientifically and reasonably, and all existing facilities are used to construct and develop the new town, without occupying fertile land. The land use indicators are determined based on the reality, and construction and development are achieved according to the actual conditions of mountain terrain, which neither pursues a large area, nor a fashionable design.

③ **Importance is attached to the eco-environmental protection of the towns, reflecting the landscape and local characteristics of mountain areas.**

In order to create good living conditions and living environment for the residents, the planning emphasizes the utilization and greening layout of the surrounding environment of the town, and gives careful consideration to the artistic layout of the urban group space and the architectural complex. In particular, the author makes a useful discussion on the main urban space level for mountain towns with complex and undulating terrain, and also conducts earnest planning and design on the hilltops, ravines and open spaces within the planning scope.

④ **Combination of the short and long terms focusing on the short term**

Priority is given to arranging recent construction projects, which is centralized and compact in the layout structure, with the development direction from the center to the periphery, leaving room for development. Hence, the planning is not only phased, but also maintains the continuity of development.

(3) **Town center design**

For a rural town in mountain areas, its function, nature, scale, content, and objective constraints of topography and landform are different from those in plain towns. In this respect, the streets of traditional mountain towns and Guandu Old Town have excellent reference value. This kind of street is not only a place where people live, but also a space for neighborhood communication, as well as a vibrant cultural and commercial area, so it has a vivid and harmonious atmosphere, which makes people feel cordial and pleasant. Standing at a new height, this traditional street market is revived to form a residential, cultural and commercial complex. According to the guiding principle of the China Spark Program, combined with the social, traditional, technological and economic conditions of Guandu Town, a new mountain town center with traditional atmosphere and mountain characteristics is created.

The dwellings are arranged on a sunny slope with high terrain in the form of low-rise and high-density layout. The residential row is arranged along the contour line, and the residential complex, combined with the terrain, is arranged along the contour line in an organic curve, forming a traditional street market. Combined with the local traditional houses and street doorways, the long snake-shaped market is appropriately divided and limited in space, so that the local characteristics and traditional atmosphere of the planned street market is highlighted (Fig. 34).

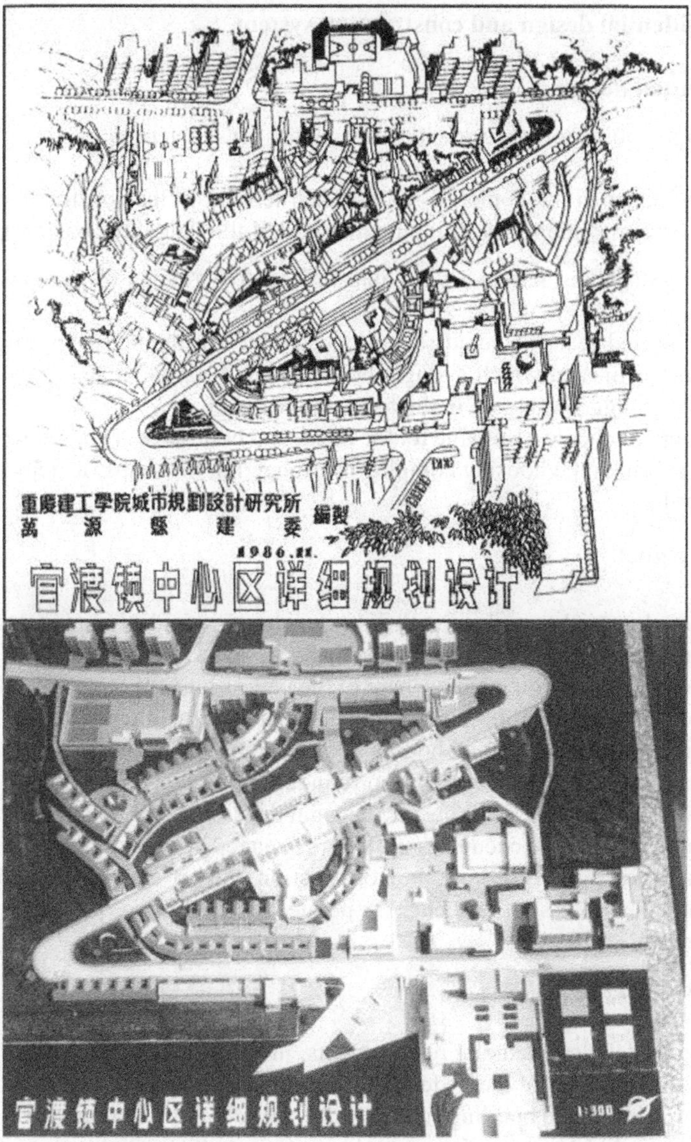

Fig. 34 Detailed Planning and Design of the Central Area of Guandu Town. *Source* Huang Guangyu. *Mountainurban Planning and Design*. Chongqing: Chongqing University Press. January 2003, p. 29

(4) Residential design and construction system

① Residential design

a. Close combination of residential design and hillside terrain

In order to adapt to the complex gradient and terrain on the construction site, the residential scheme adopts the combination of the following three basic approaches:

- The non-staggered scheme is suitable for the construction of sections with flat terrain;
- The front and rear staggered scheme is suitable for the houses arranged along the contour line, in which a single house spans the contour line longitudinally, forming a staggered layout in the front and the back;
- The stepped scheme is suitable for the houses on both sides of the steep pedestrian street, in which the housing complex is arranged in the shape of steps across the contour line;

b. Plan layout

In order to adapt to the needs of pedestrian streets, the residential plan design takes form of store in front and yard behind, or inner patio is.

This layout enhances the atmosphere of the pedestrian shopping street, facilitates the life of residents and the communication between neighbors, and reflects the characteristics of the residential houses in the towns of Sichuan Province.

Residential houses in the towns require independent courtyards to facilitate self-construction by the masses. Taking into account the different population composition of the households and the occupations they are engaged in, such as business, domestic industry and livestock breeding, the construction area of each household is about 120 m^2, and the urban staff residence is operated by a point-type scheme with 60 m^2 per household.

c. Residential style

According to detailed planning requirements, most residential houses are centrally arranged on both sides of the pedestrian commercial street, with the street width of only 4–7 m. Hence, the buildings on both sides should not be too large. Considering factors such as lighting, ventilation, and street view effect, the number of floors is mainly two, and not more than three in some localities. With the change of topographic relief, the layout is arranged in a staggered way to form a plain, but elegant, appearance with rustic style.

The residence for industrial and mining workers is arranged at the edge of the town according to the detailed design. Considering the structural characteristics of block walls, it is advisable not to exceed six floors.

② Residential structural system

The shear wall bearing system is suitable for both low-rise residence and multi-storey residence for industrial and mining workers, and energy-saving small concrete hollow brick blocks are selected as wall materials.

Guandu Town is close to the Hou River, which is very convenient to collect sand and stone, and the local area is rich in cement, which provides abundant local materials for the production of concrete blocks.

With the basic specification of 390 × 190 × 190 mm, small concrete hollow blocks can be produced by residents themselves. The block molding machine produced by the Farm Tools Plant in Wanyuan County is only 400 yuan per set, by which each person produces 100–120 blocks on average per day. A more mechanized block making machine can also be used, which is more efficient than the former one. Four blocks can be formed at a time. Each block is equivalent to eight ordinary clay bricks in volume and costs only 0.30 yuan, only half of the price of ordinary clay bricks. The price per cubic meter of small concrete blocks is 47–60% lower than that of ordinary bricks. The number of masonry blocks per cubic meter is 62.5, and that of ordinary bricks is 512. Hence, the consumption of masonry mortar can be reduced by 60% (Fig. 35).

The unit price of houses built with small blocks is only 80 yuan/m^2, while the unit price of houses built with ordinary bricks is 90–100 yuan/m^2, which can reduce the cost by 11–20%. At the same time, as the thickness of the load-bearing wall is only 20 cm, the usable floor area of each suite can be increased by 3–4%. In addition, the blocks are hollow and each weighs only 19 kg. Calculated by weight per m^2 wall,

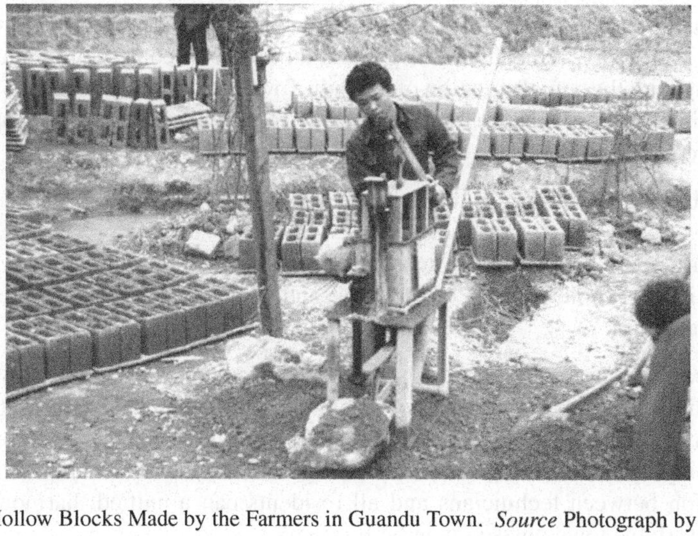

Fig. 35 Hollow Blocks Made by the Farmers in Guandu Town. *Source* Photograph by the author in 1985

Table 2 Cost estimation table

Project	Unit	Scale	Unit price (yuan/m^2)	Cost (yuan)	Remarks
Residence	M^2	16,780	90	1,510,200	Building blocks
Public construction	M^2	17,820	150	2,673,000	Concrete
Vehicle lane	M^2	13,650	50	682,500	Masonry
Street square	M^2	8,740	60	524,400	
Greening	M^2	2,4040	2	48,080	
Pipeline layout	M	2,050	80	164,000	
Outdoor excavation and filling	M^3	55,000	2	110,000	
Ground works				5,712,180	
Total					

Phase I project investment: (1986)
Unit self-raising: 700,000 yuan (loan)
Local investment: 600,000 yuan (loan)
State investment: 790,000 yuan
Total investment: 2.09 million yuan

the self weight of a 19 cm-thick block wall can be reduced by 45% than that of a 24 cm-thick ordinary clay brick wall. Replacing clay bricks with small hollow blocks can also save cultivated land and energy, and protect eco-environment, in order to achieve good social, environmental and economic benefits.

(5) **Cost estimation**

There was no ready-made unit price for the construction of towns and villages at that time. Therefore, referring to the general unit price of construction in cities and combining with the actual situation of Guandu Town as a building materials base, the facilities in the central area are roughly classified as rough estimates after appropriate downward adjustments. Refer to Table 2 for details.

(6) **Public participation in town construction**

Public participation in the town construction is the principle that we have always promote in the whole process of planning and design of Guandu. There are two specific practices:

① Since not everyone understands the importance of town planning, we should publicize the importance of town planning and design to all because strengthen the overall perspective in building houses, and understand the importance of abiding by the planning and not building randomly. Only through close cooperation between technicians and all residents can a unified, harmonious and comfortable town environment be formed.

Table 3 Basic framework of mountainurban non-development land planning and control technology

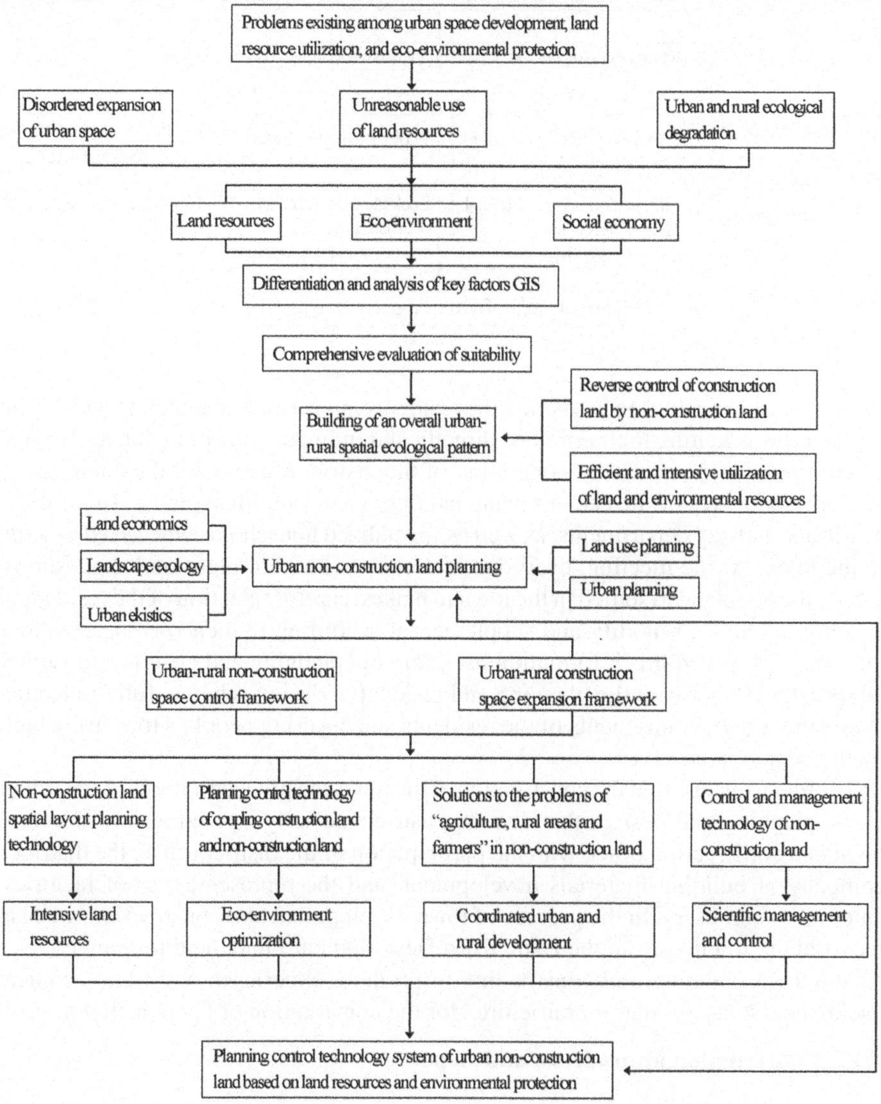

② According to the different needs of users and the different requirements of operators on housing, we should encourage users to participate in designing and modifying the design scheme together with the designer, in order to ensure that the design scheme, with various functions, can satisfy the users.

Table 4 Layout planning process of mountainurban non-development land

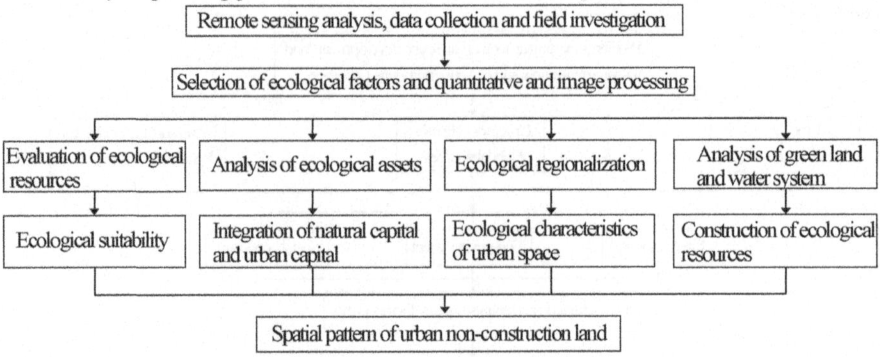

To this end, in addition to visiting the residents many times in planning and design process and soliciting their opinions directly and indirectly, the planning and design meeting was held under the cooperation of the district office, with the attendees of the representatives of the cement plant and Qingyuan Iron Plant, and the education, health and railway departments, as well as specialized households who asked to settle in the town. At the meeting, everyone actively put forward plans and suggestions, asking the people who settled in the town to make a careful selection of the residential design scheme, and modify and supplement it according to their own needs. After the meeting, we fed this information back into the planning and design, and further modified and improved the planning and residential design scheme, fully reflecting the interests and requirements of the residents and ensuring people's town truly built by the people themselves.

In order to ensure that the construction of the town can proceed smoothly according to the planning and design, the Guandu District specially formed a town development and construction office with the participation of the district office, the district's company of building materials development, and the representatives of factories, enterprises and users. In the process of town development and construction, we also received the funding from the Dadi Town Development Fund, held training courses on the town planning and construction to promote experience and train technical backbones, so as to "start a prairie fire" for the construction of Guandu Town.

(7) **Construction approaches and steps**

We consider that self-built and publicly-aided housing is an effective way to utilize local technology, give play to and respect the creativity and personality of the masses for village and town construction. The so-called self-built and publicly-aided housing means the construction of housing and living environment according to their own needs and hobbies that the central or local government provides loan funds, organizes technical forces, provides design and construction consulting guidance, and cooperates with resident family members or several households based on the blueprint for comprehensive development, overall planning, and unified construction, which has many advantages:

① Obvious economic efficiency refers to the construction of houses according to the number of residents applying for housing, which will not cause idleness;
② Residents themselves are the real owners of residential construction, who understand the quality of residential design in daily life, and know how to save materials and use land and space;
③ Residents acquire the knowledge and skills of planning, designing and building houses so that they can take care of and maintain their homes and beautify their environment in the future;
④ Residential and outdoor space are the products of mutual cooperation between neighbors. In the process of jointly creating a new life and living environment, a common sense of neighborhood will be generated, which will make the relationship between people closely integrated.

Therefore, except for large public buildings such as shopping malls and movie theaters, which are organized and constructed by the local government, others such as houses, green spaces and their embellishment facilities may be built by residents themselves according to the planning and design drawings. The steps are as follows: the county and district governments construct infrastructure in a unified way, connect roads, water and electricity to each household, complete earthwork excavation, and level the ground in accordance with the plan. At the same time, a special agency responsible for self-construction work is organized, whose task is to accept self-built applications, help self-builders select suitable sites according to the planning and undertake necessary design work, prepare loans, and provide necessary assistance in the construction process and self-builders with the planning and design files of the whole lot for the reference of residents' self-management. Before the construction starts, the self-builders begin to participate in the selection process. Therefore, we focused on the large spatial structure system in our planning, accordingly.

Due to the participation of the masses, rather than mechanically carrying out the intention of a designer, the masses have their own special interests and hobbies, with different aesthetic standards and using requirements. Under the guidance of the overall planning, it will finally present variable space layout and artistic effect. Over time, the appearance of street markets will become more diversified and harmonious, with more in-depth humane colors and greater cohesion. (More than 20 years have passed since the pilot practice of this project. Although the development of Guandu today is not what it used to be, the ecological design concepts and approaches throughout it are still meaningful and valuable for reference.)

4 Planning and Control of Mountainurban Non-development Land

(1) Meaning of planning control of mountainurban non-development land

Mountainurban non-development land is the land that must be protected and controlled for development within the urban planning area, which includes ecologically sensitive areas such as basic farmland protection areas, scenic spots, wetlands, water source protection areas, and distribution areas of underground mineral resources, as well as green land and land for river and lake water system in urban construction.

There is a coupling relationship between non-development land and construction land in mountain cities, featuring coordination of Yin and Yang, and virtual and real complement, the Yin-yang relationship in China's traditional TaiChi diagram (Huang 1994) (Fig. 36).

(2) Basic framework of Mountainurban non-development land planning and control technology

The mountainurban non-development land planning and control technology is the basic framework of urban non-development land planning and control technology on the basis of ecological suitability evaluation of urban-rural space within the scope of urban planning areas guided by the problems such as disordered expansion of urban space, irrational use of land resources, and urban and rural ecological degradation that have emerged in the rapid urban development in recent years, which integrates multiple disciplines such as urban science, ecology, geography, landscape science, sociology and economics using the 3S technology platform (Table 3).

The planning and control technology of mountainurban non-development land mainly consists of the following four aspects:

(1) **Layout planning technology of mountainurban non-development land**

Non-development land layout planning is to comprehensively identify the ecological resource factors in the urban and rural planning area by means of the analysis of

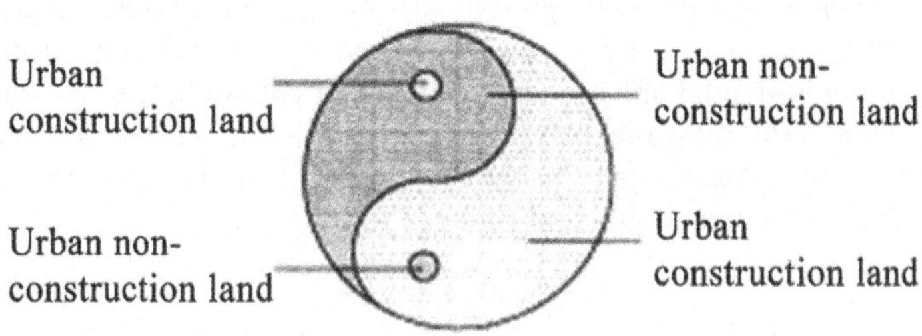

Fig. 36 Coupling Relationship between Urban Construction Land and Non-development land. *Source* Drawing by the author

satellite imagery and aerial photography of the GIS system, and then establish the spatial layout of urban non-development land based on the overall urban and rural land use pattern. The natural resources not for development and construction, such as mountains, water systems, gullies, forest lands, farmlands, wetlands and wildernesses within the planning area, together with the urban construction and development land, constitute an organic part of the overall mountainurban spatial structure.

The layout planning process of mountainurban non-development land is as follows: ecological investigation—ecological evaluation—overall ecological spatial pattern construction—non-development land form and area control (Table 4). The planning approach applies the land suitability analysis with ecological factor analysis and overlay mapping technology as the core, in order to analyze the ecological service function of non-development land and optimize the landscape ecological pattern according to the principle of landscape ecology, and construct a basic ecological pattern of mountainurban non-development land.

(2) **Planning and control technology coupling urban construction land and non-development land**

The planning approach of coupling non-development land and construction land is to focus on the establishment of the ecological spatial pattern of non-development land, coordinate the relationship between "non-development land" and "construction land", and support and guide urban construction and development with a healthy and safe ecological base system, and maintain a dynamic balance between urban development and natural evolution for reasonable construction of the spatial structure of mountainurban non-development land and construction land that are complementary to each other.

(3) **Technology of coordinating urban-rural spatial development in urban non-development land planning**

The urban non-development land planning itself aims at the urban-rural space within the urban planning area for the purpose of coordinated development of urban-rural space.

First, the coupling environmental conditions of urban-rural spatial ecology should be evaluated, in which the approach of ecological footprint can be applied to six major land uses: development land, agricultural land, grassland, forest land, water area and land for energy resource.

Based on the evaluation of urban-rural spatial ecological coupling environment conditions, the urban non-development land control and the development of rural eco-industry are combined, and the development pattern of "urban village" based on eco-industry is proposed.

Eco-industry is a kind of network, evolutionary and compound industry organized according to the laws of circular economics based on the carrying capacity of ecosystem, with a complete life cycle, efficient metabolic process and harmonious ecological function. As a new round of industrial ecological revolution, it will bring

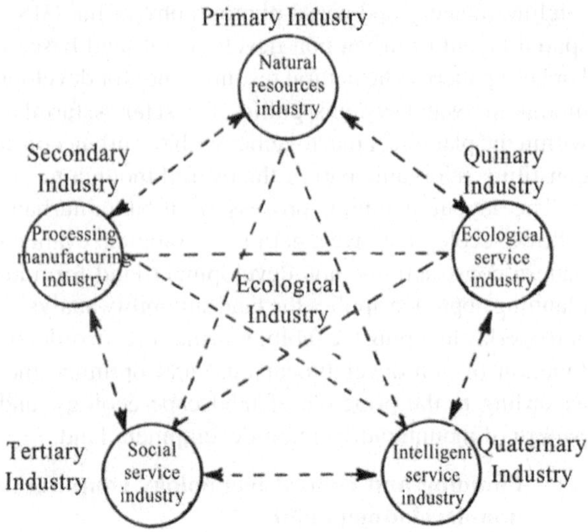

Fig. 37 Eco-Industry Classification. *Source* Wang Rusong. "Industrial Ecology Method of Circular Economy Construction" in *Compound Ecology and Circular Economy*, Beijing: China Meteorological Press. October 2000, p. 158

about the reduction of the disparities between industry and agriculture and between urban and rural areas.

In circular economy, eco-industry is divided into five categories: primary industry: the natural resource industry for the purpose of producing photosynthetic resources and mineral resources; secondary industry: the tangible processing industry for the purpose of manufacturing material and energy products; tertiary industry: the human ecological service industry for the purpose of providing social services; quaternary industry: the intellectual services industry for the purpose of research, development, education and management; quinary industry: the natural ecological service industry for the purpose of material reduction, environmental protection and ecological construction (Fig. 37). The five major industries are the integration of primary, secondary and tertiary industries in the traditional sense, that is, the urban-rural space built on it is the relationship of harmony and symbiosis.

The urban spatial system based on eco-industry is no longer a simple urban space, but an integrated urban-rural spatial system. For example, in the pilot urban villages of Chengdu City, the settlement scale and development form can be diversified, instead of necessarily adopting a scale standard and a development pattern. The settlement scale can be large or small, ranging from thousands of people to tens of thousands of people, generally not less than 3,000 people, with the operating area radius of about 2 km. The existing scattered natural rural settlements are moved, combined, and concentrated in urban villages to improve the living conditions of farmers, increase the area of cultivated land and conduct intensive management. At the same time, the residents in urban villages can be engaged in "industries, agriculture and businesses" in the ecological industrial chain, which is in a transition to agricultural modernization, ecologicalization and urban-rural integration. The following measures can be taken in actual construction:

① According to the principle of "integrated development, priority to efficiency", we should take the road of industrial agglomeration development, at the same time, promote the tertiary industry in the regional center towns where conditions allow, and pay attention to industrial transformation and structural upgrading. The site selection of new rural communities should be combined with the agricultural industrialization projects, close to the secondary and tertiary industry projects with a certain scale, so as to meet the requirements of urban areas supporting rural development and nurturing agriculture by industry.

② In the vast rural areas, we should adjust the agricultural structure, optimize the industrial layout, and vigorously develop characteristic agriculture on the premise of ensuring food security. Through the industrialization of agriculture, we should extend the operation chain and value chain of the primary industry before, during and after production, and transform the production mode and allocation mechanism of agriculture through the large-scale socialized production of characteristic agriculture.

③ We should guide farmers to transfer the land use rights for some agricultural and forestry industries or cash crops with comparative advantages according to the principle of model guidance and farmers' willingness in the form of "company + farmer household" or "association + farmer household". In the process of transfer, the centralized land management can not only achieve scale benefits, but also enable farmers to obtain rent in transferring land use rights, and ensure the normal living income level, promoting farmers to transform into large-scale operation or agricultural workers of agricultural industry companies.

④ According to the new urbanization approach of "urban villages" and the planning standards of urban residential areas, we should reasonably determine the number, spatial layout and time sequence of construction of rural communities, make overall arrangements for and set up public service and municipal public facilities with the new communities in rural areas as the center, and strengthen the protection of non-development land such as agricultural and forestry ecological land.

(4) Control and management technology for mountainurban non-development land

According to the ecological service function of urban non-development land, we put forward the control approach of "hierarchical control, classified protection and step-by-step implementation" of urban non-development land, and the hierarchical, classified and step-by-step standards, formulate the control plan of urban non-development land, and strengthen the control and management of non-development land, so as to provide a set of scientific management approach system for the management and control of non-development land.

Urban non-development land planning is based on the control technology of land resources and environmental protection, which overcomes the traditional planning concepts and approaches that emphasize construction land, development and cities, and neglect non-development land, environmental protection and countryside, and combines land use planning with overall urban planning. The two different levels of planning of land and resources department and construction department are better coordinated and connected through urban non-development land planning to optimize the reasonable allocation of urban and rural land resources, give full play to the comprehensive ecological service function of land resources, and achieve efficient and intensive use of land resources so as to achieve the purpose of adapting measures to local conditions and making the best of land resources.

(3) Practice cases of Mountainurban non-development land planning and control

In 2004, we selected Yunyang New County with complex mountain terrain as the pilot for urban non-development land planning and control. Considering the completeness of basic data in the research process, we selected evidence-based independent factors with complete data and high operability and measurability as analysis factors: topography, gradient, elevation, landscape elements, geological hanzards, thermal environment, historical context, landscape line of sight, to make the land suitability evaluation of each single factor respectively. The comprehensive suitability zoning of land is obtained through holographic superimposition, and the non-development land is optimized using landscape pattern optimization approaches. With natural mountains and waters such as Longji Mountain, Tiefeng Mountain, Mulong Mountain, the mountains on the South Bank of the Yangtze River, the Yangtze River and Pengxi River as ecological matrixes, and mountain ridges, gullies, water systems, historical and cultural landscape corridors and urban wind porches as ecological corridors, urban construction land patches are coupled with non-development land networks to establish the dendritic basic ecological pattern of non-development land of "Two Rivers, Four Mountains and Twelve Veins" with Longji Mountain as the backbone (Fig. 38).

According to the overall urban planning of Yunyang County and the observation and simulation analysis of thermal environment in typical areas (Fig. 39), the thermal environment conditions and influencing factors of Yunyang are summarized, and the suggestions on thermal environment planning and control are proposed to guide the development and utilization of urban construction land and the protection and control of non-development land. In the main urban area, as the dominant wind blows into the whole city along the mountain terrain, that is, along the direction of the Yangtze River, we should appropriately widen the main road junctions, plan the air flue parallel to the dominant wind direction, build few buildings at the entrance of the air flue on both sides of the road, and demolish or rebuild some buildings that severely block the wind field, such as densely populated residential areas by the river, in order to guide the river wind into the city, and ensure the unimpeded opening of the windward side of the air flue. The waterfront area with a low altitude is not only the main ventilation landscape corridor of a city, but also a good place for leisure and entertainment of citizens.

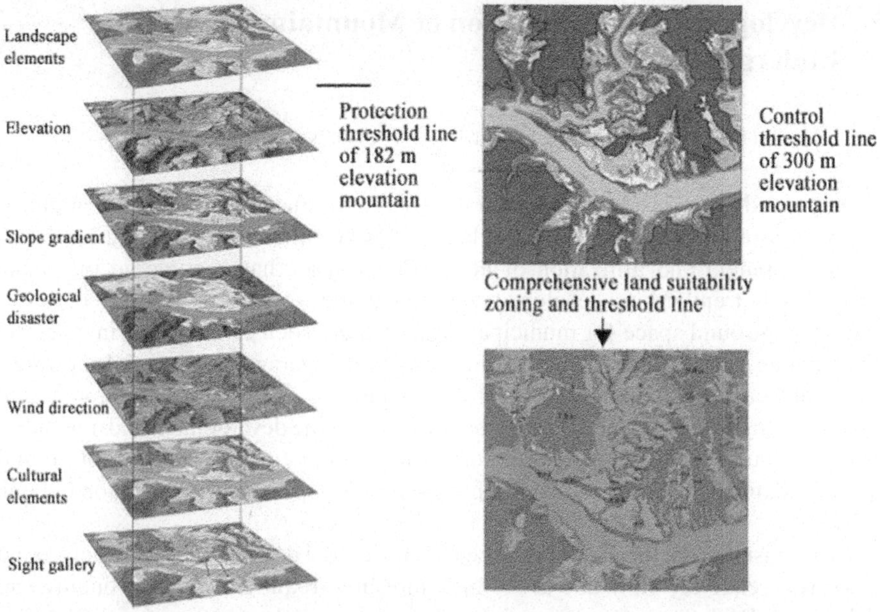

Fig. 38 Constructive Process of Ecological Pattern of Non-development land in Yunyang New County. *Source* Text of Master Plan of Yunyang County (Moderator: Huang Guangyu). 2005

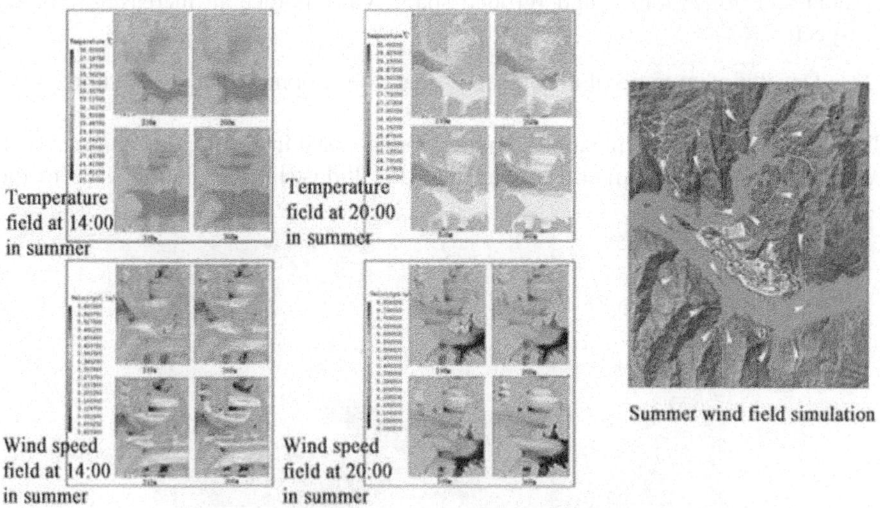

Fig. 39 Simulation and Analysis of Urban Thermal Environment in Yunyang New County. *Source* Text of Master Plan of Yunyang County (Moderator: Huang Guangyu). 2005

5 Development and Utilization of Mountainurban Underground Space

(1) Necessity of development and utilization of urban underground space

Underground space is a precious natural resource for mankind. As the development and utilization of ground resources in large cities become more and more saturated, the development and utilization of underground space has become an inevitable development trend. Many developed countries in the world have extensively developed underground space for municipal infrastructure such as the following: sewers and pipeline sharing ditches; subway, highway traffic, garage, commercial and entertainment facilities; facilities of medical and scientific research laboratory; storage and aviation refuge facilities, etc. Practice has proven that the development and utilization of underground space plays an important role in mitigating the shortage of ground space land, traffic congestion, environmental pollution and safety protection in large cities.

In the past, except for a few developed countries, the development of underground space was relatively slow due to the high threshold requirements of economy and technology. With the aggravation of global environmental problems, the rapid development of economy, science and technology, as well as the improvement of people's understanding of the development and utilization of underground space, the development and utilization of underground space will embrace an increasingly broad prospect.

(2) Development trend of underground space development and utilization

The development and utilization of underground space in developed countries such as Europe and North America began with the Industrial Revolution. Due to the

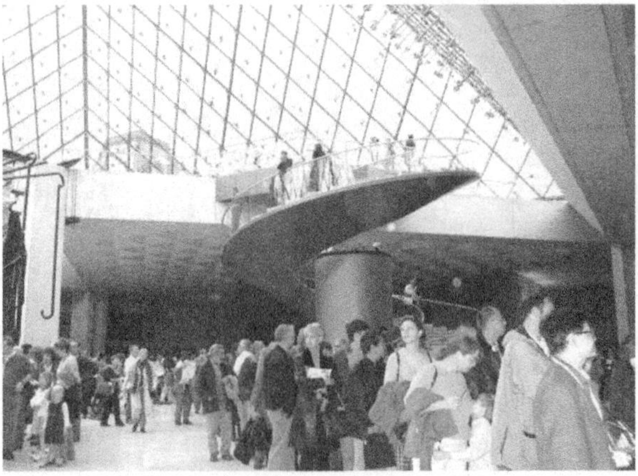

Fig. 40 Underground Entrances and Exits of Louvre. *Source* Photograph by the author in 2005

needs of defense and environmental protection, the development and utilization of underground space in countries, such as Northern Europe, Switzerland, Finland and Britain, France and Germany, are mostly developed and utilized for air defense facilities, subway traffic, water-supply lines and sewers, sewage treatment and refuse transportation. Air defense facilities can usually be used for storage, parking, entertainment and sports, which include the world's first London subway built in 1863, as well as underground facility complexes such as the sewers in London and Paris, the underground parking lot and underground street in Louvre (Fig. 40), the underground or semi-underground residences built to overcome severe weather in hot areas such as Minnesota and Ohio in the United States, and the underground network of underground commercial streets, subway stations and underground sidewalks built to overcome the harsh climate and create a comfortable environment in Toronto and Montreal in Canada (Fig. 41).

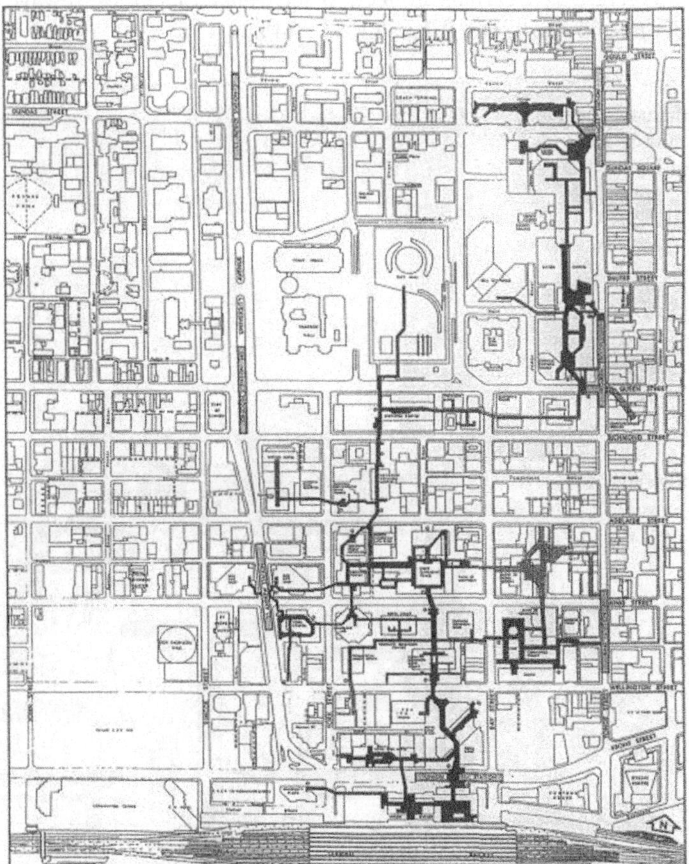

Fig. 41 Underground Space Network in Toronto. *Source Underground Metropolis*. Tokyo: SEIBUNSHA. August 1989, p. 87

Japan has advanced development technology of underground space. Due to a large population and less land, more mountains and less cultivated land in large cities such as Osaka, there is a high density of population, buildings and transportation, and the development and utilization of underground space has been highly valued. They have not only developed and constructed underground power plants, subways, underground tunnels and commercial streets in the hard bedrocks of mountain areas, but also vigorously built subways, underground commercial streets and parking lots on the general soft soil foundation of coastal floodplains and in local medium-sized cities. And the development and construction has gradually deepened into a deeper space, such as within the range of 50 m. There are many types of underground facilities in Japan, such as subways, underground highways, tunnels, underground parking lots, underground power stations, gas supply and distribution systems, communication systems, common trenches for various lines, sidewalks, underground commercial streets and plazas, as well as underground culture, sports facilities, storage facilities, underground oil reserves, garbage collection and transfer facilities, sewage treatment plants, underground dams, etc. (Figs. 42, 43 , 44 and 45).

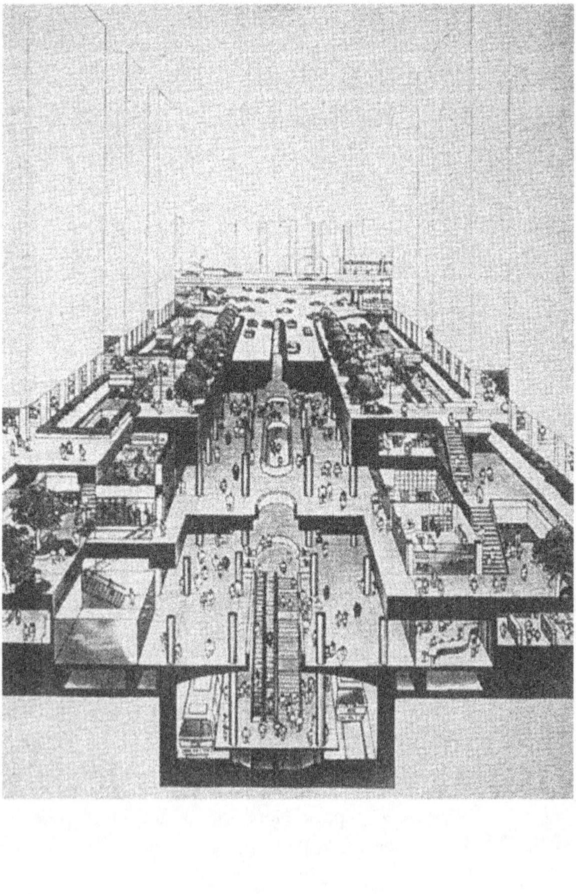

Fig. 42 Schematic Diagram of Underground Space Utilization in Japan. *Source Underground Metropolis*. Tokyo: SEIBUNSHA. August 1989, p. 12

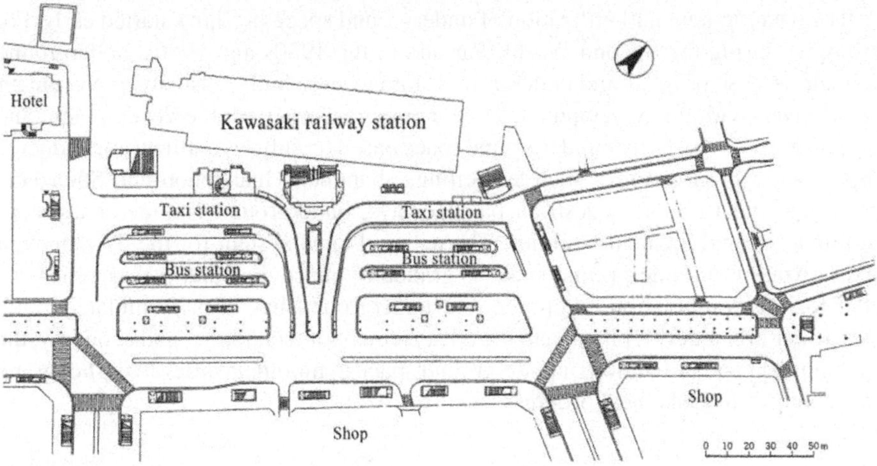

Fig. 43 Floor Plan of the Railway Station in Kawasaki, Japan. *Source Underground Metropolis.* Tokyo: SEIBUNSHA. August 1989, p. 146

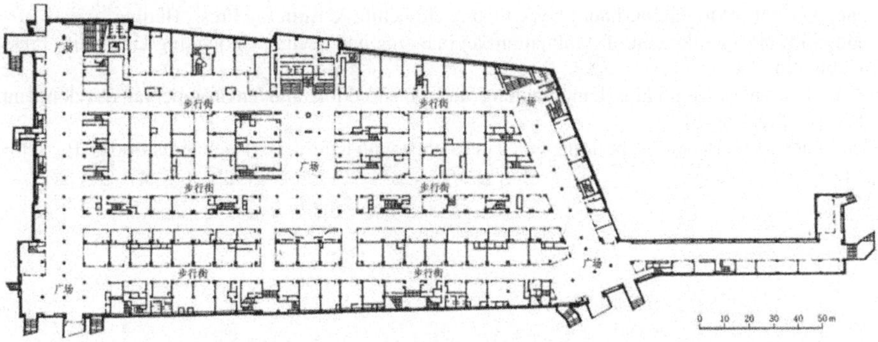

Fig. 44 Floor Plan of the First-floor underground Railway Station in Kawasaki, Japan. *Source Underground Metropolis.* Tokyo: SEIBUNSHA. August 1989, p. 146

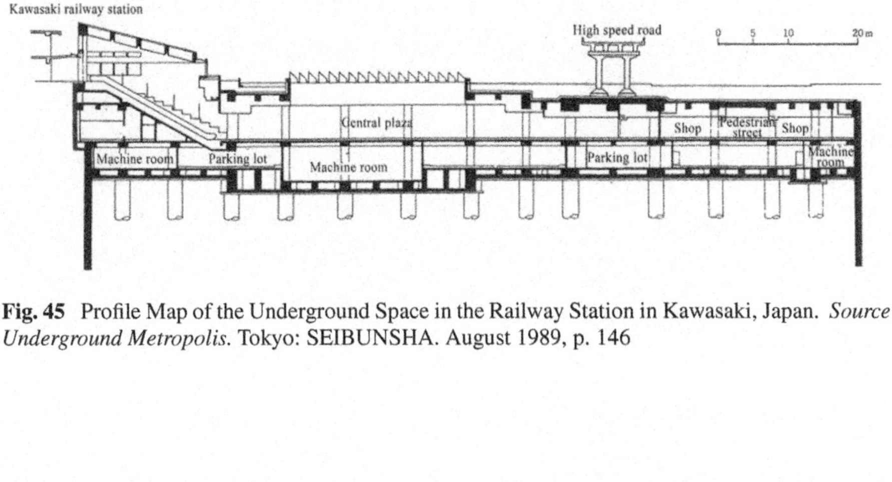

Fig. 45 Profile Map of the Underground Space in the Railway Station in Kawasaki, Japan. *Source Underground Metropolis.* Tokyo: SEIBUNSHA. August 1989, p. 146

The development and utilization of underground space in China started early. For example, during the Second World War and in the 1950s and 1960s, underground civil air defense projects and underground storage were built passively to prevent air raids. Later, with the development of economic construction, the development and supporting construction of underground space based on subway traffic was conducted successively in large cities such as Beijing, Shanghai, Guangzhou and Shenzhen, including underground pedestrian passageways, underground commercial streets, and underground medical facilities. The national legal system for the development and utilization of underground space is gradually improving, and a comprehensive plan for the combination of ground and underground has been formulated. With the further urban development and the advancement of technology and economy, the development and utilization of underground space in mountain cities like Chongqing will continue to make new progress.

References

Huang G (2002) Mountainurbanology. China Architecture & Building Press, Beijing

Huang G (2003) Collection of mountainurban planning and design. Chongqing University Press, Chongqing

Huang G (1994) Mountainurban planning and construction and environmental development. Science Press, Beijing

Richard R (2003) Ecocities: building cities in balance with nature. Berkeley Hills Books, Berkeley, California

Chapter 6
Design

In modern urban design, we should not only pay attention to the space, axis, visual corridor and continuity formed by a square and a street in a city, but also more focus on the improvement of environmental quality and people's psychological and physiological experience to achieve the goals and values of sustainable development it pursues through the formulation of a variety of management policies. Modern urban design is also a continuous decision-making process, which should be part of public policies.

1 Characteristics of Mountainurban Design

(1) Complexity and limitation of land use

The complex natural geographical conditions and eco-environment of mountain cities, such as terrain changes, geological conditions, climatic characteristics, biological and landscape diversity, as well as the constraints of mountains, rivers and elevation, bring a series of complex and varied factors, influences and constraints for urban design, which need to be considered comprehensively.

(2) Multi-dimensionality and richness of space

The rich and varied topography and landforms of mountain cities are precious natural resources, which serve as the natural basis for the multi-dimensional mountainurban spatial changes. For example, the areas of Chongqing are named after "Gang", "Ling", "Liang", "Gu", "Ya", "Po", "Ping", "Chang", "Kan", "Bei", "Ya", "Ao", "Gou", "Wan", "An", "Xia", "Tuo", "Tan", "Hui", "Dao" and "Zui" according to various topographic features and characters. These different topographical and geomorphic features are also important factors for the formation of rich and varied landscapes and microclimates in mountain cities, which provide a variety of changing natural environmental conditions for mountainurban design (Guangyu 2003).

© The Author(s), under exclusive license to Springer Nature Singapore Pte Ltd. 2021

G. Huang, *Theory of Mountainurbanology*,
https://doi.org/10.1007/978-981-16-0819-3_6

2 Basic Principles of Mountainurban Design

(1) Ecological priority

Ecological priority is an important guiding principle for modern urban design, which is particularly important in mountainurban design. In terms of ecological priority, the principle of respecting nature in design should be set up first to realize the harmony and unity between man and nature. In the design process, we should make full use of all favorable conditions of nature to achieve economic, practical and aesthetic purposes, such as how to combine natural climate, terrain, water, local building materials, and excellent traditional construction techniques and approaches. Mountain cities, with more natural conditions, should further give play to the design creation with the integration of planning, architecture and gardens.

(2) Interactive with water

Humans like to be close to water. In landscape cities, rivers and lakes constitute an important material basis for the spatial pattern of landscape cities. In the mountainurban design, the natural water system should be used as much as possible to make the urban spatial structure full of anima. And there should be certain water surface space in urban residential areas to increase infinite temperament and interest to citizens.

Su Dongpo, Ouyang Xiu and Yuan Shen, the famous writers in the Chinese history, attached great importance to the use of water in their positions as local officials, such as the exploitation of the West Lake in Hangzhou City, the West Lake in Huizhou City and the Slender West Lake in Yangzhou City, which not only bring endless benefits to the urban residents, but also add unlimited scenery to the cities. These have all have become world-famous tourist destinations Fig. 1.

(3) Slope selection

Slope land is a decisive factor in the mountainurban design. The size of slope land affects the selection of land use, overall planning layout and road traffic organization, and more importantly, the design, architectural layout and construction mode of the city.

Different urban construction land has different requirements for gradient and slope direction, and residential land is generally required to be arranged on a sunny slope below 30%. When the gradient is less than 5%, the building can be flexibly arranged, which is basically not affected by gradient. If the gradient exceeds 5%, it is necessary to adopt special arrangements such as faulting of slab ends, dislocation, stilt and climbing with parallel, vertical or oblique contour lines to meet the requirements of different terrains Fig. 2.

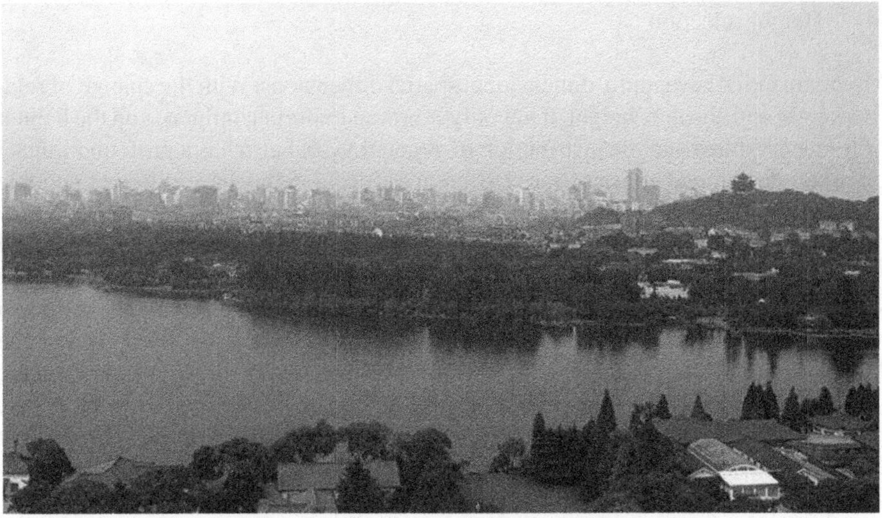

Fig. 1 West Lake in Hangzhou. *Source* Photograph by the author in 2004

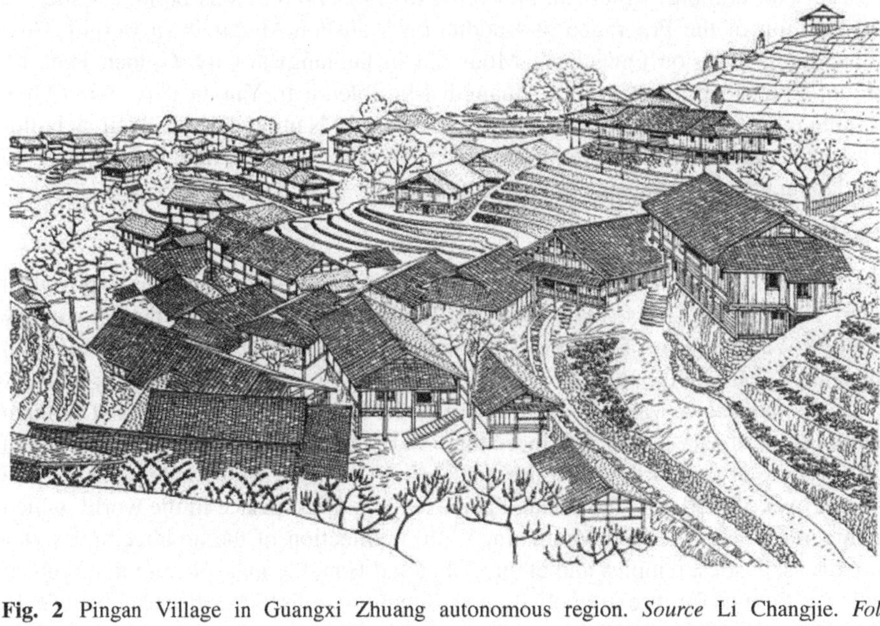

Fig. 2 Pingan Village in Guangxi Zhuang autonomous region. *Source* Li Changjie. *Folk Architecture in Northern Guangxi.* Beijing: China Architecture & Building Press. May 1990. p. 31

(4) Height selection

Mountain cities have multi-dimensional spatial dimensions. With the change of relative height and absolute height, it not only exerts an important influence on the habitat of living creatures and the distribution of vegetation, but also has a profound impact on urban design and construction. Different vertical heights have different suitable design and construction approaches. For example, public buildings with large flow of people or industrial buildings with high water consumption and large demand for transportation should be arranged at a lower altitude near river. In contrast, the scenic or memorial buildings that can be ascended to enjoy a broad, unobstructed and distant view should be arranged in higher places, such as the Stork Tower, the Yellow Crane Tower, the Yueyang Tower, the Tengwang Pavilion and other scenic spots. As a famous ancient poetry of Wang Zhihuan, the *Climbing the Stork Tower*, goes, "……To see a thousand miles, one should ascend another story", and the Yellow Crane Tower, Tengwang Pavilion and Yueyang Tower are all excellent scenic spots to inspire people's emotional world.

In addition, there are also some buildings with special significance and use requirements, such as the Potala Palace in Lhasa City and the Acropolis of Athens in Greece with combined political and religious powers, religious buildings such as the Pavilion of the Fragrance of Buddha on Wanshou Mountain in Beijing City, Shangqing Palace on Qingcheng Mountain in Dujiangyan City, Golden Peak on Mount Emei in Emeishan City, Huangdi Mausoleum in Yan'an City, Sima Qian Temple in Weinan City, Sun Yat-sen Mausoleum in Nanjing City, as well as buildings with special functional requirements, such as Purple Mountain in Nanjing City and Dongchuan Observatory in Kunming City.

Potala Palace

In 633 A.D., Songtsan Gambo, the King of the Yalong Tribe, established a powerful Tubo Dynasty in Lhasa, built Jokhang Temple and Ramoche Temple to marry Princess Wencheng into Tibet, and constructed the Potala Palace on the Red Mountain in center of Lhasa Fig. 3.

The Potala Palace was built on the top of the Red Mountain at an altitude of 3,700 m in the center of Lhasa, which is more than 370 m long from east to west, more than 100 m wide at the north and south, and 117.19 m high, with a total construction area of over 57,700 m^2. The Potala Palace is the highest palace in the world, which is integrated with the Red Mountain. With a collection of the architectural styles of palaces, castles, temples and ethnic Tibet and Han, the majestic and magnificent palace is famous for its extremely high architectural, religious and artistic values. In 1961, it was rated as the first batch of key cultural relics protection units in China, and listed as a world cultural heritage by UNESCO in 1994.

Acropolis of Athens

Parthenon Temple, a jewel in the crown of the Acropolis, was built on the base of three steps Fig. 4. The columns on both sides are slightly inclined inward to correct the visual deviation. The combination of Doric order and Ionic order is a

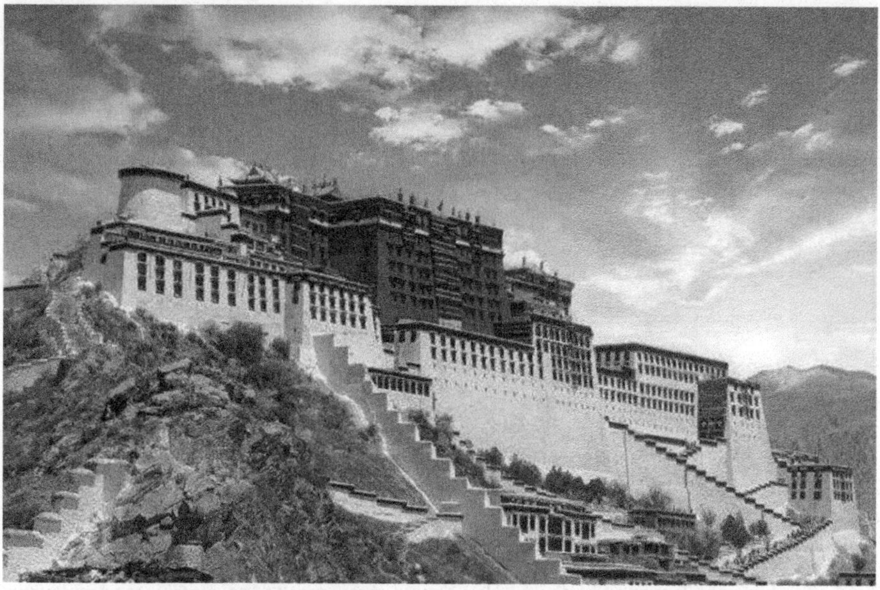

Fig. 3 Panorama of Potala Palace in Lhasa. *Source* Photograph by the author in 1986

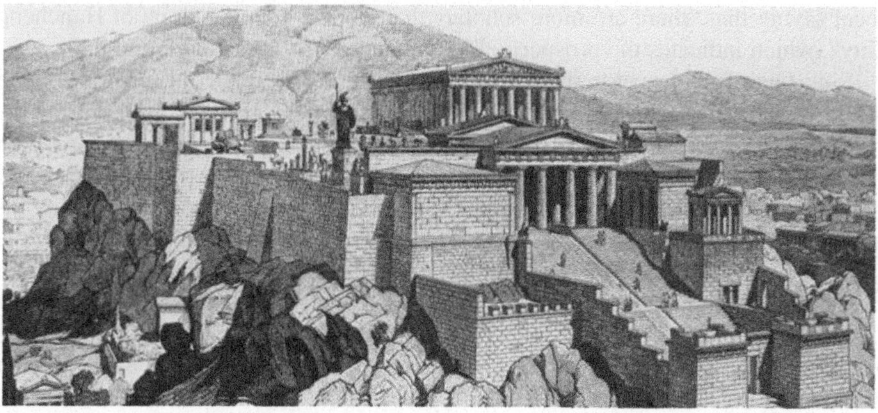

Fig. 4 Restoration map of Acropolis. *Source* Website of Ancient Greek Culture http://www.ntgs.com.cn

typical representative of ancient Greek classical architecture, which has an important influence on the architecture of later generations.

Sima Qian Temple

Sima Qian Temple was built on the slope of Hancheng City on the bank of the Yellow River in Shaanxi Province, the hometown of Tagaung Shi Figs. 5 and 6. There is a

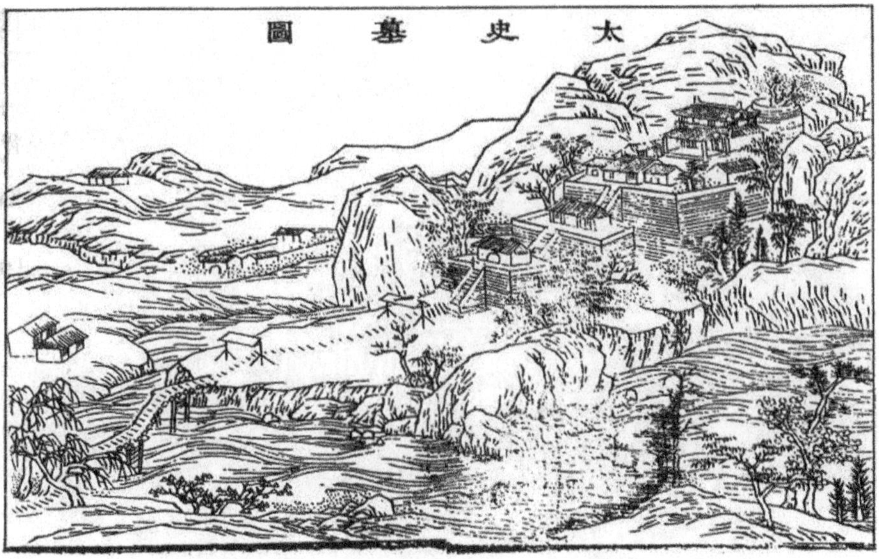

Fig. 5 Tomb map of Tagaung Shi. *Source* Hou Youbin. *Aesthetics of Ancient Chinese Architecture*. Harbin: Heilongjiang Science & Technology Press. April 2000, p. 81

local saying that "there are more scholars than donkeys on the slope of Hancheng City", which indicates the prosperity of this famous historical and cultural city.

Sima Qian Temple was first built in 307 A.D. (during the reign of Emperor Yongjia in the Western Jin Dynasty), and was expanded from 1068 to 1077 A.D. (during the reign of Emperor Zhengping in the Song Dynasty). The ancestral hall was built with bricks and stones, with a 4-m-high colorful statue of Sima Qian. Around the ancestral hall, there are 59 inscriptions recording Sima Qian's life experience and praising his achievements. Behind the temple is the Eight-trigram Tomb of Sima Qian. After the liberation of China, the ancestral hall was renovated. The hall was built on a small overlooking mountain ridge of Hancheng City, which ascends layer by layer from the ancestral gate of "Sun of Mountains and Rivers", the archway of "Beholding a High Mountain with Awe" to the Eight-trigram Coffin Chamber, with a magnificent momentum Fig. 7.

In 1958, the Chinese litterateur Guo Moruo wrote a poem that reads "Longmen, a picturesque place with mountains and rivers, cultivated many outstanding persons like Sima Qian who was very knowledgeable, surpassing his predecessors. His essays can be regarded as master pieces of many ages. However, the poor genius suffered intolerable torture. He wrote the Historical Records with anger and resentment, which was as glorious as a rainbow. The achievements of Sima Qian can be compared with those of Confucius. As an outstanding historian, he has been remembered through the ages", to praise his great spirit of enduring humiliation and finally completing the monumental historical works under the humiliating punishment of the royal court.

Fig. 6 Floor plan of Sima Qian Temple. *Source* Hou Youbin. *Aesthetics of Ancient Chinese Architecture.* Harbin: Heilongjiang Science & Technology Press. April 2000, p. 81

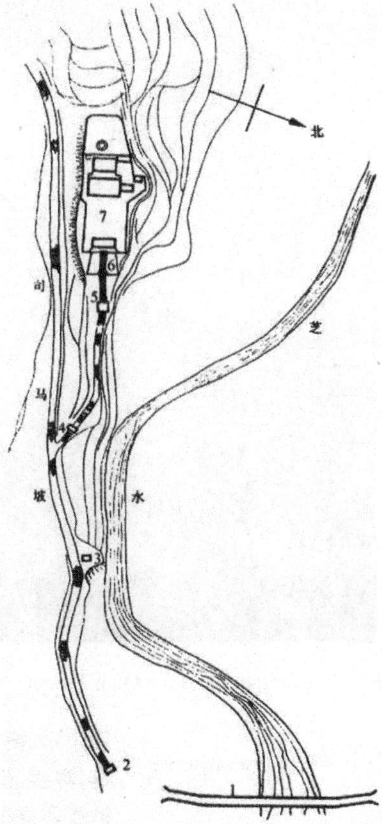

(5) Top retention

There are many mountain ridges and hills distributed in mountain cities in most cases, taking a commanding and prominent position. These mountain ridges and hilltops are important scenic spots or landscape sites in the cities, and also good destinations for citizens to climb, overlook and appreciate mountainurban scenery, which should be used as green scenic spots under normal circumstances to give fully play to the service function of natural ecology and be enjoyed by the public, rather than being occupied by individual institutions. As Wright (1992) put it well in a conversation about "Taliesin", a house from his own design and construction, "I don't think it should be built directly on the top of anything. If you build a house on a hilltop, you will lose the hill. If you build a house on the side of a hilltop, you will get the hill and the highland you desire…". Later, he applied the same concept of "Taliesin" to the design of "water villa" Fig. 8.

The city of San Francisco has protected many hilltops as a leisure green space and viewing space for citizens Figs. 9 and 10.

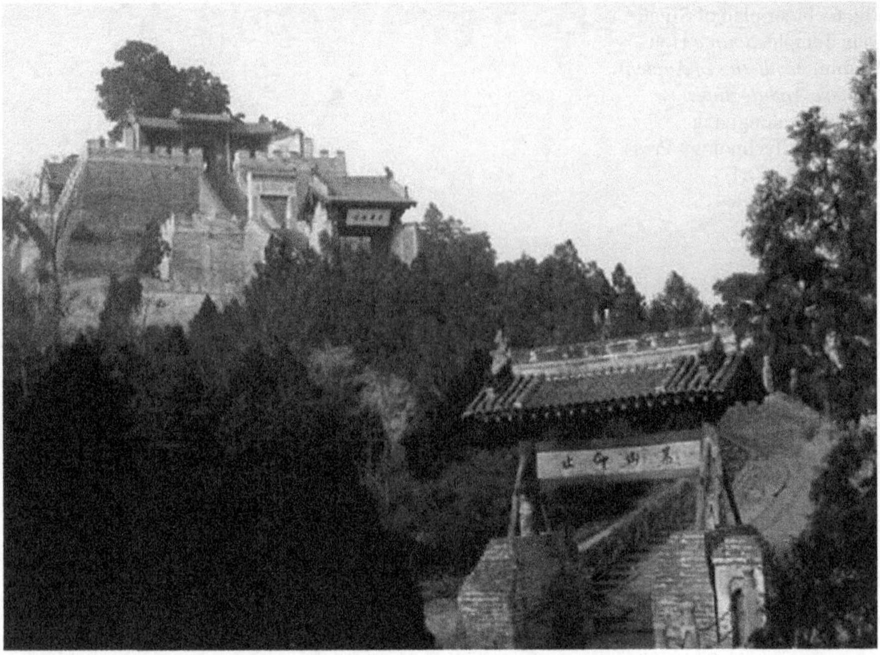

Fig. 7 Panorama of Sima Qian Temple. *Source* Photograph by the author in 2003

(6) Edge occupying

In traditional Chinese architecture, the concept of axis symmetry and building centering is often applicable to solemn and respectful royal cities, palaces and important monumental buildings. However, under the condition of complex terrain, the edge occupying principle of leaning to one side is generally applicable to a large number of buildings, so as to form a larger outdoor activity space and make the building layout freer and more flexible. In traditional Chinese painting, the layout, blank space and title are often the same Fig. 11.

(7) Multidimensional concentration

Due to the shortage of land available for construction and high ecological sensitivity in mountain areas, it is emphasized to create a multi-dimensional and intensive mountainurban space in combination with terrain and environment, including architectural space, urban traffic space and open activity space, to strive for living space and compactly save land, so as to minimize the disturbance and occupation of terrain and achieve the intensive spatial effect. The intensification of space brings about the intensification of time accordingly, which also improves the efficiency of urban activities Figs. 12, 13 and 14.

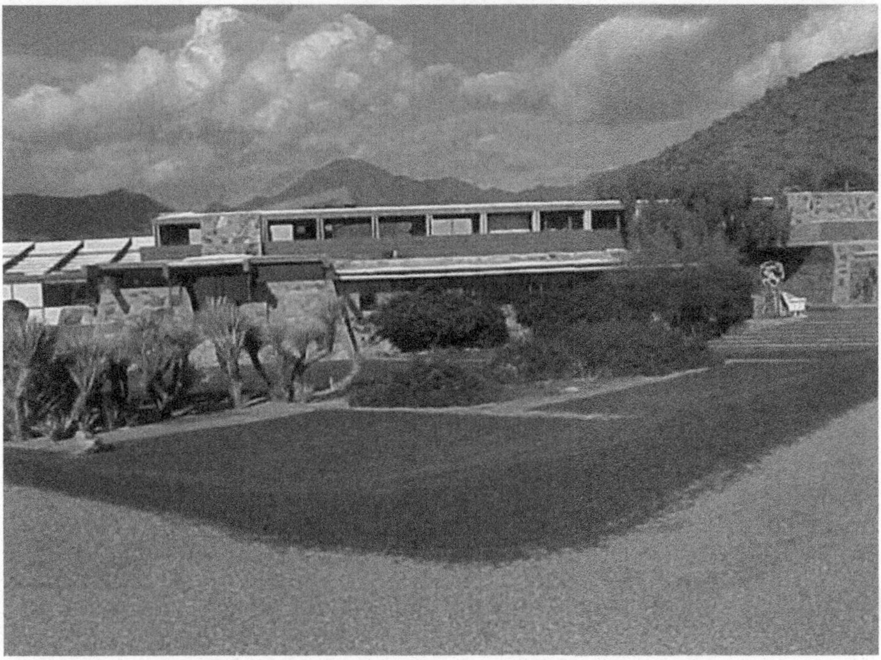

Fig. 8 Residence of Taliesin. *Source* http://www.franklloydwright.org

(8) Overall coordination

Due to the complexity of terrain and the diversity of space, mountain cities are generally different from flat cities and present rich and varied urban contour lines and skylines. The overall coordination principle of mountainurban design is that in the design of each local section or block, overall coordination should be achieved as much as possible, in order to achieve the effect of unity in change and change in unity.

3 Methodology of Mountainurban Design

The methodology of mountainurban design is to comprehensively apply the above design concepts in combination with mountains, water systems, topography, climate and current conditions of a city to create new cities, new settlements and new buildings that adapt to ecology, in order to bring new vitality into play in the cross integration of ecology, urban science and architectural science.

Lao Zi, a famous philosopher in the Spring and Autumn Period, said, "The existence of everything in the world is named and visible, but the named and visible things must take the nameless and invisible "Tao" as the root." It was recorded in his

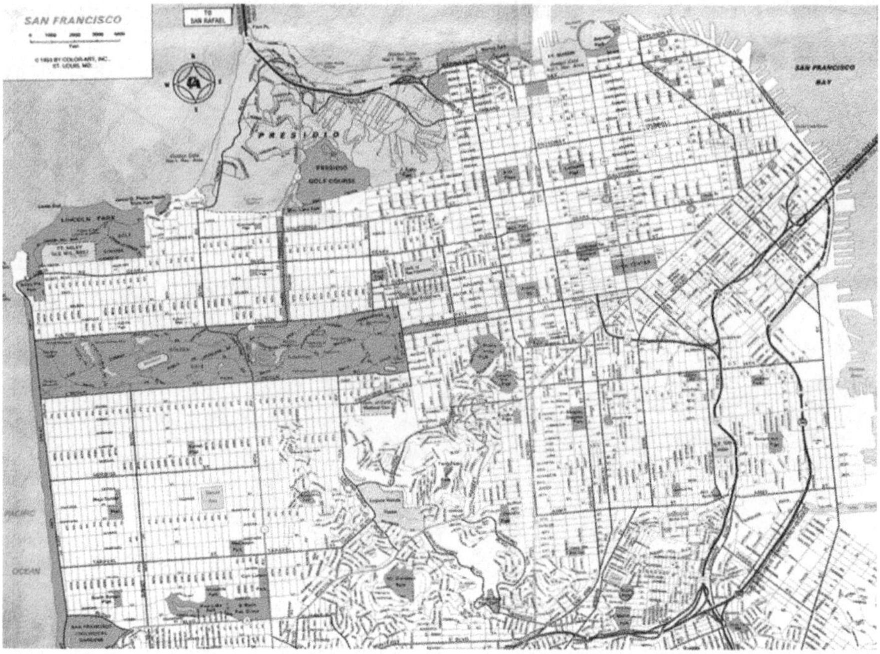

Fig. 9 Map of San Francisco Preserving Mountains as Urban Green space. *Source* Repainting based on the map of San Francisco collected by the author in 1996

famous works the *Tao Te Ching* that "Thirty spokes radiate from a hub. When there is nothing in the hub, the wheel can roll. Clay is kneaded to mould a vessel. The vessel is useful only because of the space within. Doors and windows are cut out to form a room, but it is on the interior vacancy that the utility of the room depends. What is useful will do good, and what seems useless may become useful". (*Tao Te Ching Chap.* 11) These words of Lao Zi are often taken as a guideline for design and creation by architecture teachers and students. Lao Zi emphasized the importance of space from the function of wheels to a gharry, from potting to pottery, and from building a house with windows to the function of a house. It is said that the American architectural design master F·L·Wright's design concept of flowing space and organic unity was once influenced by the thought of Lao Zi. We use his words to understand the design of urban space, which is the same. It shows that the creation of a vibrant urban spatial environment is extremely important for the "giant" city to give play to its comprehensive functions. The core is to be used and needed by people. The premise is to establish a harmonious and friendly relationship with the natural environment and make use of nature without damaging it.

There are many natural conditions that we can take advantage of in mountain areas and cities. For example, in valley cities, the difference of topographic height between the valley bottom and the valley slope can be used to enhance the air convection between the top and foot of a mountain and between daytime and nighttime,

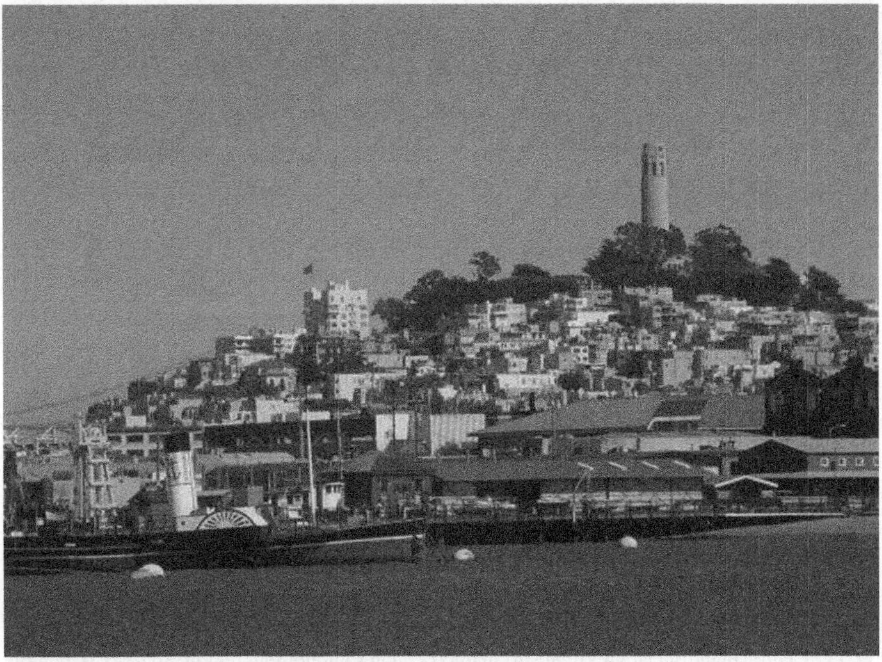

Fig. 10 Waterfront architectural complex of San Francisco complex built around Hillsides. *Source* Photograph by the author in 2005

so as to mitigate the hot environment. In waterfront mountain cities, the temperature difference between land and water can be used to introduce cool breeze into urban streets and valleys, in order to improve the urban microclimate and sanitary conditions. It is also possible to improve the local climate and microclimate through building construction, plane layout, facade retraction and strewn distribution of different heights. For example, Yantai, a coastal mountainurban, makes use of its geographical advantages and the altitude difference between the mountains and the sea to connect the green core of Nan Mountain and the bay through the ventilation corridors of the city, allowing the mountains and the sea to be connected, which has a good effect on the ventilation and cooling of the hot environment in summer. Chongqing is located in the upper reaches of the Yangtze River and the parallel ridge valley at the eastern margin of the Sichuan Basin. The lush forests of Nanshan Mountain, Gele Mountain and Tieshanping play an important role in improving the urban thermal environment and providing leisure tourism for citizens. However, due to the high-intensity development and construction of residential areas on the mountains in recent years, the forest vegetation has decreased year by year, which greatly weakens the cooling effect of convection between the cold airflow and the warm airflow on the top and at the foot of a mountain.

In order to cope with the grim situations of the global environmental crisis, climate warming, increasing urban population and building density, sharply decreasing water

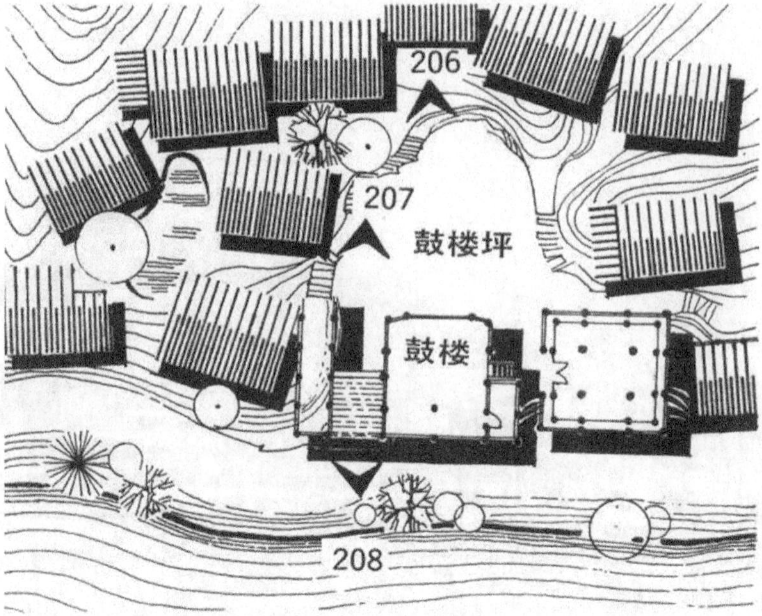

Fig. 11 Central plane of Datian Village in Guangxi Zhuang autonomous region. *Source* Li Changjie. *Folk Architecture in Northern Guangxi.* Beijing: China Architecture & Building Press. May 1990. p. 123

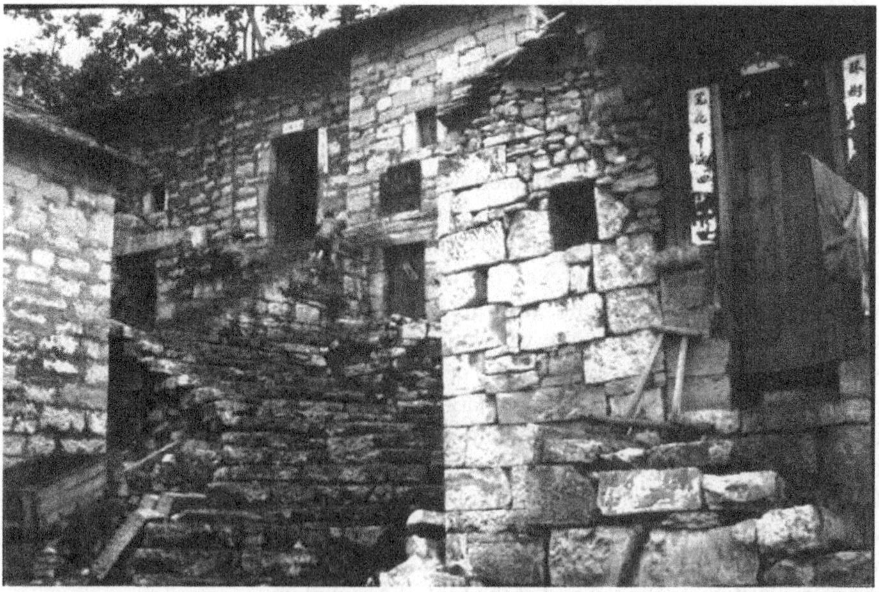

Fig. 12 Multi-dimensional intensive space in Shitou Village, Guizhou City. *Source* Photograph by the author in 1998

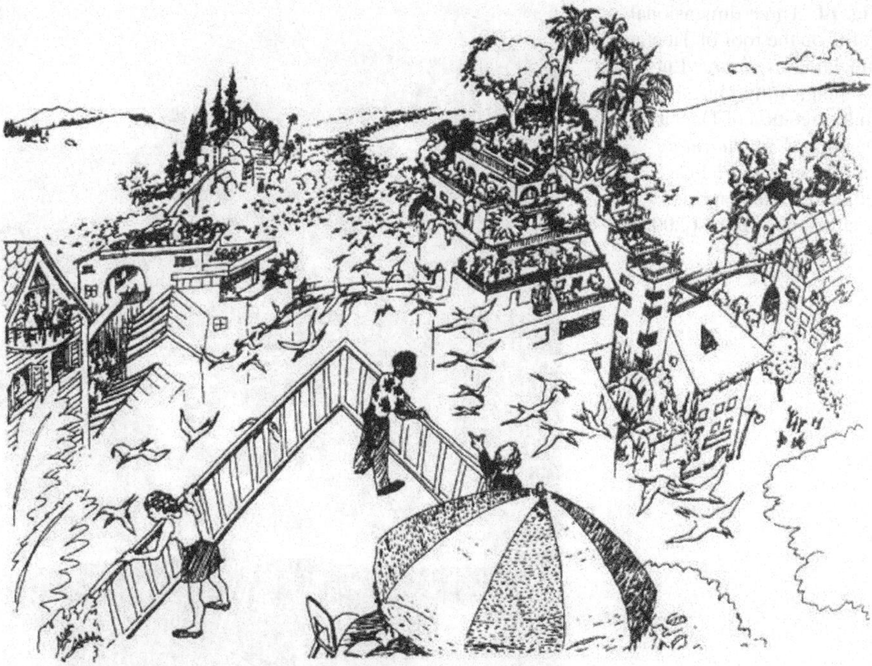

Fig. 13 Roof space in Rogersart's imagination. *Source* (US) Richard Register. *Eco-cities, Beikeley Hills Books*. 2002, p. 182

resources and intensifying environmental pollution, many cities have begun to reflect on the past foolery of reclaiming land from lakes, filling lakes for housing and filling rivers for roads, which has led to the deterioration of urban environment and the sharp decline of biodiversity and landscape diversity, and have implemented ecological landscape restoration measures and wide-area urban environmental policies, such as Seoul in South Korea and Beijing City, Wuxi City and Hangzhou City in China.

Wu Liangyong, Professor of Tsinghua University, created the concept of "quasi courtyard" through the protection and renewal of traditional quadrangle residences in Beijing, and made meaningful practice and exploration in improving building density and saving urban construction land on the basis of preserving local traditional residential culture and original courtyard environment and trees.

Chen Qigao, Professor of Chongqing University put forward the concept of "geomantic field and environment" through the research on traditional courtyard residences, and proved through experiments and calculations that a traditional quadrangle courtyard is actually a geomantic field and environment, which has the functions of absorbing fresh air, purifying, storing and automatically adding oxygen, and can often automatically keep fresh air in the courtyard space to make up for the shortage of oxygen at night.

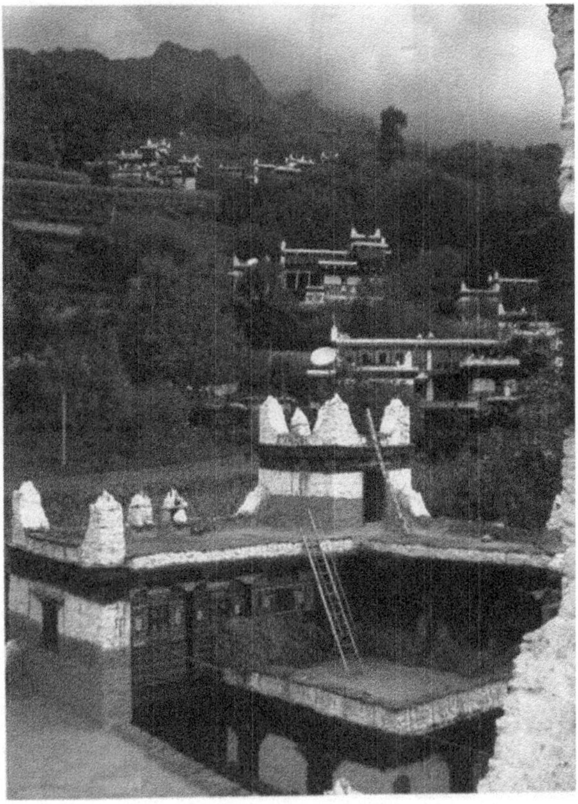

Fig. 14 Three-dimensional traffic on the roof of Tibetan Blockhouses. *Source* Luo Jianping, Xun Ping. "Interpretation of Danba Dwellings" in *Huazhong Architecture*. Periodical Office of Huazhong Architecture. August 2004, p. 74

The American ecologist and architect R. Reigest has been engaged in the practice of eco-city and eco-architectural design with a successful pilot in Berkeley.

Through the study of tropical regions and architecture in India, the Indian architect C. Correa put forward how to use open space (such as under the shade of trees) to develop "open-air architecture" and "tubular housing", that is, "tubular housing" designed for dry and hot areas by using shaded outdoor or semi-outdoor activity spaces or natural ventilation systems Figs. 15, 16 and 17.

The Malaysian architect K. Yeang applied ecological concepts in the design of high-rise buildings in combination with the principles of biological climatology, and adopted natural ventilation or multiple design techniques such as walls, roofs, floors and greening, to solve the problem of energy conservation, achieving obvious results Fig. 18.

Guizhou Huaxi Guest Hotel is located in the Huaxi scenic spot 17 km southwest of Guiyang City. Its designer Luo Deqi fully combined mountains, waters and trees, conducted special mapping on the trees in the site, retained 718 arbors with a stem of more than 10 cm, and kept clear the trees in the architectural layout, in order to organically combine the planning, architecture and garden, creating a building with rich regional cultural characteristics and personality Figs. 19, 20 and 21.

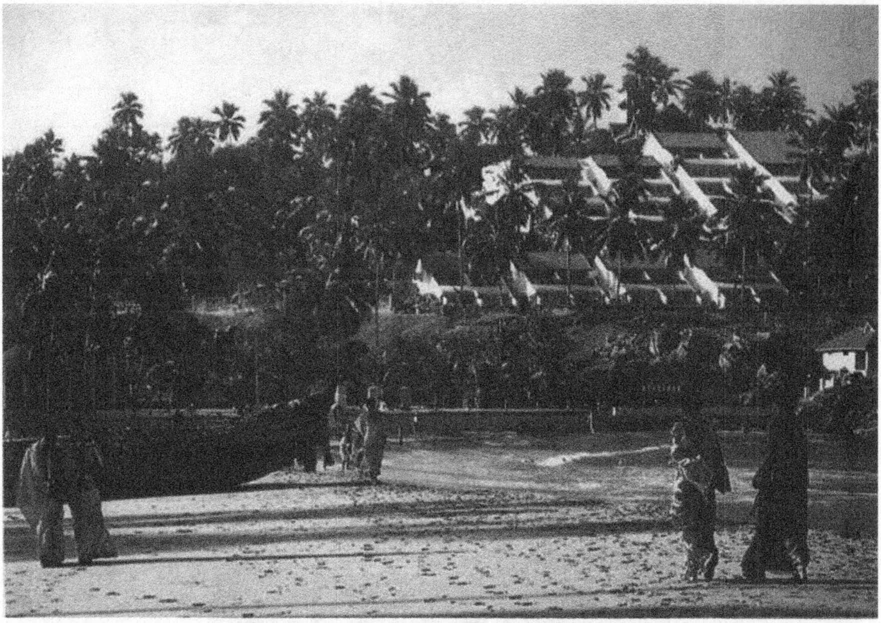

Fig. 15 Kovalam Beach Hotel in India. *Source* Lu Jiwei, Wang Haisong. *Mountain Architecture Design*. Beijing: China Architecture & Building Press. February 2001, picture on the inside cover page

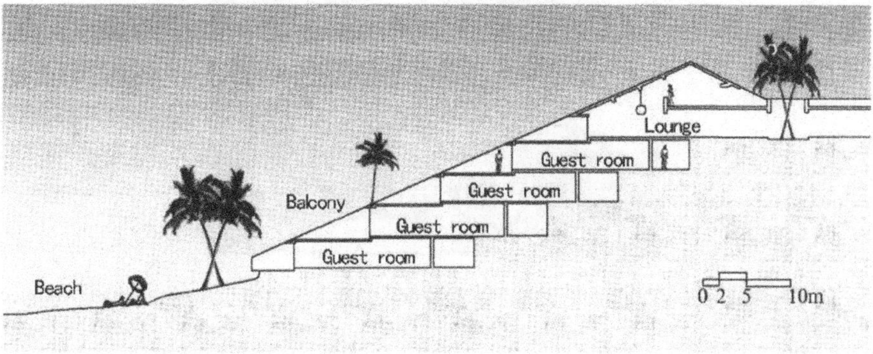

Fig. 16 Profile map of Kovalam Beach Hotel in India. *Source* Wang Fang. *Charles Correa*. Beijing: China Architecture & Building Press. August 2003, p. 113

The Apple Primary School was built in Darchen, Ali Region, in 2004, with a construction area of 1,850 m^2. The building fully combines the natural climate characteristics of high altitude, plateau drought and anoxia in Tibet, and the local terrain, and adopts the layout of low-rise courtyard Figs. 22, 23 and 24.

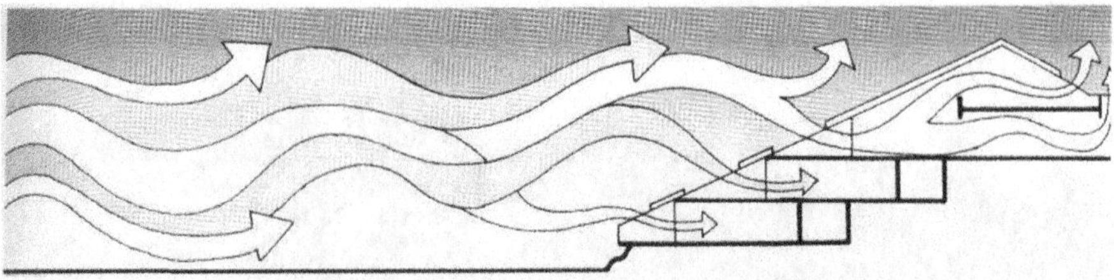

Fig. 17 Airflow analysis diagram of Kovalam Beach Hotel in India. *Source* Wang Fang. *Charles Correa*. Beijing: China Architecture & Building Press. August 2003, p. 113

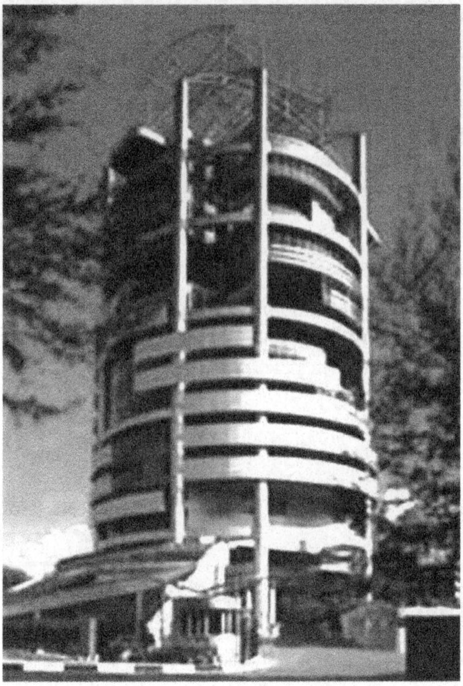

Fig. 18 Menara commercial building in Malaysia designed by K. Yeang. *Source* ABBS Architecture BBS http://abbs.com.cn/

4 Accessible Design in Urban Environment

(1) Concept of accessible urban environment

In the past, accessible design was more considered to provide convenience for people with mobility difficulties, but it is not enough. Accessible design should be extended to urban environments of various public activities and communication venues such

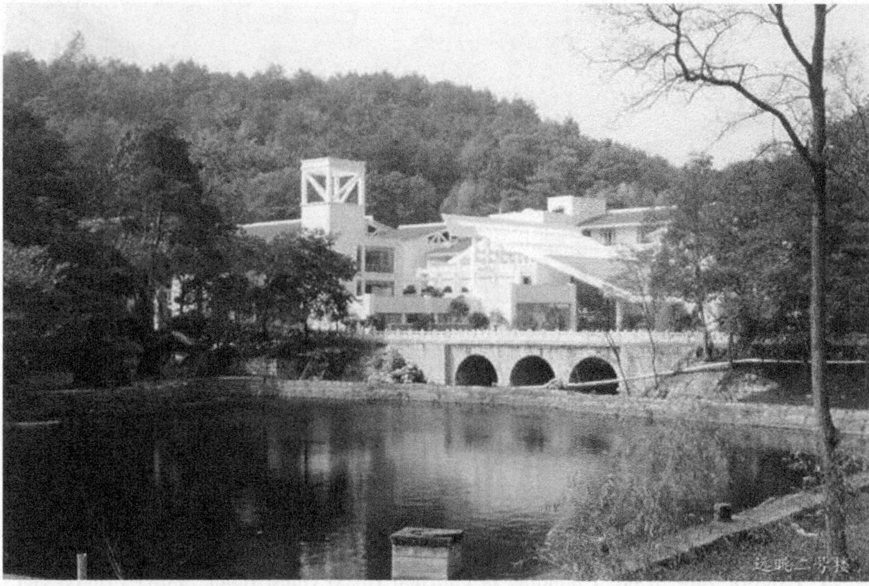

Fig. 19 Panorama of Guizhou Huaxi Guest Hotel. *Source* Luo Deqi. "Huaxi Guest Hotel: An Architectural Design Respecting Geographical Environment" in *Architectural Creation*, Periodical Office of Architectural Creation, December 2004, p. 68

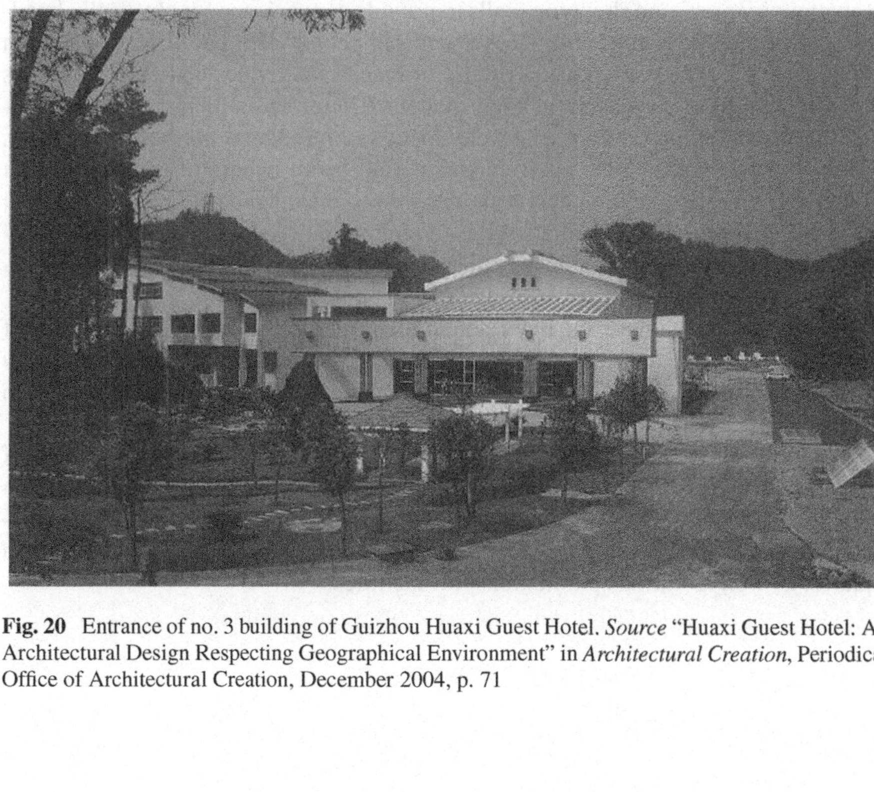

Fig. 20 Entrance of no. 3 building of Guizhou Huaxi Guest Hotel. *Source* "Huaxi Guest Hotel: An Architectural Design Respecting Geographical Environment" in *Architectural Creation*, Periodical Office of Architectural Creation, December 2004, p. 71

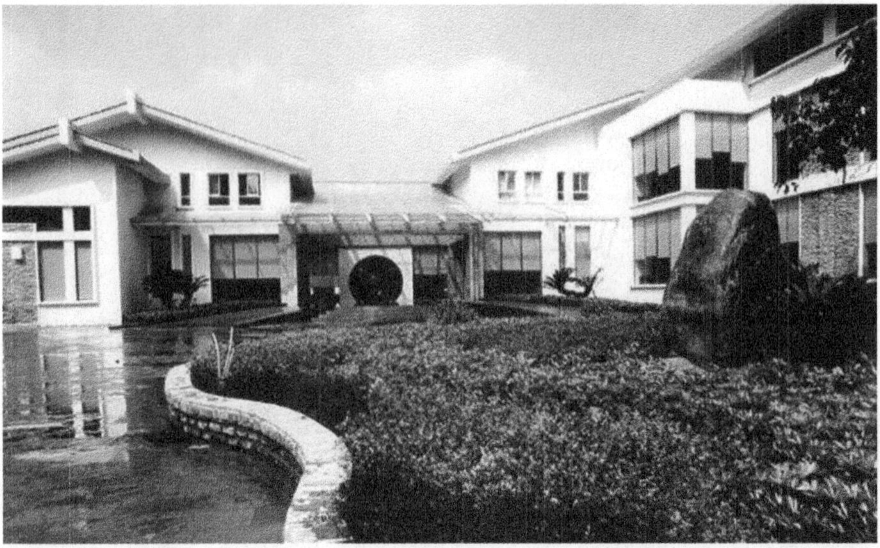

Fig. 21 Exterior view of no. 1 building of Guizhou Huaxi Guest Hotel. *Source* "Huaxi Guest Hotel: An Architectural Design Respecting Geographical Environment" in *Architectural Creation*, Periodical Office of Architectural Creation, December 2004, p. 67

as community parks and shopping malls to provide convenience for physically handicapped people. There are more than 80 million people with mobility problems in China, who should equally participate in all kinds of activities and enjoy all fruits of civilization in urban society in a fair manner. Therefore, in the process of implementing the scientific concept of development and building a harmonious society, building an accessible urban environment should be an important part that cannot be ignored in urban planning and design and environmental construction. In recent years, various cities have gradually given attention to accessible design, and accessible facilities have been installed in main streets, transportation facilities, and important public buildings. However, regardless of the scale and content of the installation, it is far from meeting the needs. In the actual construction of many cities, the needs of people with mobility difficulties have not been carefully considered, and even many sidewalks for the blind exist in name only. For example, blind people are rarely seen walking on the sidewalks.

According to the construction experience of developed countries, the design of accessible urban environment should be comprehensive in the places where urban people interact intensively, including public places such as roads, community environment, various public buildings, medical institutions, parks, scenic spots, and urban squares, as well as various public transport facilities.

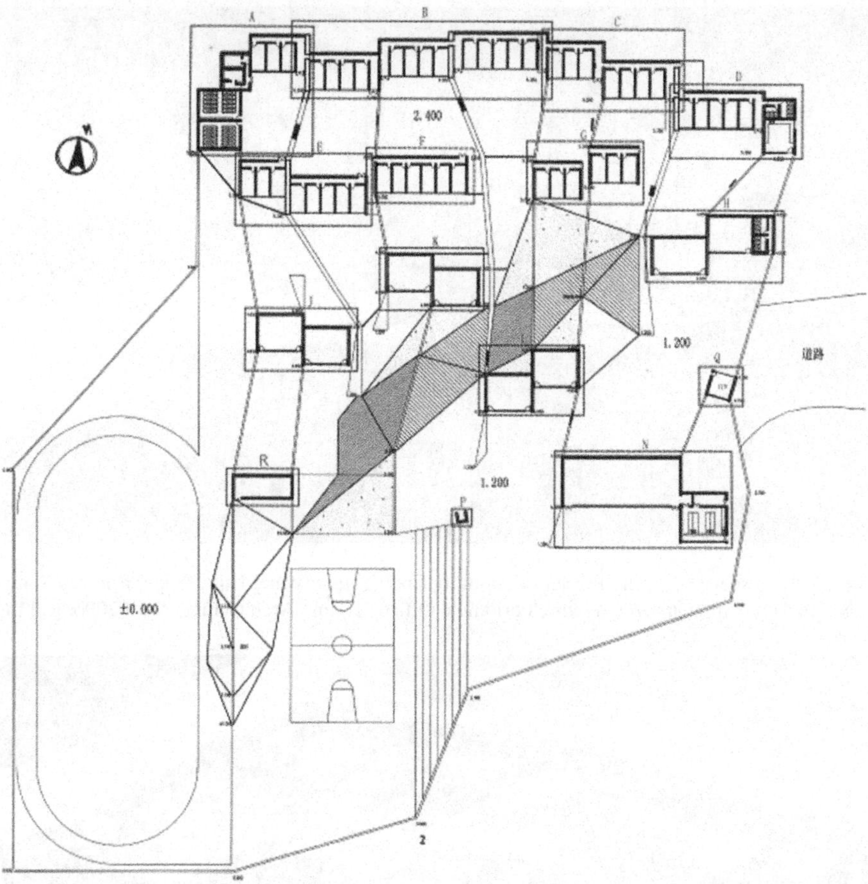

Fig. 22 Floor plan of Tibet Apple Primary School. *Source* Wang Hui. "Apple Primary School in Ngari, Tibet" in *Time Architecture*. Periodical Office of Time Architecture. April 2006, p. 115

(2) Example of accessible urban environment construction in Taipei City

In the 1980s, Taipei City promulgated to implement the "Disability Welfare Law", and begun to promote the construction of accessible environment. In 1995, the Taipei Municipal Government established the "Accessible Environment Promotion Committee", which held meetings regularly to formulate and promote policies for the construction of accessible environment in the city. In 1998, the "Chap. 10 Facilities for Disabled Persons in Public Buildings" was added to the *Building Technical Regulations—Building Design and Construction*, to implement the construction of accessible environment in public buildings and venues, which have been gradually completed after long-term construction. At present, a relatively complete accessible urban environment has been formed in Taipei City (Table 1).

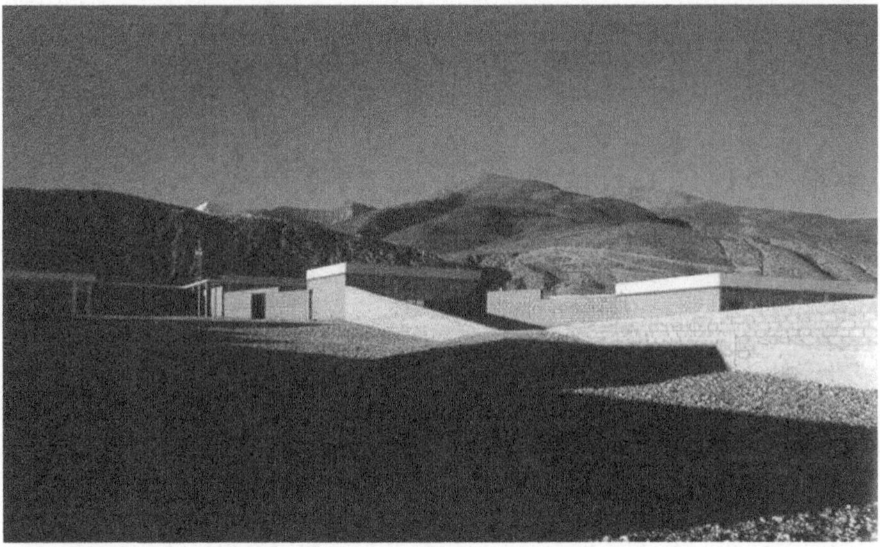

Fig. 23 Real scene of Apple Primary School in Tibet. *Source* Wang Hui. "Apple Primary School in Ngari, Tibet" in *Time Architecture*. Periodical Office of Time Architecture. April 2006, p. 114

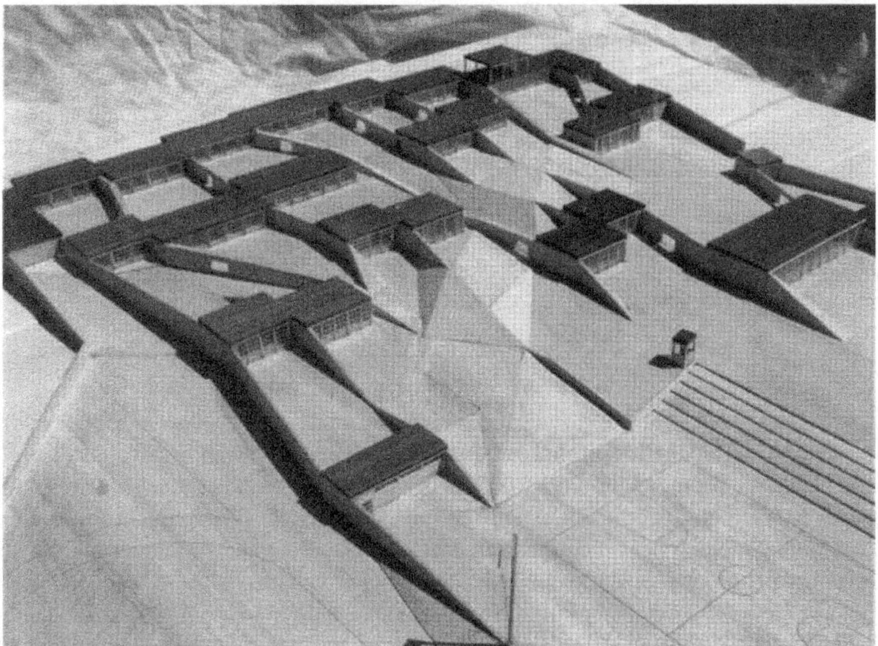

Fig. 24 Model of Apple Primary School in Tibet. *Source* Wang Hui. "Apple Primary School in Ngari, Tibet" in *Time Architecture*. Periodical Office of Time Architecture. April 2006, p. 115

Table 1 Basic framework for the construction of accessible urban environment in Taipei city

Type of construction	Construction content	
Point construction	Public building	Such as government agencies, hospitals, maintenance institutions, stations, schools, department stores, etc
	Public place	Such as squares, large parks, Duli Park, Riverside Park, etc
Line construction	Sidewalk	–
	Arcade	–
	Avenue	–
	linear park	–
	Special footpath	Such as hiking trails, leisure trails, etc
Surface construction	Community	–
	Specific area	–
	Underground street	–
	Bicycle lane system	–
Traffic construction	Personal transport tool	Such as three-wheeled motor tricycles, personal vehicles, etc
	Niche transit system	Such as taxis, small air-conditioned cars, etc
	Mass transit system	Such as general bus, low-chassis bus, MRT system, etc

5 Renovation and Historical Conservation in Traditional Districts of Mountain Cities

(1) Reasonability and legitimacy of the existence of traditional districts

The traditional blocks and old buildings in most mountain cities in China are gradually formed in the long-term urban development, with profound historical and cultural accumulation, small building volume, narrow streets, pleasant scale, harmonious neighborhood relations, rich local social and cultural life atmosphere, and obvious traces of the old city and profound historical memories, leaving obvious traces of traditional districts and decorous historical memory. Many historical events have taken place here, and many moving stories have been staged here. Thus, things flow away, but the future can still be redeemed. Later generations who live in the urban environment with historical and cultural accumulation may be touched by the scenery and reminded of the history and culture when they visit the traditional blocks, old city walls, old buildings, former residences of celebrities, old trees and ancient wells. A city without a history is like a city without a memory, soul, or culture. "Diversity is the nature of a city," said Jane Jacobs. She advocates keeping small-scale blocks and small shops on the streets to increase the chance of people meeting each other in life, so as to enhance the sense of security and intimacy of the streets. Some old streets in Ciqikou Ancient Town in Chongqing Municipality, Dali City and Lijiang City in Yunnan Province, and Yangshuo County, Guilin City in Guangxi Zhuang Autonomous Region have attracted people's interest because of their small-scale street culture. Therefore, in terms of our attitude towards historical buildings and cultures such as old cities and traditional blocks, we should first acknowledge the reasonability and legitimacy of their existence, strengthen their protection and utilization, and then try to improve the unreasonable part that does not meet the needs of modern life, instead of totally repudiating and starting over. In 2004, the Ministry of Construction issued the *Measures for the Administration of City Purple Lines* and the *Guiding Opinions on Strengthening the Planning and Protection of Excellent Modern Buildings in Cities*, emphasizing the requirements for the protection of historical buildings, modern buildings and former residences of celebrities. The 15th International Council on Monuments and Sites (ICOMOS), which was held in Xi'an in October 2005, adopted the *Xi'an Declaration* on the protection of ancient buildings, ancient ruins and the surrounding environment of historic areas was adopted, and clearly indicated that "The environment of historical buildings, ancient ruins or historic areas is defined as a direct and extended environment, that is, as or constituting an integral part of its importance and uniqueness. In addition to the physical and visual meanings, the environment also includes the interaction with the natural environment, the utilization or activities of intangible cultural heritage such as past or current social and spiritual activities, customs and traditional knowledge, as well as other forms of intangible cultural heritage, which create and shape the environmental space and the current, dynamic cultural, social and economic context". The concept and scope of protection have been further expanded, which has important theoretical and practical guiding significance Fig. 25.

(2) Main problems in the traditional districts of mountain cities

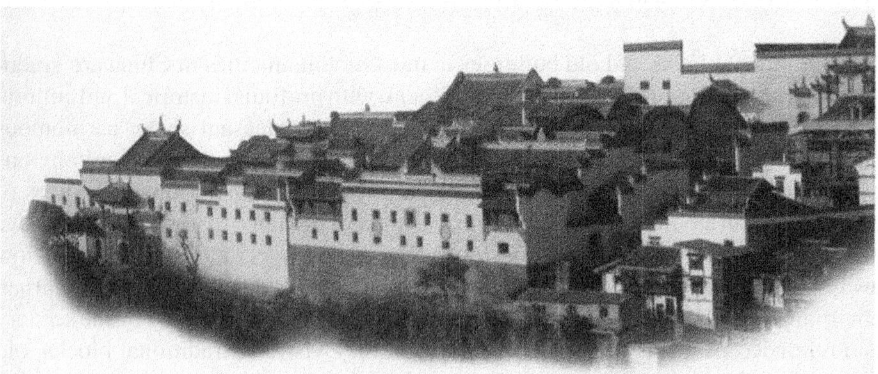

Fig. 25 Panorama of Huguang Guild Hall of Chongqing after Restoration. *Source* Chongqing Municipal People's Government. City · Planning · Art. Shanghai: Academia Press. October 2005, p. 15

Generally speaking, the traditional blocks in mountain cities have the following problems:

(1) Densely crowded buildings and poor fire fighting conditions;
(2) Scarce green lands and open spaces, with poor environmental quality;
(3) Narrow and unsystematic roads;
(4) Lack of public service facilities such as culture and entertainment facilities;
(5) Imperfect pipeline network systems; no centralized water supply and drainage systems in some traditional districts.

(3) Main measures for the renewal of traditional districts and historical and cultural protection

With the development of urbanization, mountain cities have further opened up, the living standards of residents have been continuously improved, and the number of motor vehicles is increasing day by day. The contradiction between urban renewal and historical and cultural protection is also becoming increasingly prominent, which needs to be dealt with carefully. In protecting historical and cultural relics, it is necessary to improve the municipal infrastructure and eco-environment of traditional districts. The main measures are, as follows (Guangyu and Yong 2002):

(1) To delimit the protection scope for traditional blocks and historic buildings with historical and cultural values, as well as their surrounding environment, and strictly strengthen the protection and management in accordance with the Law on the Protection of Cultural Relics and relevant regulations.
(2) To improve the road network system, lead out the transit traffic, increase the fire rescue channel, and improve the pedestrian traffic system.
(3) To demolish the dilapidated houses without protection value, increase roadside green space and square vacant land, and improve the ability of disaster reduction and prevention.
(4) To improve indoor and outdoor water supply and drainage conditions of neighborhoods and residences, and the system of municipal facilities such as water supply and drainage, electricity, communications, heating, and gas supply.
(5) To increase the conditions of cultural, entertainment and commercial services in traditional districts, improve community management, and raise the level of comprehensive public cultural service facilities.

6 Vertical Design and Outdoor Engineering of Mountain Cities

(1) General requirements for mountainurban vertical design

In order to meet the convenience and comfort requirements of outdoor activities, the technical and economic requirements of various engineering facilities and pipeline laying, and prevent and reduce the occurrence of ground disasters, it is necessary to do a good job in the vertical design and the comprehensive work of engineering pipelines in the planning and construction of mountain cities.

The mountainurban vertical design includes the comprehensive requirements for construction site preparation, earthwork engineering balance, road vehicular and pedestrian traffic system, determination of the design elevation of outdoor square, ground drainage, reasonable determination of the plane position and elevation of various engineering pipeline networks, outdoor site layout of buildings and landscape control of multi-dimensional and three-dimensional space of mountain cities. On the premise of meeting the requirements of outdoor activities and engineering technology and economy, the damage degree to natural terrain and the amount of outdoor earthwork should be reduced as much as possible.

The mountainurban vertical design and engineering pipeline integration is a highly professional, technical, economic and comprehensive task, which needs to coordinate and deal with various complex contradictions in engineering technology, economy and time–space construction timing sequence. For example, there are many kinds of engineering pipelines in modern cities, including water supply pipelines, rainwater pipelines, reclaimed water pipelines, sewage pipelines, gas or natural gas pipelines or thermal pipelines, information and communication lines. These pipelines have different technical performance and requirements. Similarly, there are also specific requirements for the plane and spatial location of the urban land, which must be implemented in land use through the vertical design of the site and pipeline engineering integration. In order to save land, facilitate engineering maintenance and leave room for future expansion and development, the better way is to set up a public pipe trench, in order to allow most pipelines to be arranged in the common pipe trench in a unified way. Therefore, multi-disciplinary and multi-task cooperation are needed to select the best implementation scheme through a comprehensive technical and economic comparison of multiple options.

(2) Outdoor pedestrian stairway system

Pedestrian stairway is the most common means of transportation in the outdoor pedestrian transportation system in mountain cities. The design of the pedestrian stairway system takes up less space than that of the vehicle road traffic system, with flexible and free layout in combination with the terrain. The layout and design of pedestrian stairways should be people-oriented, and meet the walking needs from the physiological and psychological feelings of pedestrians. In addition to meeting the basic needs of transportation, the walking system should be regarded as an organic part of the outdoor space in mountain residential areas, which is often connected with outdoor venues such as streets, squares and building courtyards, becoming an organic composition that reflects the mountainurban landscape characteristics.

Based on the physiological needs of pedestrians, the reasonability of functional organization, comfort and safety of the mountain walking system should be considered. From the general requirements of an adult, the height of stairway step should be slightly lower than that of indoor steps, and the width should be a little wider, which should be 140 × 350 mm generally. Furthermore, a rest platform should be set up in an appropriate step.

The gradient of outdoor stairways should be controlled within 1/12 in general, and the skid-proof requirement must be considered in ground pavement. Starting from people's psychological needs, the setting of stairways should make people feel psychologically safe and comfortable, such as the requirements for appropriate height segmentation. When the stairways reach a certain height, guardrails and handrails should be set up. In terms of the detailed treatment of stairways, the characteristics of mountain terrain and landform should be combined to enrich the spatial characteristics of mountain areas. Greening and sketch treatment should be utilized to enrich landscape changes and reduce the fatigue of pedestrians. For the connecting stairways set up in leisure and open activity space such as mountain parks and riverside green belts, the extended rest platform of the stairways shall be equipped with facilities, such as benches and chairs when necessary to meet the needs of pedestrians for resting, overlooking, resting and talking.

When a pedestrian stairway is set up in a place where pedestrian traffic is concentrated and the height difference is large, an elevator should also be installed as much as possible to meet the walking needs of different groups of people.

In mountain cities and settlements, the design requirements for accessible environment must be considered to meet the needs of people with mobility difficulties.

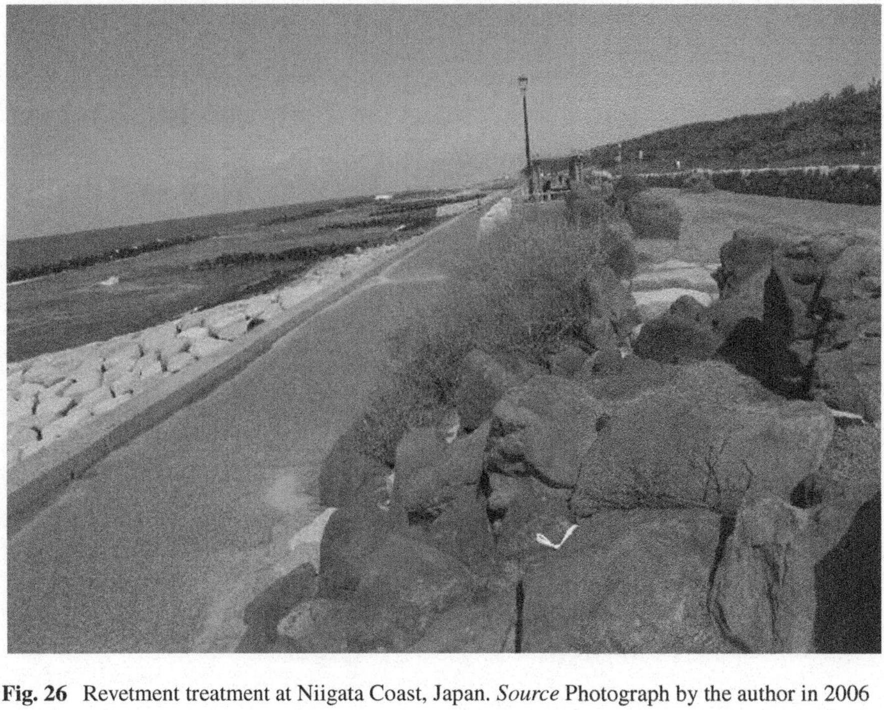

Fig. 26 Revetment treatment at Niigata Coast, Japan. *Source* Photograph by the author in 2006

For this reason, ramps or vertical elevators need to be set up for them along the stairways on hillsides in public places with large crowds.

(3) Revetment

Revetment is formed due to the elevation difference among buildings, between buildings and roads or outdoor environment caused by the changes in mountainurban terrain. A gentle slope surface is formed between the top and the foot of a revetment, and treated with greening by lawns or shrubs and trees to form a beautiful outdoor environment Fig. 26.

(4) Retaining wall

Retaining wall is a common outdoor project in mountain cities and buildings. It is a wall structure set up to withstand the lateral pressure on the slope surface, which has different structural types according to its characteristics, such as gravity retaining wall, anchored retaining wall, crib retaining wall and reinforced retaining wall Fig. 27. The setting of retaining wall can greatly reduce the distance between the top of a mountain slope and the foot of the slope compared with a natural slope or revetment, in order to save the site and improve its utilization efficiency. If the height difference is large and the retaining wall is very high, the set-back model should be adopted, in which each step should be greened and beautified to reduce the depression of space Figs. 28 and 29.

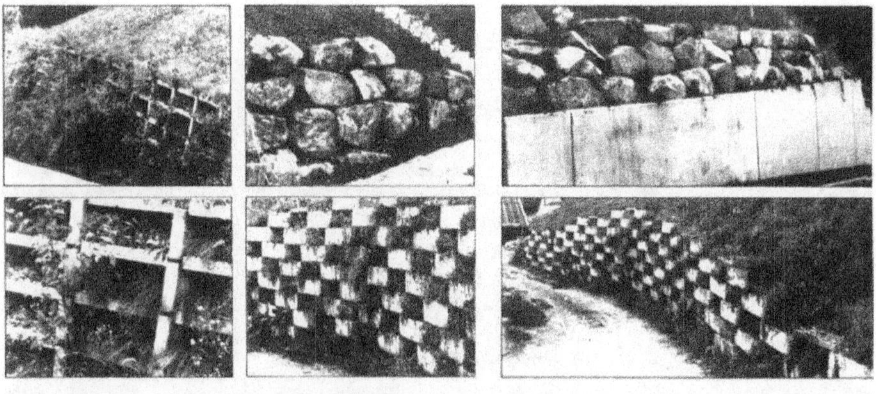

Fig. 27 Various treatment forms of retaining walls. *Source* Rolf Ridk. *Handbuchsiedlungsokolgische Eckwerte Zum Bebauungsplan*. p. 202

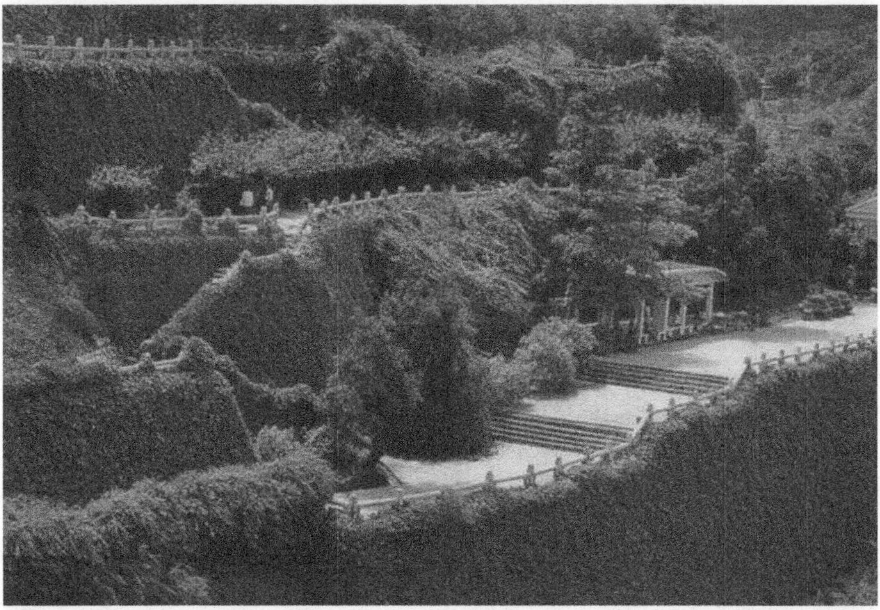

Fig. 28 Green retaining wall of Xiangshan Park in Zhuhai City. *Source* Division of Urban Construction of Ministry of Housing and Urban-Rural Development of the P.R.C. *Landscape Garden Cities in China*. Beijing: China Architecture & Building Press. September 1999, p. 37

(5) Flood intercepting ditch

Flood intercepting trench is a drainage ditch between the foot of the mountain ridge, retaining wall or revetment and the road to prevent rainwater from scouring. There are two kinds of materials for flood intercepting trench: masonry and concrete pavement. The size of its section depends on the catchment area of the mountain or revetment.

7 Case Analysis of Mountainurban Design

(1) Sunshine Grand Stairway in downtown Panzhihua

The grand stairway, built in the 1980s, won the National Design Award. In 2005, the second phase of expansion was completed, and the expanded grand stairway was updated Figs. 30 and 31, with the design theme of integrating business, leisure, culture, art and tourism. The stairway extends from the downtown square to the waterfront of the Jinsha River, which is connected with the green space of the Binjiang Park. The grand stairway has a total length of 660 m from north to south, a width of 40 m and a height difference of 110 m. There are 648 steps in total, with an average gradient of 15%. The grand stairway is arranged in sections, including waterfalls, fountains,

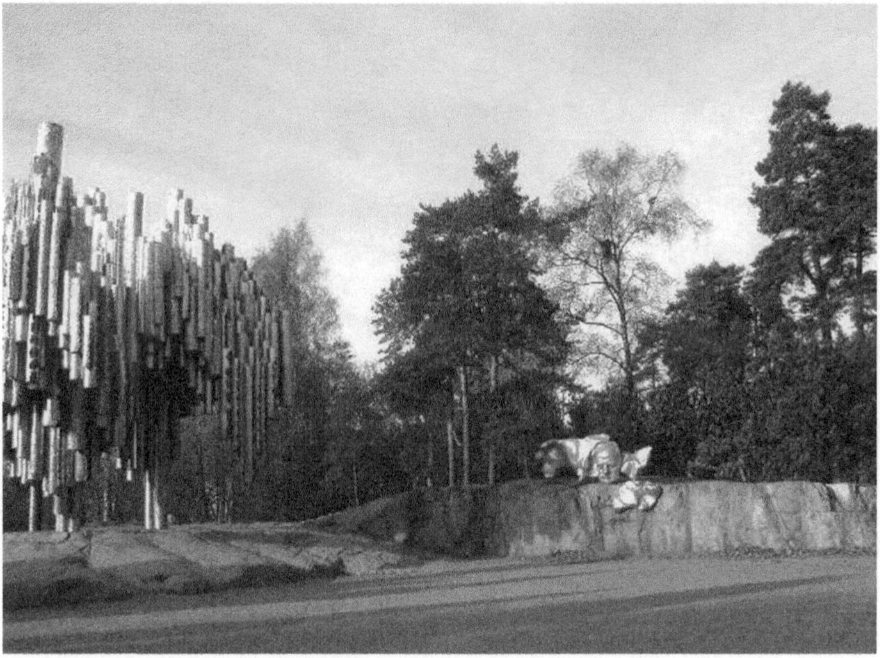

Fig. 29 Artistic treatment of the natural Fortress in Sibelius Park, Finland. *Source* Photograph by the author in 2005

waterscapes, flyovers, sculptures, cultural walls, public performance stages, leisure communication platforms, underpasses, etc. On both sides of the grand stairway are the main commercial shops and tourist service shops. The strong mountain-urban cultural atmosphere, humanized natural water and ecological landscapes, as well as pleasant, flexible, diverse and well-arranged spatial treatment of the city are refreshing, natural, friendly and popular among the citizens, which makes the only young mountainurban named after a flower more brilliant Figs. 32, 33 and 34.

(2) Urban Design of Yuzhong Peninsula in Chongqing Municipality

In 2002, Chongqing Municipality launched the collection of urban design schemes at home and abroad in Yuzhong Peninsula, among which the "leaf vein" scheme of the French AS Design Company was the most creative. It compares the vein ecology of nature with the human ecology of the Peninsula, and takes the ancient poem of "A leaf floating on the land of Bazi State" as a visual creativity with novel conception. After the comparison and study of the design schemes by six Chinese and foreign design institutions for many times, 10 main elements of the urban design of the Peninsula are summarized:

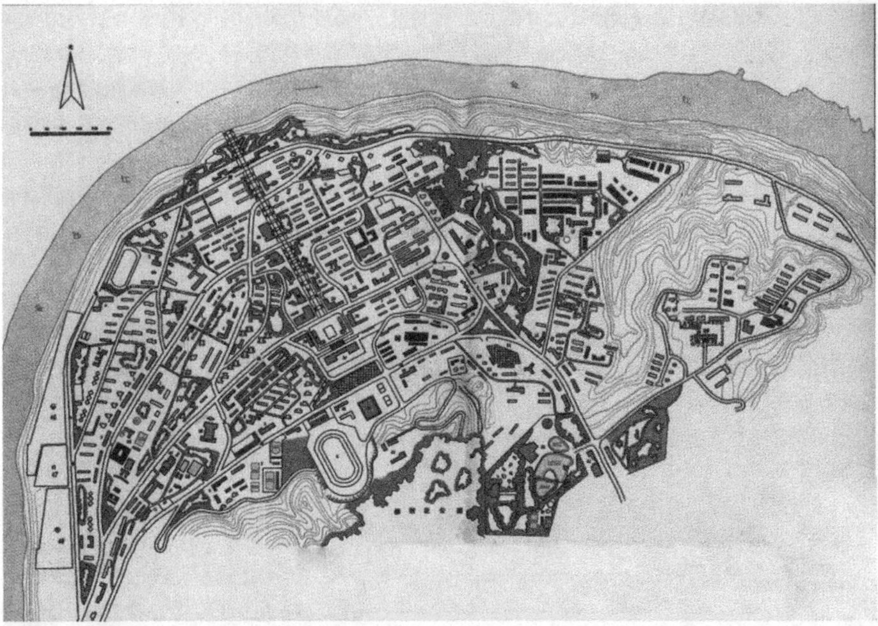

Fig. 30 Floor plan of Bingcaogang central area in Panzhihua City. *Source* Bao Shixing, Chen Jiayun. *Forty Years of Blossom of Panzhihua*. Beijing: China Architecture & Building Press. December 2005, p. 76

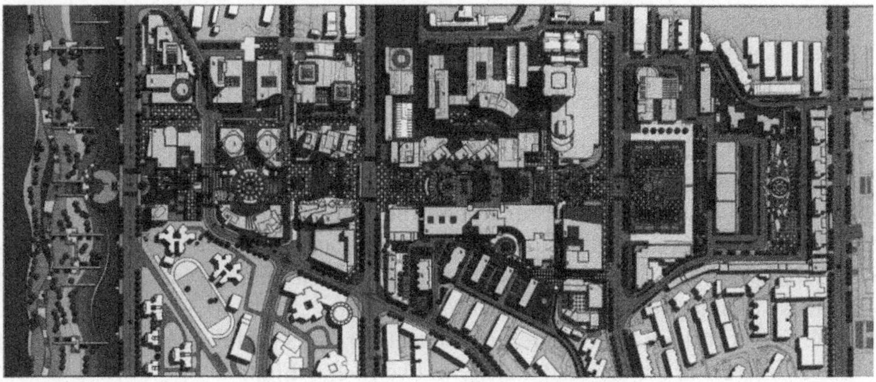

Fig. 31 Floor plan of the grand Stairway in Bingcaogang central area in Panzhihua City. *Source* Bao Shixing, Chen Jiayun. *Forty Years of Blossom of Panzhihua*. Beijing: China Architecture & Building Press. December 2005, p. 111

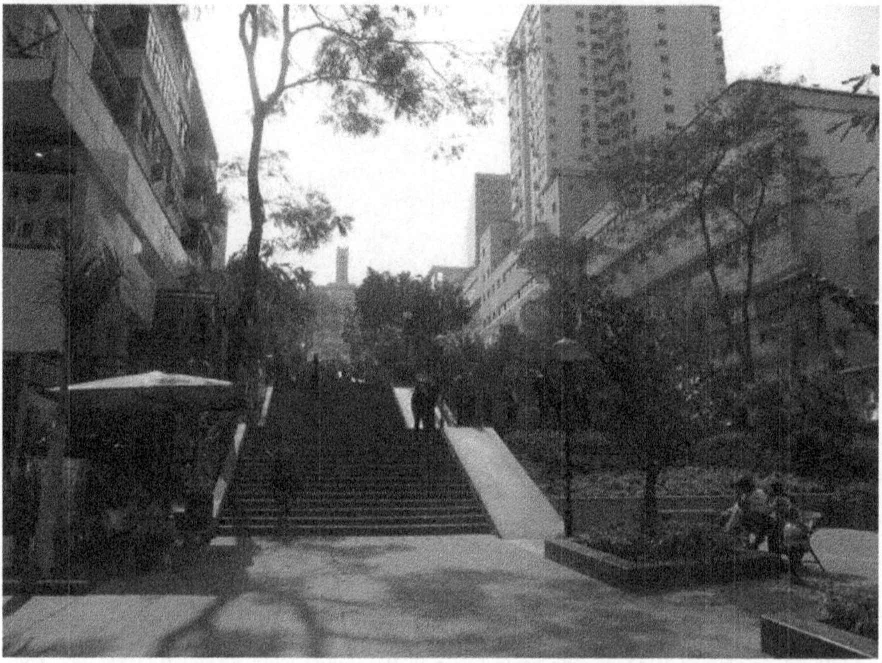

Fig. 32 Scene I of civic activities on the grand Stairway. *Source* Photograph by the author in 2006

(1) Urban crown: It enriches the cityscape mold with mountainurban characteristics and the skyline of the Peninsula;
(2) Peninsula gate: It emphasizes the city entrance of mountainurban identification;
(3) Landscape loop: It connects the elements of urban design in series to form a tree-lined landscape loop full of the characteristics of mountainurban street scene and the interest of street space Fig. 35.
(4) Urban balconies: These form a platform facing the two rivers using the steep rocks and slopes of the Yuzhong Peninsula, and become a place for people's leisure activities. The walkway also connects these urban balconies, making these an integral part of the pedestrian system.
(5) Green corridors: Seven green ecological corridors realize the sight gallery of "two rivers in the city and the city in two rivers" Fig. 36.
(6) Leisure waterfront: This shapes the leisure space along the Yangtze River and the Jialing River using different ways, according to the characteristics of Binjiang Road along the rivers.
(7) Urban context: It is the accumulation of urban historical development and the source of shaping urban future characteristics. The urban context is divided into three levels:

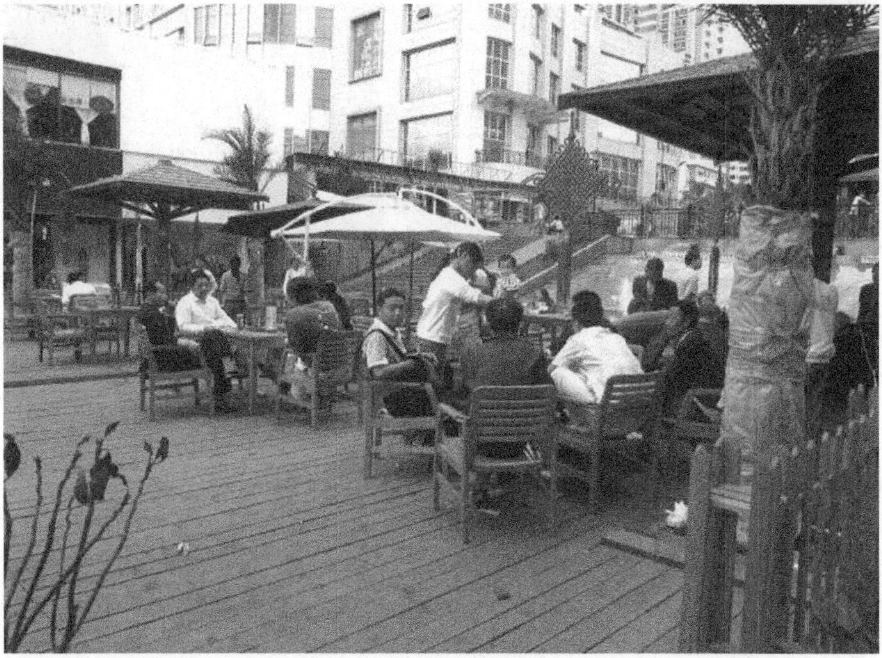

Fig. 33 Scene II of civic activities on the grand Stairway. *Source* Photograph by the author in 2006

① Macro level—urban pattern.
② Meso level—traditional block.
③ Micro level—cultural relics and historic sites.

(8) Chongqing International Financial Street: It is a crisscross pedestrian street located around the Chongqing People's Liberation Monument.

(9) Mountainurban walkway: There are nine walking paths, including eight from the southwest and one from the southeast. The walkway and venue connect the most attractive public space and key buildings on the Yuzhong Peninsula, combine green corridors and urban balconies to connect the pedestrian staircases in the upper and lower parts of the city, and assist to solve the current situation of inconvenient traffic, forming a pedestrian system with complete structure across the Peninsula. At the same time, it connects different historical fragments of the mountainurban and becomes a "time tunnel" to experience the urban evolution and development Fig. 37.

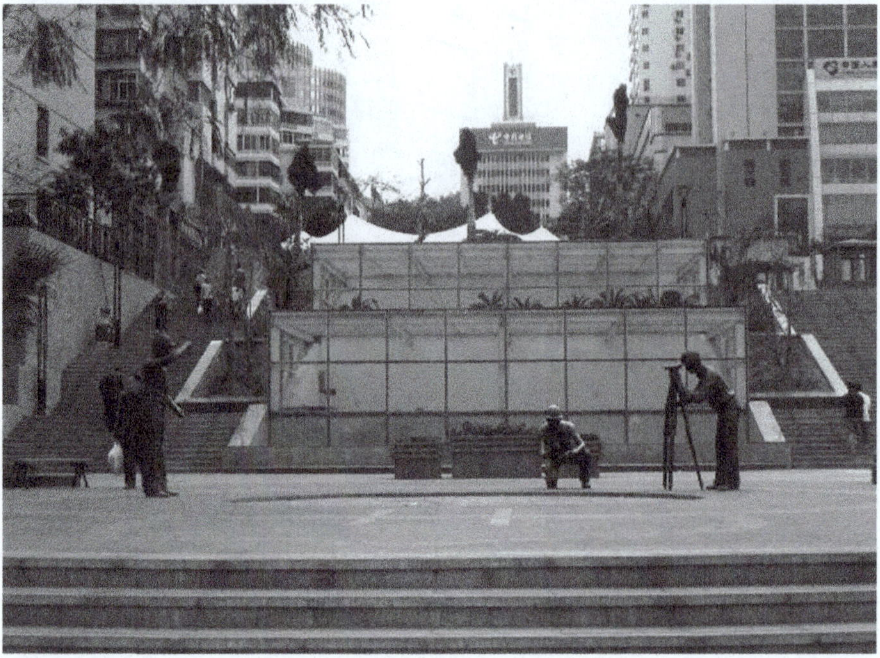

Fig. 34 Scenes III of civic activities on the Grand Stairway. *Source* Photograph by the author in 2006

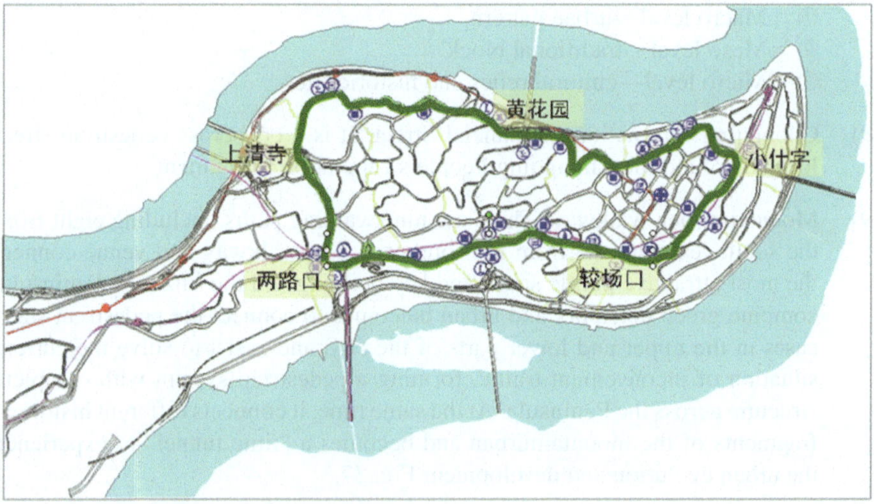

Fig. 35 Schematic diagram of Mountainurban landscape loop planning. *Source* Text of Master Plan of Metropolitan Area in Chongqing in 2005

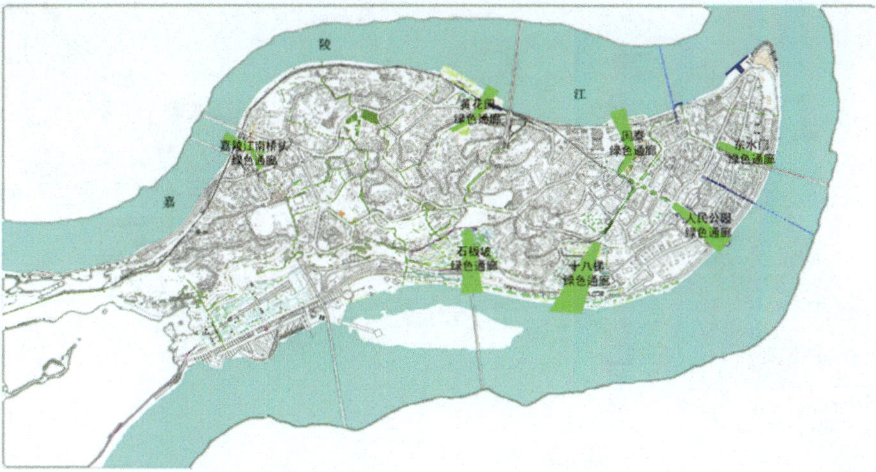

Fig. 36 Schematic diagram of Mountainurban Green Corridor planning. *Source* Text of Master Plan of Metropolitan Area in Chongqing in 2005

(10) Rail node: It solves the traffic in the areas where buildings and people are concentrated. The overall urban design greatly improves the mountainurban characteristics of the Yuzhong Peninsula. At the same time, the overall environmental capacity of the Yuzhong Peninsula is controlled to create vacant lot and increase greening. The population scale will be adjusted from the current 480,000 to 450,000. In addition, the objectives of phased implementation on planning and design are put forward from 2003 to 2030.

(3) Design plan for the new campus of Chongqing University

The new campus of Chongqing University, which is located at the east foot of the Jinyun Mountain National Nature Reserve in Chongqing Municipality, is an organic part of the western new city of Chongqing. The planning combines the natural terrain and landscape characteristics of the site, with a free and flexible layout, which integrates the artificial construction with the natural environment Fig. 38.

(4) Planning and design of Canberra as the new capital

The planning of Canberra, the capital of Australia, exemplifies another typical case of integration of urban design with nature. The city is located on undulating hills, surrounded by mountains in the east, south and west. The Molonglo River flows through the west. A dam is built in the west of the city, forming an artificial lake, which divides the city into two parts. The southern part of the city is dominated by

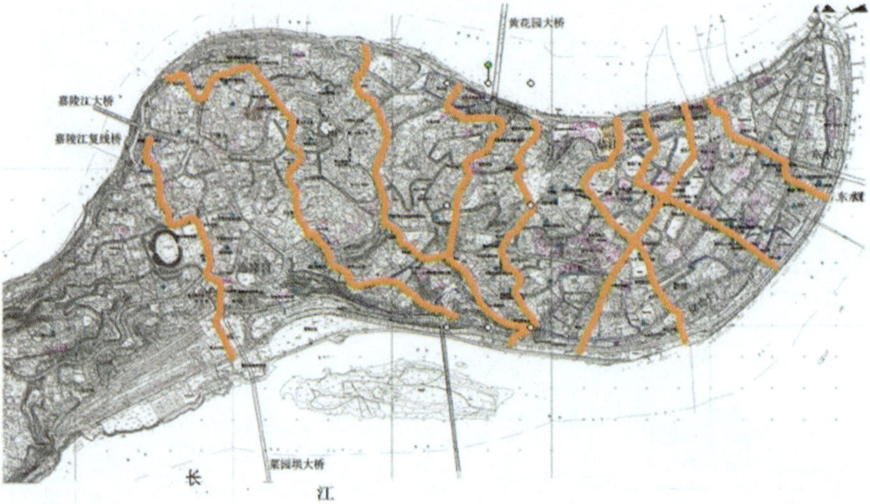

Fig. 37 Schematic diagram of Mountainurban walkway planning. *Source* Text of Master Plan of 23 Metropolitan Area in Chongqing in 2005

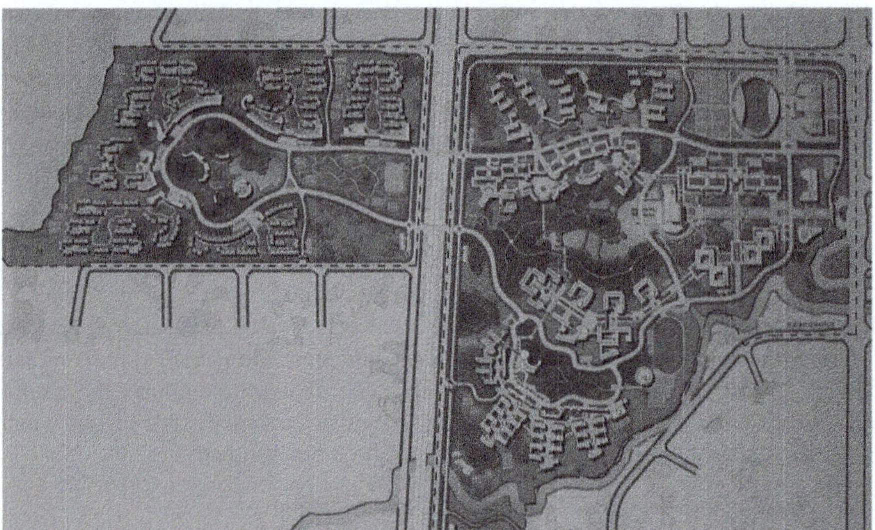

Fig. 38 Design scheme of new campus of Chongqing University. *Source* Text of Planning and Design Scheme of Huxi New Campus of Chongqing University (Moderator: Huang Guangyu). 2004

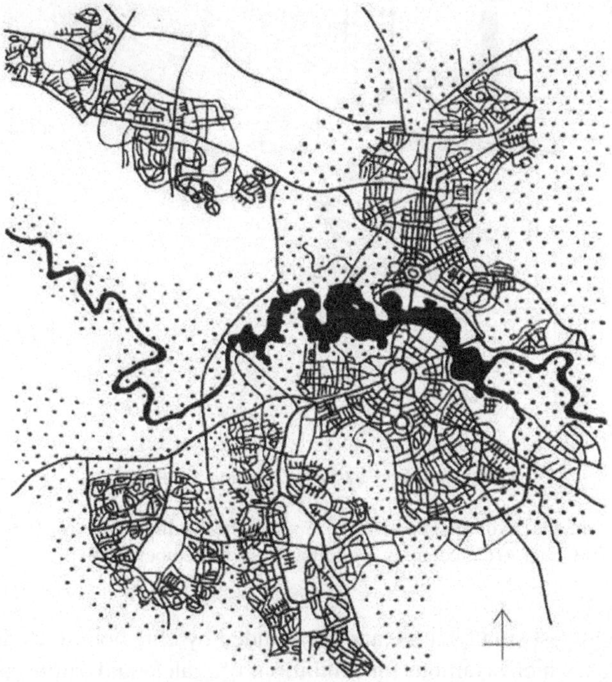

Fig. 39 Schematic diagram of the spatial structure of Canberra, the capital of Australia. *Source* Li Dehua. *Principles of Urban Planning (Third Edition)*. Beijing: China Architecture & Building Press. June 2001, p. 260

administrative offices. In the south of the city, administrative office is the main body, and the north part of the city is mainly for living, of which both are connected by the central square and the park. The urban spatial structure adopts multiple clusters of hexagonal and octagonal geometric figures, and connects each spiderlike cluster with three radial axes to form a beautiful landscape of "garden city" surrounded by mountains, waters and lakes, which add radiance and beauty to each other. However, this overly geometric composition is not suitable for complex terrain environments.

The scheme for the implementation of the new capital Canberra was designed by the architect W·Bri Griffin selected in an international competition organized by the Australian government in 1911 Fig. 39.

(5) Urban Design of the Piazza di Spagna in Rome

Built in the first half of the eighteenth century, the Piazza di Spagna has always been the tourist center of the ancient Roman city, with typical characteristics of mountain-urban square. The square consists of a set of giant artistic steps, including 137 stone steps connecting the upper and lower streets, which are built on a trinity hill, with spectacular momentum Fig. 40. The Fontana della Barcaccia, created by the father of Giovanni Lorenzo Bernini, the most famous Baroque artist of seventeenth century, is

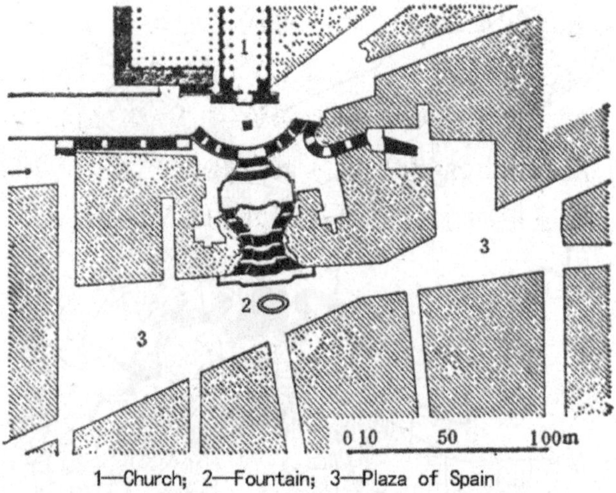

Fig. 40 Plane of the Piazza di Spagna in Rome. *Source* Shen Yulin. *History of Urban Construction Overseas*. Beijing: China Architecture & Building Press. December 1989, p. 85

located in the lowest small square and surrounded by chic boutiques, jewelry shops and Greek cafes, which is famous for exhibitions of azaleas and paintings in the spring and fashion shows in the summer and autumn. In the past, it was a gathering place for visiting intellectuals and artists from Italy and abroad. Many celebrities, such as Byron, Soest, Goethe, Stendhal and Balzac, lived in the neighborhood. Lingering tourists appreciate the rich and multi-dimensional sense of space and time formed by the combination of three-dimensional space and long history, providing infinite artistic reverie Figs. 41 and 42.

(6) Urban design of the Stepped Garden Hall in Fukuoka, Japan

Stepped roof greening of the Fukuoka International Congress Center in Japan.

From the view of the park, the stepped roof courtyard connecting the Fukuoka International Congress Center and Tenjin Central Park is designed like a mountain landscape. The greening technology of the high-rise large-scale courtyard is adopted, reflecting the space and greening composition of changing seasons. In terms of drainage, a basic natural drainage system modeled on a hillside is adopted, with rainwater from the roof floor (the 13th floor) penetrating into the soil and transmitting to the lower stepped courtyard through the water channel, and finally reaching the square. The building was designed by the Japanese architect Narifumi Murao Figs. 43 and 44.

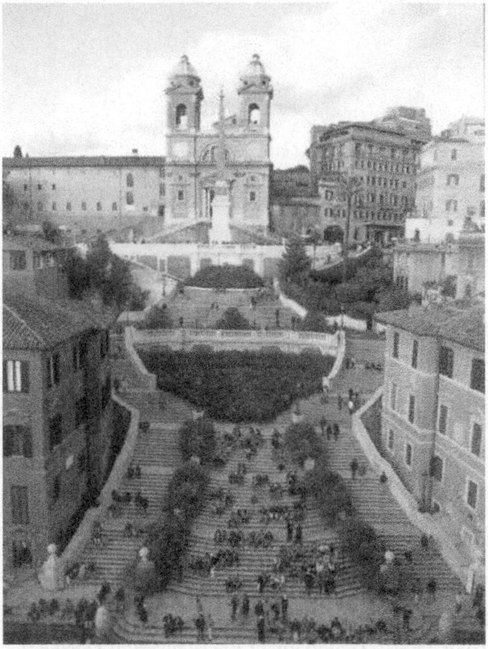

Fig. 41 Panorama of the Piazza di Spagna in Rome. *Source* Chen Zhihua. *Essays of Ancient Italian Architecture*. Hefei: Anhui Education Press. December 2004, p. 26

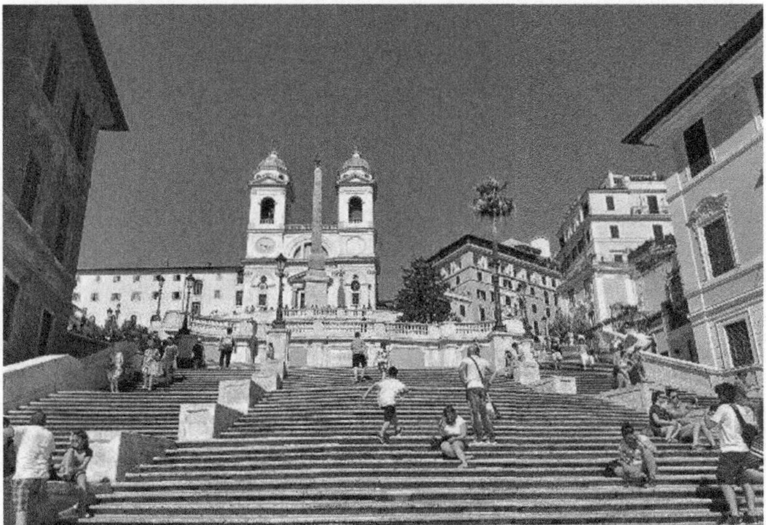

Fig. 42 Grand Stairway of the grand royal palace in Barcelona, Spain. *Source* Photograph by the author in 2005

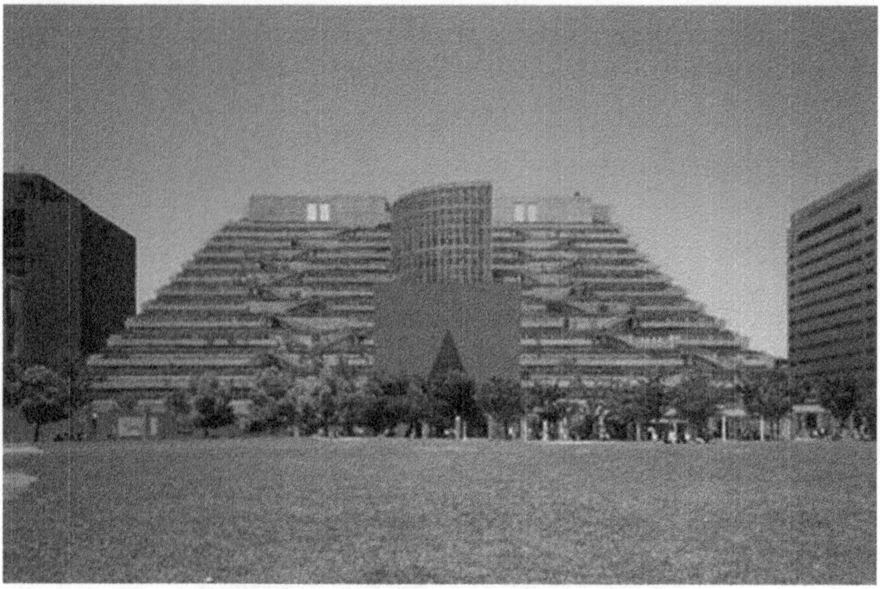

Fig. 43 Facade of Fukuoka international congress center, Japan. *Source* Architectural Institute of Japan. (Translator: Li Biyu, et al.). *Handbook of Environmental Design (Comprehensive Edition)*. Beijing: China Architecture & Building Press. February 2003, p. 501

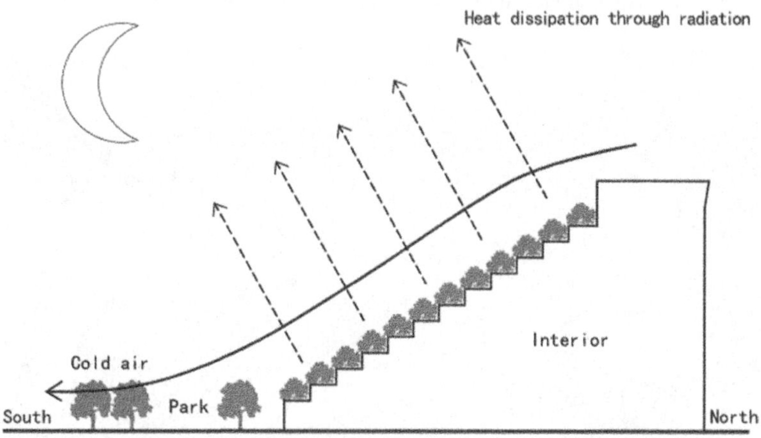

Fig. 44 Schematic diagram of the heat dissipation at night in Fukuoka international congress center. *Source* Drawing by the author

References

Guangyu H (2003) Collection of mountainurban planning and design. Chongqing University Press, Chongqing

Guangyu H, Yong C (2002) Ecocity: theory and design approach. Science Press, Beijing

Wright FL (1992) (Translator: Weng Zhixiang). Future of architecture. China Architecture & Building Press

Chapter 7
Architecture

Mountain architecture is an important part of the artificial ecosystem of mountain cities, a physical entity that meets the needs of living and production, and an important material element to constitute the urban style and characteristics (Tang 1995). In mountain cities, the natural landscape patterns, geomorphic features and vegetation forms are the natural basis and landscape matrix that constitute the mountainurban features, and mountain architecture is an inlay embedded in this natural matrix Fig. 1. Therefore, the design and construction of mountain architecture must adapt to the general background of the regional characteristics of mountain natural ecosystem.

1 Characteristics of Mountain Architecture

(1) Regional characteristics of mountain architecture

Different mountain areas have different geographical, climatic and topographic characteristics. For example, the Yangtze River Basin and the Yellow River Basin have distinct regional characteristics, and different ethnic groups have different local customs, practices and life styles. Even in a river basin, the natural and social conditions of the upper, middle and lower reaches are also obviously different. These differences will be reflected in the architecture, which makes the mountain architecture reflect strong regional characteristics, such as stilted buildings in Chongqing Municipality, Hunan Province and Hubei Province, the buildings of stone carving and bole fences in Sichuan, Yunnan and Guizhou provinces, the bamboo buildings in Xishuangbanna City, earth buildings in Wuyi Mountain, Fujian Province, the cave buildings and earth buildings in the Loess Plateau, etc.

(2) Coordination characteristics of mountain architecture

Mountain architecture is the architecutre built in a specific mountain area, so the regional characteristics must inevitably reflect the coordination with nature.

© The Author(s), under exclusive license to Springer Nature Singapore Pte Ltd. 2021
G. Huang, *Theory of Mountainurbanology*,
https://doi.org/10.1007/978-981-16-0819-3_7

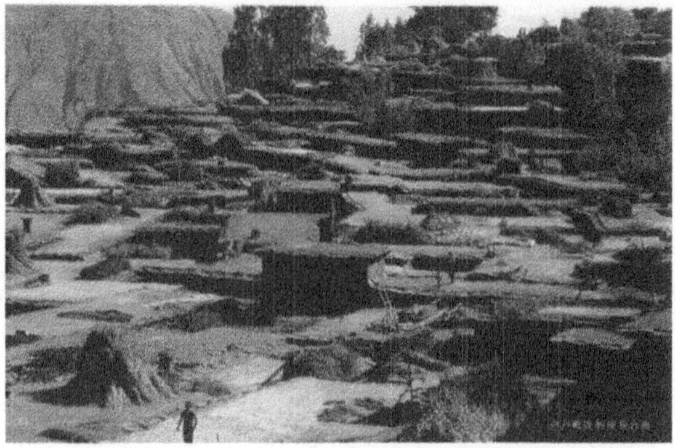

Fig. 1 Yidula Village of the Yi Nationality in Yunnan Province. *Source* Zhou Wenhua. *Villages and Buildings in Mountain Cities of Yunnan Province.* Kunming: Nationalities Publishing House of Yunnan. March 2004, p. 84

In its design, consideration must be first given to the requirements of energy conservation and land saving, and the combination with the local characteristics of topographic relief, climatic environment and building materials, in order to achieve a harmonious and symbiotic relationship between architectural design, group combination, plane layout, facade modeling, building materials and color treatment and local natural conditions. Thus, a variety of ecological buildings were adapted to the complex changes of mountain morphology and climate emerge, achieving the ideal effect of adjusting measures to local conditions (Tang 1995). Some buildings have planning and design drawings in advance, and some wonderful local buildings are not based on design drawings at all, but are on the actual terrain. These are buildings "growing" on mountains, which are also called, "barefoot" buildings, such as the carved buildings, cave dwellings and earth buildings mentioned below.

2 Design and Construction of Mountain Architecture

The design of mountain architecture should be adapted to local conditions and combined with terrain, in order to save land, energy and resources, and organically integrate buildings, gardens and urban environment, which is widely reflected in traditional and modern mountain architecture.

(1) Cluster type

In the mountainurban design, the residential units are arranged on slopes suitable for construction to make more compact use of hilly lands, and reserve more open spaces for public use, green land and recreation parks, in order to reduce the occupied area, and the destruction and change to natural slopes (Guangyu 2003). This is conducive to the formation of well-spaced mountain residential landscape, such as the Shibanpo Residential Complex and Wangjiapo Cluster Residential Complex in Chongqing Municipality Fig. 2.

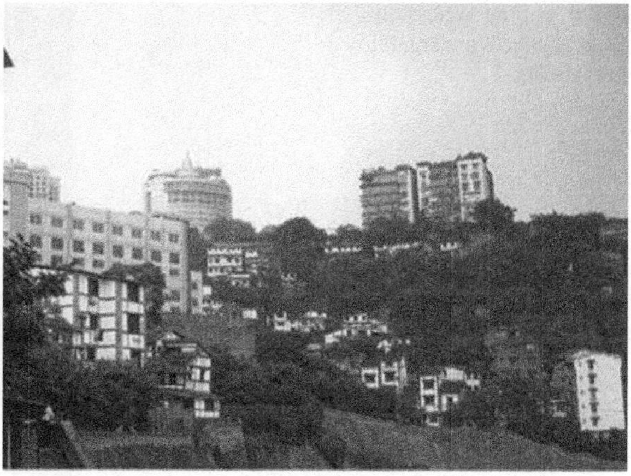

Fig. 2 Shibanpo residential cluster in Chongqing municipality. *Source* Photograph by the author in 2005

(2) Stilted type

It is a land-saving layout form to protect slopes from human disturbance and loss of balance, and avoid soil erosion caused by land filling and excavation, such as stilted dwellings in Tujia-Miao Autonomous Prefecture of Xiangxi Fig. 3.

(3) Balcony type

The balcony layout is adopted to save land under the conditions of complicated terrain and limited construction site Fig. 4.

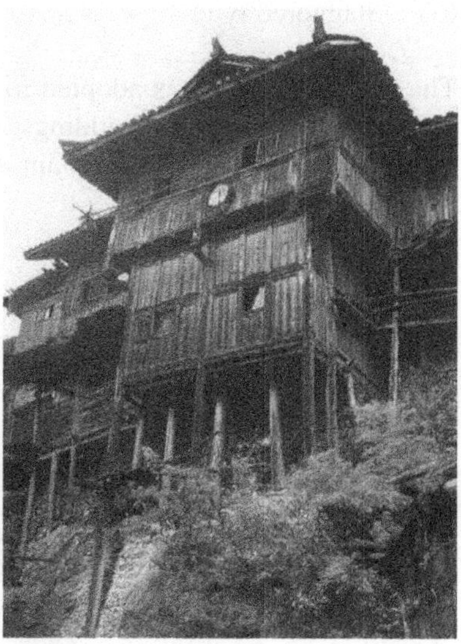

Fig. 3 Stilted Buildings in Huangli Village, Guangxi Zhuang Autonomous Region. *Source* Li Xiankui. Bole-fence Residential Buildings of the Miao Nationality. Beijing: China Architecture & Building Press. June 2005, picture on the inside cover page

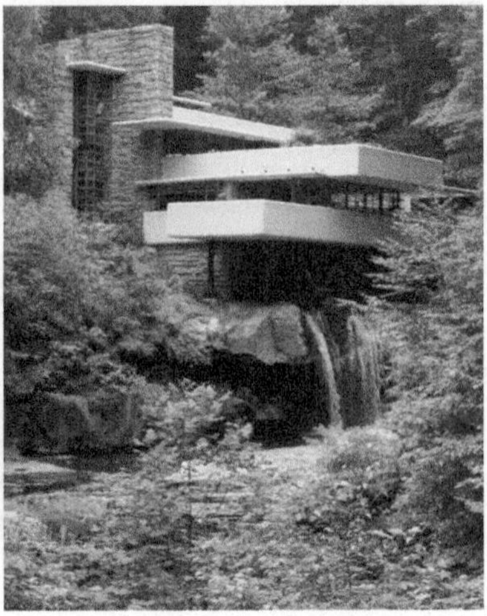

Fig. 4 Falling Water Villa. *Source* http://www.franklloydwright.org

(4) Set-back type

The outdoor steps are extended to the indoor to save the outdoor space, overcome the shortage of the outdoor sloping field, and strengthen the spectacular architectural style Figs. 5, 6, 7 and 8.

(5) Descending type

A building descends along the slope into a stepped layout along a vertical contour line, forming a strong sense of rhythm Figs. 9, 10 and 11.

(6) Staggered type

The staggered layout is adopted to adapt to the topographic relief under complex terrain, integrating the buildings with the natural environment, such as Wuyi Mountain Villa in Wuyi Mountain, Fujian Province Fig. 12.

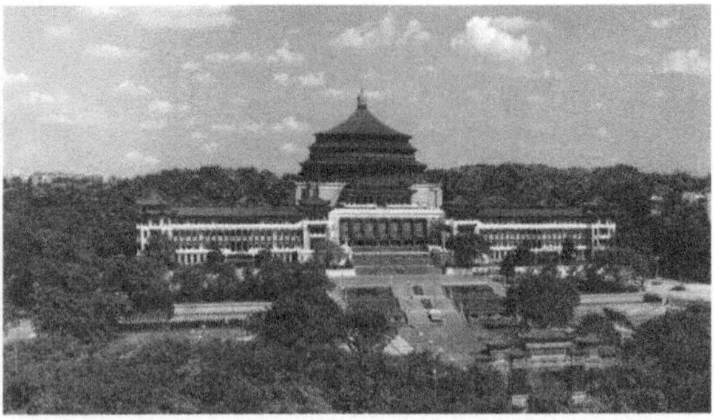

Fig. 5 Panorama of Chongqing people's auditorium. *Source* Chongqing Municipal People's Government. *Fifty Years of Construction in Chongqing*. Chongqing: Chongqing Publishing Group. October 1999, p.12

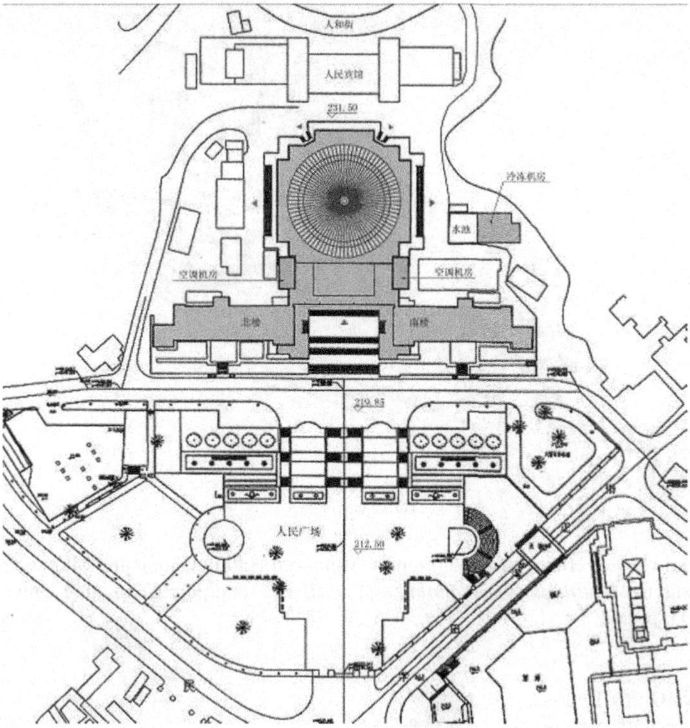

Fig. 6 Floor plan of Chongqing people's auditorium. *Source* Information Provided by Professor Liang Dingsen

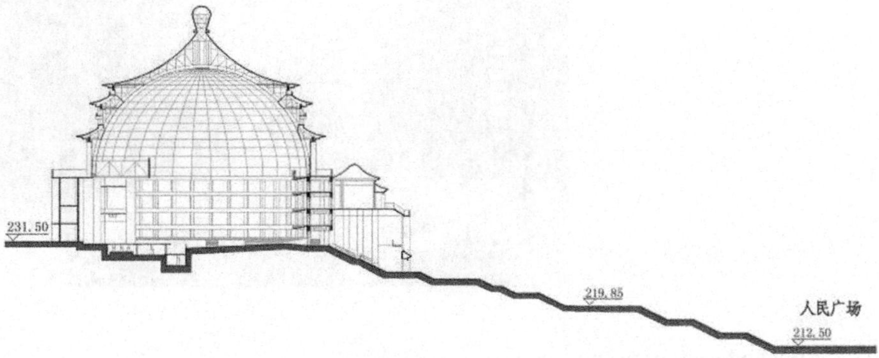

Fig. 7 Profile map of Chongqing people's auditorium. *Source* Repainting based on the information provided by Professor Liang Dingsen

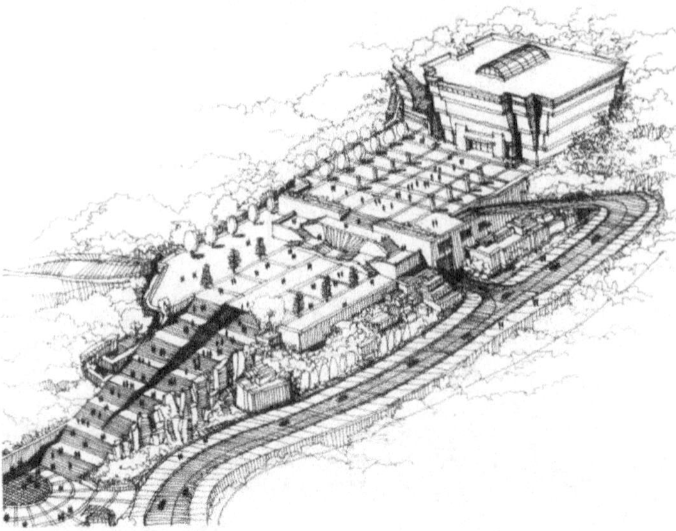

Fig. 8 Aerial view of Hongyan Revolutionary Memorial Hall in Chongqing Municipality. *Source* Huang Guangyu. *Mountainurban Planning and Design*. Chongqing: Chongqing University Press. January 2003, p. 102

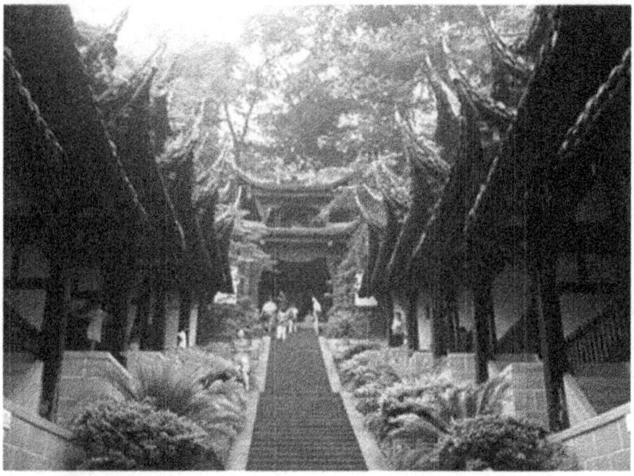

Fig. 9 Town God's Temple of Dujiangyan City. *Source* Photograph by the author in 2005

(7) Attached-wall type (mosaic type)

The buildings of the attached-wall type rise from the ground and form a vertical inlay on the steep rock wall, forming a unique mountain architectural landscape Fig. 13, or the buildings suspended and attached to the rock wall of the mountain form a mountain architectural wonder Fig. 14.

The Xuankong Temple in Shanxi Province, an architectural spectacle built in the Northern Wei Dynasty, has a history of more than 1,500 years. The building site

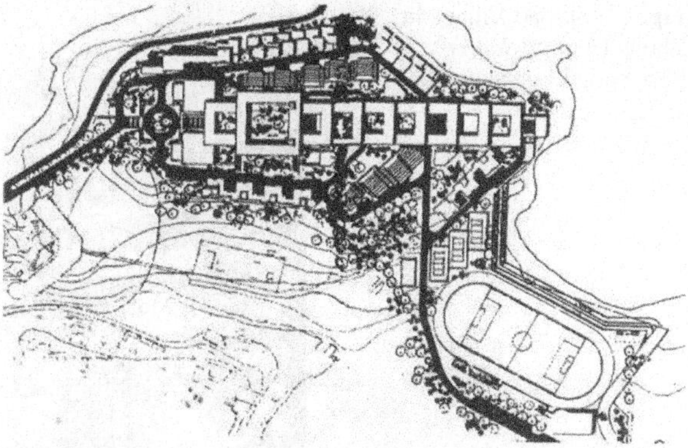

Fig. 10 Floor plan of East Asia College, University of Macau. *Source* Lu Jiwei, Wang Haisong. *Mountain Architecture Design*. Beijing: China Architecture & Building Press. February 2001, p. 132

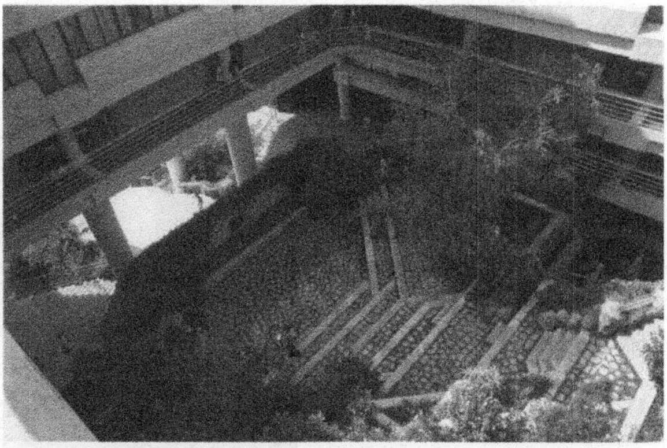

Fig. 11 Courtyard of East Asia College, University of Macau. *Source* Lu Jiwei, Wang Haisong. *Mountain Architecture Design*. Beijing: China Architecture & Building Press. February 2001, picture on the inside cover page

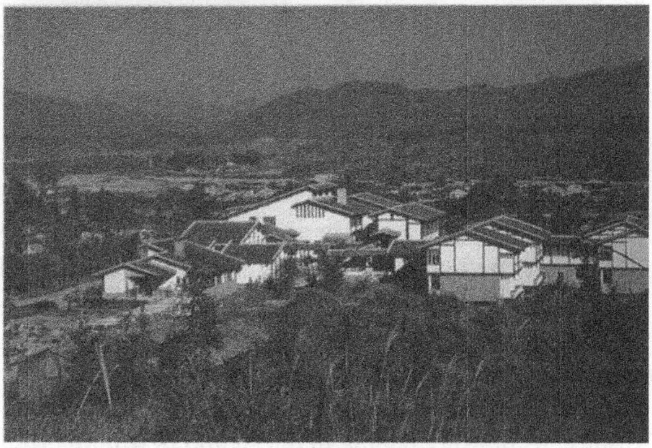

Fig. 12 Wuyi Mountain Villa in Fujian Province. *Source* Lu Jiwei, Wang Haisong. *Mountain Architecture Design*. Beijing: China Architecture & Building Press. February 2001, picture on the inside cover page

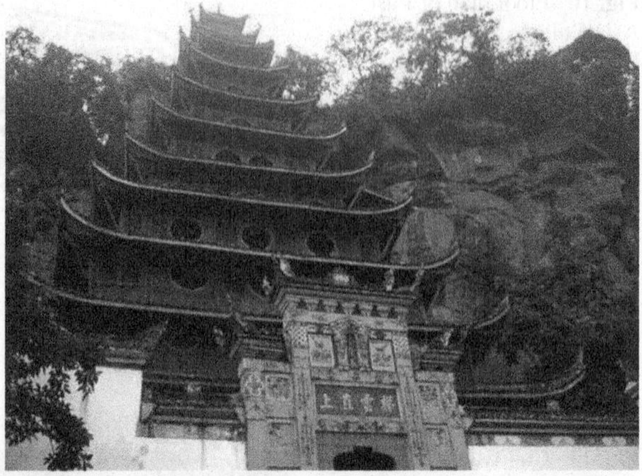

Fig. 13 Shibao Village in Zhong County. *Source* Photograph by the author in 1995

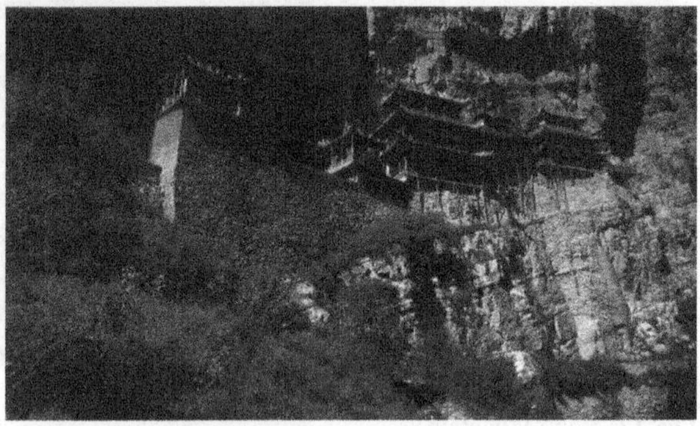

Fig. 14 Xuankong Temple in Shanxi Province. *Source* Lu Jiwei, Wang Haisong. Mountain Architecture Design, Beijing: China Building Industry Press, 2001.2, inside cover page

is located in the middle of Heng Mountain, with a river valley of more than 90 m underneath. As the riverbed is constantly silted up and piled high, it is still 28 m high. Due to the meticulous site selection, the building is located in a place with little wind and rain erosion in the mountain. It can block the north wind in winter and the monsoon in the south in summer, which makes the temple intact after more than 1,500 years without being damaged by natural forces. Local people call this "the best in the world", because "it cannot be exposed to wind, rain, flood, the sun and rocks".

All the famous Buddha statues, sculptures or buildings that have been passed down for a long time in China should be considered to avoid the damage of wind, rain and other natural forces as much as possible. For example, the site of the Dazu Stone Carvings is selected in the secluded Buddha Bay, and is more than 30 m behind Leshan Giant Buddha. This not only avoids the erosion of wind and rain, but also provides a foundation for the protective buildings covered in the body of the Buddha.

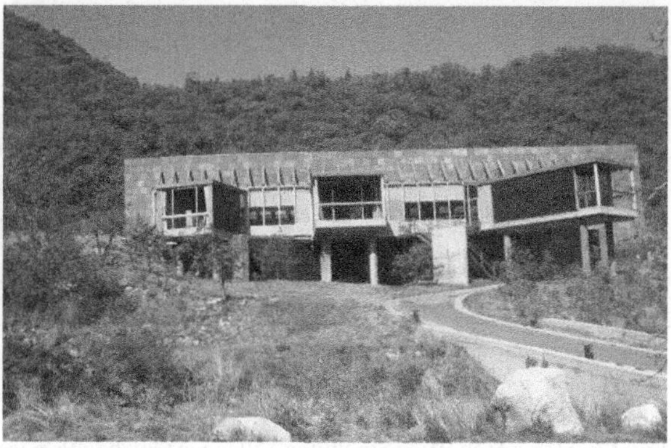

Fig. 15 Overhead Buildings at the Foot of the Great Wall. *Source* Photograph by the author in 2005

(8) Overhead type

In an overhead building, the superstructure is used to set up corridors to solve traffic problems Fig. 15.

(9) Masonry type

In areas where limestone, granite and other stone mountains are developed, people often use local natural rock materials to build residential and other civil buildings, production buildings, bridges and reservoirs, such as the buildings in Taoping Qiang Village on the upper reaches of Min River, and Shitou Village in Guizhou City Figs. 16 and 17.

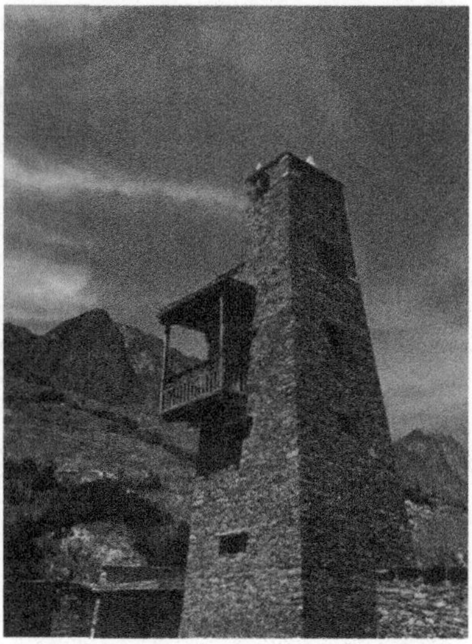

Fig. 16 Watchtower buildings of Taoping Qiang Village. *Source* Photograph by the author in 2002

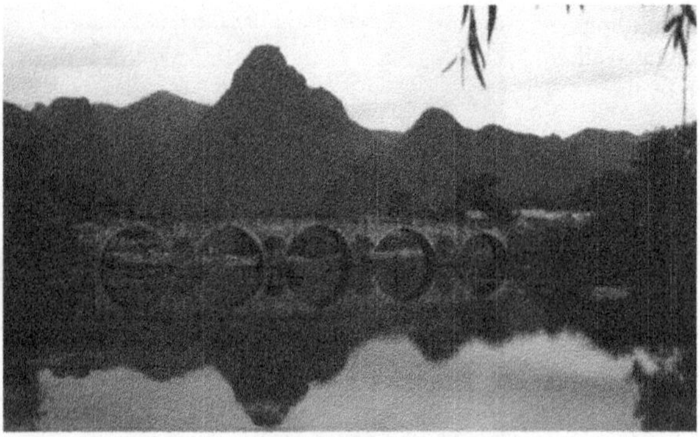

Fig. 17 Stone Arch Bridge in Shitou Village in Guizhou Province. *Source* Photograph by the author in 1998

Taoping Qiang Village is located in the Aba Tibetan and Qiang Autonomous Prefecture in Sichuan Province, and in the high valley and gorge of Zagunao River, a tributary of the upper reaches of Min River. Its buildings, built against mountains, descend along the contour line, which are closely connected, overlapping and scattered at random, and forming a highly dense spatial form with irregular layout and chaotic, but orderly intensive space. The roadway, which is only 2–3 m wide, is connected with the truncated roof from top to bottom. The buildings have the overall architectural characteristics of clusters in the style of ancient castles, giving people strong sense of local and regional styles.

This kind of extremely irregular layout, which originated from folk craftsmen and developed gradually with their rich construction experience and skills, is based on local materials, adapts to local conditions, and make the best use of the circumstances, reflecting the amazing fit between architecture and nature. According to the survey, Taoping Qiang Village was built in the Yuan Dynasty with a history of more than 700 years. This kind of ancient village architecture is very representative in the mountain areas of the Tibetan and Qiang ethnic groups in the upper reaches of Min and Jinsha river basins, which needs to be explored, summarized, and carried forward.

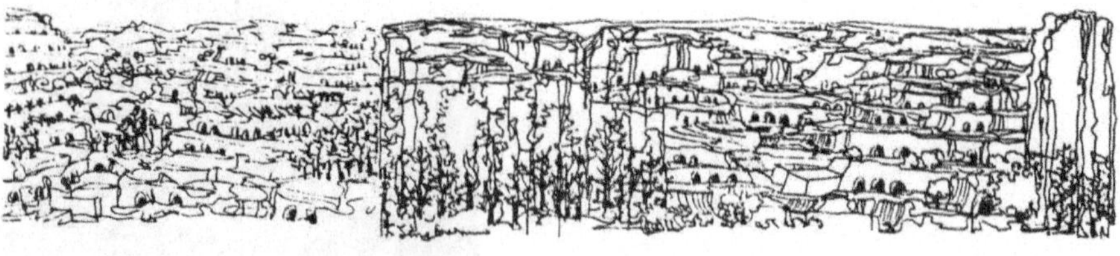

Fig. 18 Scenery of Cave complexes on the Loess Plateau. *Source* Hou Youbin. *Aesthetics of Ancient Chinese Architecture*. Harbin: Heilongjiang Science & Technology Press. April 2000, p. 215

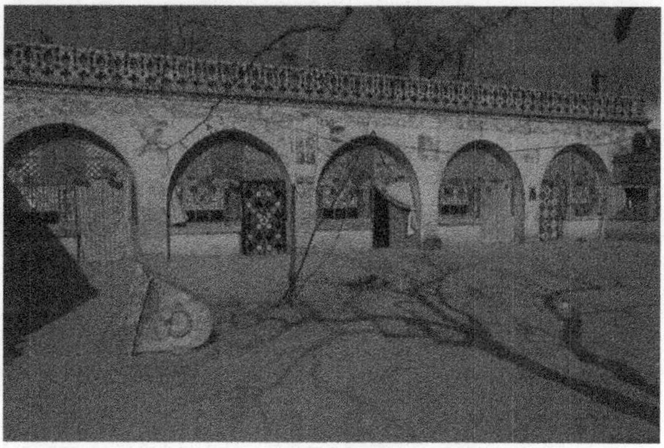

Fig. 19 Cave buildings in Yan'an City. *Source* Photograph by the author in 2003

(10) Cave type

In gully areas such as Shaanxi, Gansu and Shanxi provinces on the Loess Plateau, caves are excavated along the foothills to form earth cave buildings that are warm in winter and cool in summer, and energy- and land-saving Figs. 18 and 19.

(11) Earth-sheltered type

Buildings are buried underground or semi-underground to form an earth-sheltered structure to protect the ground vegetation, or adopt an excavation construction approach. After the construction is completed, topsoil is covered over the buildings to plant trees and grasslands, and form a green surface Figs. 20 and 21.

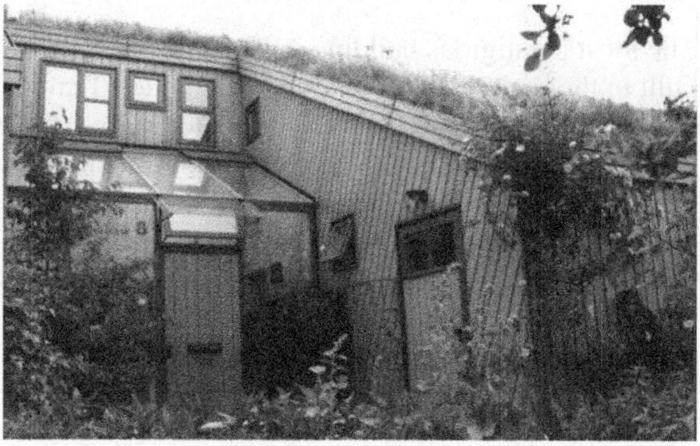

Fig. 20 Earth-sheltered Buildings in Hanover, Germany. *Source* Photograph by the author in 1987

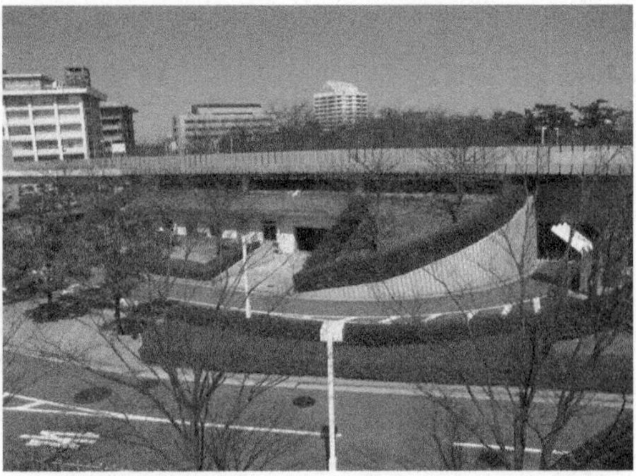

Fig. 21 Earth-sheltered Buildings in the Center of Niigata, Japan. *Source* Photograph by the author in 2006

(12) Earth type

Earth Nest Complex

Built in Xiaoshiba, Kunming and designed by artist Luo Xu, the complex is a hand-made "art craft", which covers an area of 9,720 m^2 and a construction area of 2,636m^2. It is Luo Xu organized craftsmen to build the site according to the actual situation, and guided the construction while conceiving. After completion, he mapped the plane, facade and profile according to the actual construction. The architectural conception is unique and creative Figs. 22, 23 and 24.

"Rock Church" in Finland

The most prestigious building in Helsinki, Finland is a modern underground church built in the rock - "Rock Church". It is called Temppeliaukio Church, located in the Tampere Rock Square in the urban center, which was built on February 14, 1968 and completed on September 28, 1969, with a total volume of 13,760m^3 and a project cost of 3,850,000 deutsche marks. The entire square is covered by a huge undulating rock, which is 8–13 m higher than the street next to it, and is naturally preserved, and surrounded by residential buildings of different ages Figs. 25 and 26.

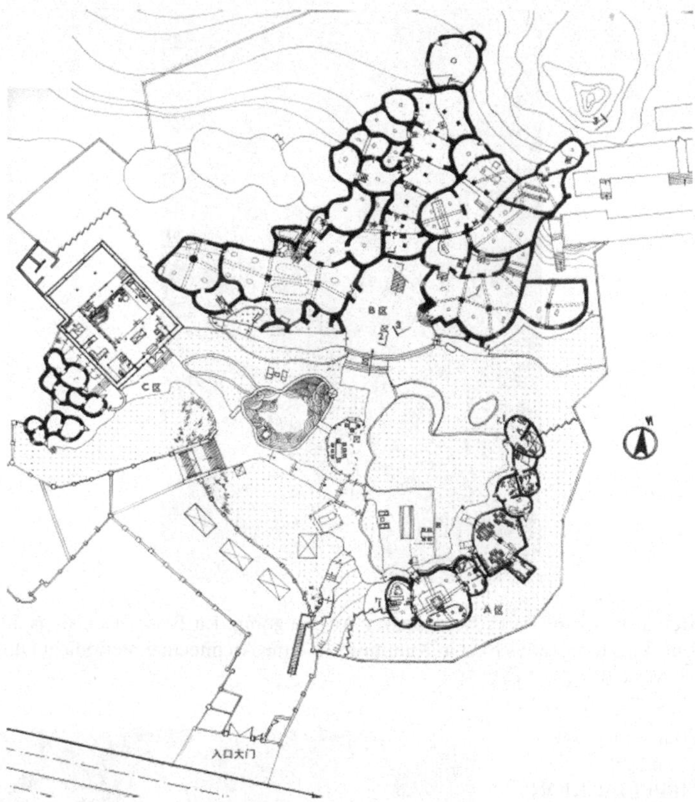

Fig. 22 Mapping plan of Earth Nest. *Source* Ye Yongqing, Lu Biao. "Delusions and Unusual Behaviors-Luo Xu's Indigenous Nest in Kunming" in *Time Architecture*. Periodical Office of Time Architecture. April 2000, p. 141

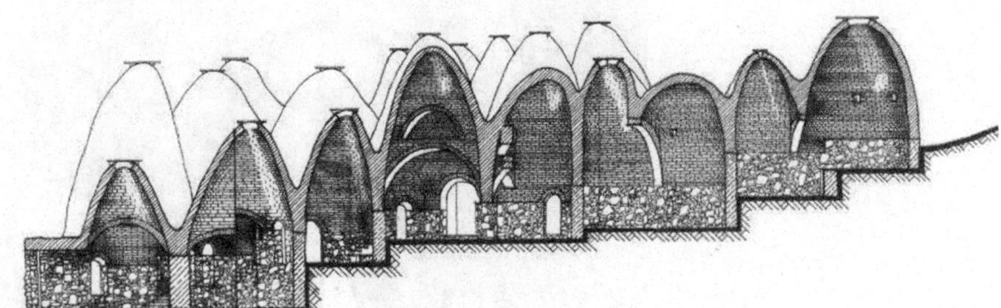

Fig. 23 Mapping profile of Earth Nest. *Source* Ye Yongqing, Lu Biao. "Delusions and Unusual Behaviors-Luo Xu's Indigenous Nest in Kunming" in Time Architecture. Periodical Office of *Time Architecture*. April 2000, p. 143

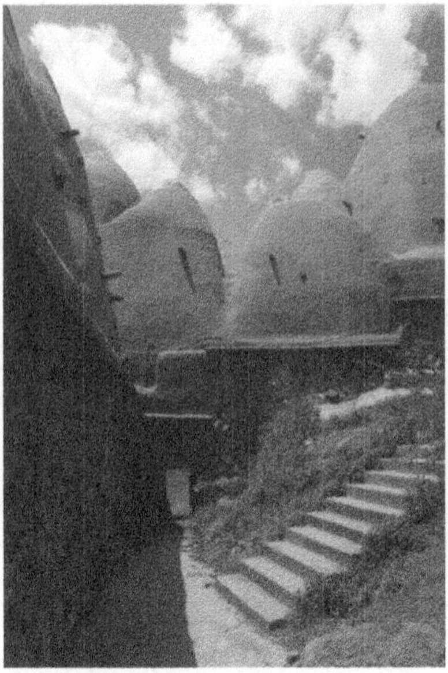

Fig. 24 External image of Earth Nest. *Source* Ye Yongqing, Lu Biao. "Delusions and Unusual Behaviors-Luo Xu's Indigenous Nest in Kunming" in Time Architecture. Periodical Office of Time Architecture. April 2000, p. 142

Fig. 25 Aerial view of the Rock Church in Finland. *Source* TEMPPELIAUKION KIRKKO, p. 3

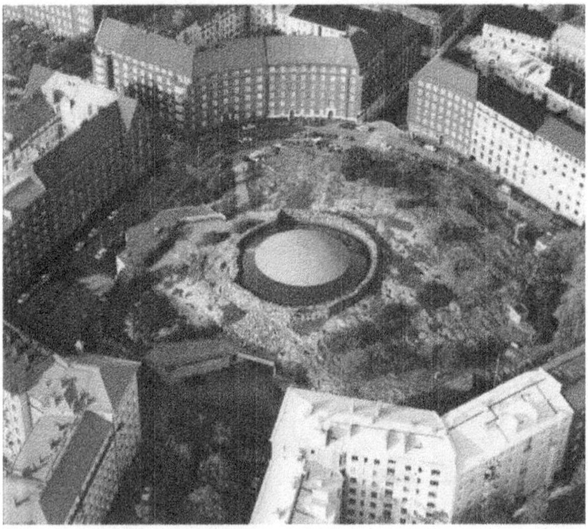

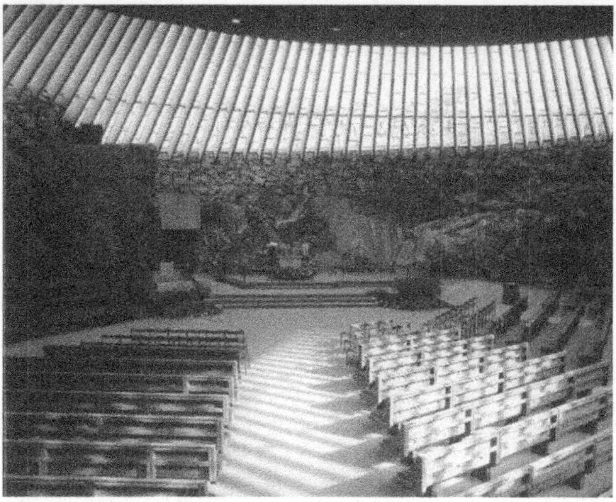

Fig. 26 Interior of the Rock Church in Finland. *Source* TEMPPELIAUKION KIRKKO, p. 5

References

Guangyu H (2003) Collection of mountainurban planning and design. Chongqing University Press, Chongqing
Tang P (1995) Mountain residential architecture. Science Press, Beijing

Chapter 8
Transportation

1 Characteristics of Mountainurban Transportation

In order to maintain the normal order of production and living in a city, there must be a large flow of people and goods traffic. Various economic, social and cultural communication activities in a city have created a demand for transportation. Therefore, transportation plays an important role in guiding and supporting the organization of various activities, such as normal production and life as well as urban spatial structure and development.

The urban transportation modes of people include walking, bicycle, public transportation, private car, rail, air and water transportations, and the transportation urban traffic modes of goods flow include truck, train, aviation and shipping transportation. There are strict requirements for site selection in a city in terms of traffic occurrence and attraction points, intricate flow of people and vehicles, especially the commercial centers, stations, ports, wharfs, airports, industries and warehouses that attract a large flow of traffic. At the same time, the urban traffic flow and direction are often complex and changeable, with uneven spatial and temporal distribution. In terms of time distribution, the four seasons, a week and a day are not the same. In terms of spatial distribution, the flow distribution on each road is also very uneven.

The reasonable organization of mountainurban transportation is to analyze and study the interactive relationship among the mode of land use, spatial layout structure and environmental impact of different types, functions and scales of mountain cities, comprehensively study the rules and characteristics of various mountainurban traffic flows, as well as the mountainurban rules for the flow of people and vehicles, characteristics of traffic organization, and its relationship with the surrounding environment (Huang and Chen 2002).

The traffic organization of mountain cities is closely correlated to land use and urban spatial layout structure. Urban transportation is the projection of urban population and logistics on the urban spatial layout plane. The mode of land use and spatial layout structure of mountain cities fundamentally determine the flow and direction of urban traffic. Whether the mountainurban layout structure is reasonable, one of

© The Author(s), under exclusive license to Springer Nature Singapore Pte Ltd. 2021

G. Huang, *Theory of Mountainurbanology*,
https://doi.org/10.1007/978-981-16-0819-3_8

its important signs is to determine whether it can reasonably distribute the flow of people and goods traffic among the functional groups of mountain cities, shorten the average spatial distance of their flow as much as possible, minimize unnecessary traffic flow in mountain cities, and reduce the total traffic flow to the minimum, in order for the mountainurban road network system, and the grade and density of primary and secondary truck roads to be reasonably organized.

The mountainurban industrial layout and residential distribution form are closely correlated to the number of commuting workers and the direction and flow of traffic. Therefore, the industrial production and working place in mountain cities should be arranged on the same elevation plane as living nearby as much as possible, in order to make commuting within a convenient walking distance.

When arranging some large industrial projects, excessive concentration of traffic in one direction during peak hours should be avoided.

In addition to the unbalance and complexity of the spatial and temporal distribution of the general urban traffic flow, mountainurban traffic highlights the diversity and three-dimensional characteristics of the terrain under the influence of obvious changes.

There are many factors that affect the organization of road traffic systems in mountain cities, the most important of which is based on the urban traffic characteristics, ways of internal and external connection, overall layout structure, and the historical status and basis under stable local natural environment. The mountainurban spatial structure formed under specific mountain natural environment affects the distribution of the flow of people and goods traffic, and road traffic organization of mountain cities, which in turn affects the layout structure characteristics of mountain cities. Therefore, it is impossible for mountain cities to lay out a grid-type road system like plain cities, and more of these adopt the decentralized group structure combined with terrain and the flexible and free road organization system (Figs. 1 and 2).

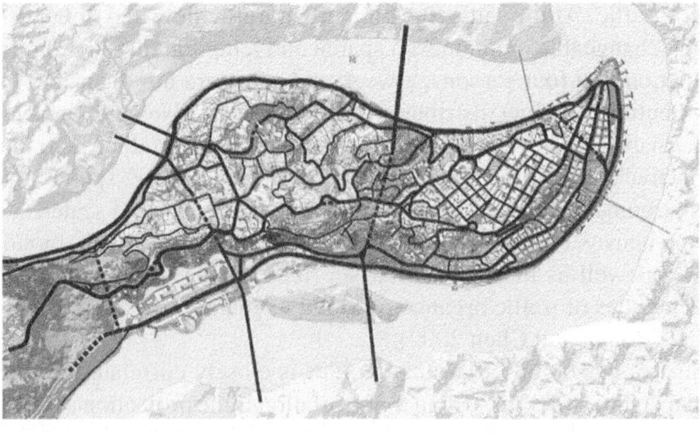

Fig. 1 Road planning system of the Yuzhong Peninsula in Chongqing. *Source* Drawing by the author

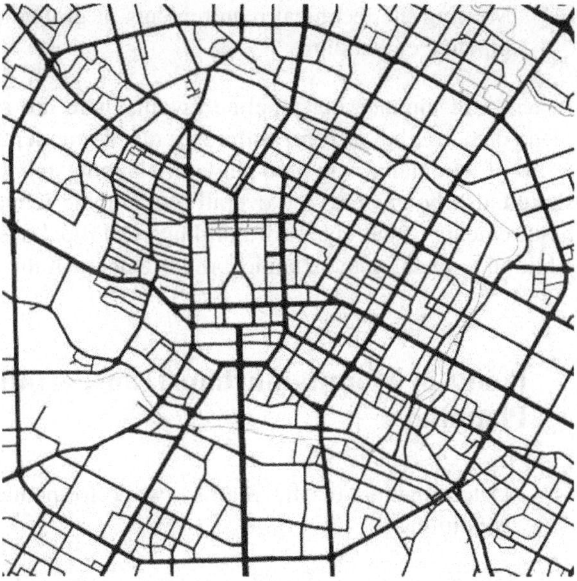

Fig. 2 Road planning system of the downtown area in Chengdu. *Source* Drawing by the author

2 Planning Requirements for Mountainurban Road Traffic System

(1) Reasonable determination of the organizational form of road traffic

In order to meet the basic requirements of safety, convenience and smoothness, the organizational form of the mountainurban road traffic system should be adapted to local conditions and combined with the mountain terrain and climate characteristics of different regions as much as possible.

(2) Reasonable determination of the function and division of roads

The mountainurban traffic routes should be as simple and clear as possible so that pedestrians and vehicles can be easily identified.

(3) Energy, land and investment saving

In combination with the land use and geological conditions of mountain areas, earth and stone excavation should be reduced, and the density of road network and the cross section form and width of primary and secondary roads should be reasonably determined.

(4) Meeting the technical requirements of ground drainage, building layout and pipeline embedding

In terms of engineering pipeline layout, there are requirements not only for the plane distance, but also for different embedding depths and longitudinal slopes. In terms of mountainurban road network planning and technical design, consideration should also be given to the overall topographic relief characteristics and technical requirements of the layout of buildings and pipelines on both sides of the road, in addition to meeting the technical requirements of the road itself.

3 Points of Mountainurban Transportation System Planning

(1) Determination of the road network form and density according to local conditions

China has a vast territory with distinct natural conditions in mountain cities in the north, south, east and west. In southwest mountain cities, the general terrain is very undulating, with high mountains, steep slopes, deep gullies and narrow valleys, and the plane, vertical and cross sections of roads are greatly constrained by terrain. In terms of geology, there are many adverse geological phenomena in mountain areas due to the thin soil layer, deep and thick rock strata, complex attitude of bed and geological structure changes, affecting the layout and design of roads. For example, in the road planning of Yunyang new county in Chongqing Municipality, a free-form road layout system is formed along the contour line in combination with the complex terrain conditions (Fig. 3). In the planning of Chongqing Longfor Blue Lake Villa Cluster, the road and building layout are organized according to the natural terrain conditions (Fig. 4).

In terms of climate, there are many rainstorms and torrential floods in southern mountain areas, where the water level of rivers and streams varies greatly. In northern mountain cities, the winter is cold, and these are covered with snow and ice. Therefore, the influence of winter climate should be considered in the planning and design of road alignment.

The direction of urban roads has a great influence on the ventilation and landscape ecology of mountain cities. In hot and humid mountain cities in the south, the truck roads should be close to the direction parallel to the dominant wind direction in summer or perpendicular to river channel, allowing the cool summer monsoon or valley wind to blow to urban residential areas or streets. In the arid or cold northern areas or areas with heavy wind and sand, the truck roads should maintain a certain skew angle with the direction of winter wind, with a corresponding shelter forest belt arranged on the urban edge, in order to avoid the attack of snow or sand.

As the natural environmental characteristics, regional population distribution density, urban distribution density and urban population size of mountain areas

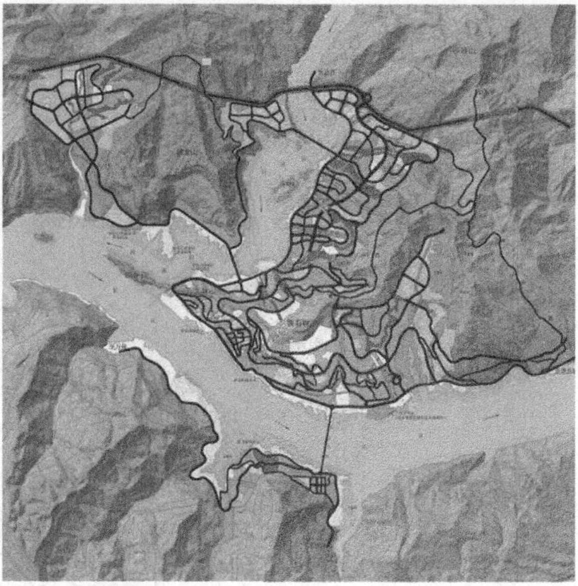

Fig. 3 Road system planning of Yunyang New County. *Source* County (Moderator: Huang Guangyu) 2005

are generally lower than those of coastal plain areas, and at the same time, mountain areas are generally located on the windward land and the upper reaches of the country, and areas with sensitive and fragile eco-environment, there is still a long way to go generally in the maintenance, restoration and reconstruction of the eco-environment. Therefore, we are faced with the arduous task of ecological conservation and economic development at the same time. To this end, it is more necessary to attach importance to the organization of mountain road transportation, which is energy saving, land saving and ecological.

In determining the mountainurban road network density and road system form, we must proceed from the natural environmental conditions and the actual situation of current foundation, and avoid formalism, determining the form and density of the road network subjectively and unilaterally, or blindly applying some forms to cause unreasonable long-term use (Huang 2003). For example, in order to pursue the axis of a road, some cities slash a truck road diagonally on the original regular trunk road network, and form a complex acute angle intersection with the original primary and secondary trunk roads, which results in traffic problems that are difficult to root out, and increases the complexity of architectural layout and uneconomical land use. Some other cities cut off the national highway transportation links connecting several provinces and regions without technical and economic demonstration, in order to open up pedestrian commercial streets and central squares, which undermines the normal traffic function of the original trunk road network, causes artificial traffic chaos, and finally takes "remedial" measures that are not worth the candle, paying high ecological and economic prices.

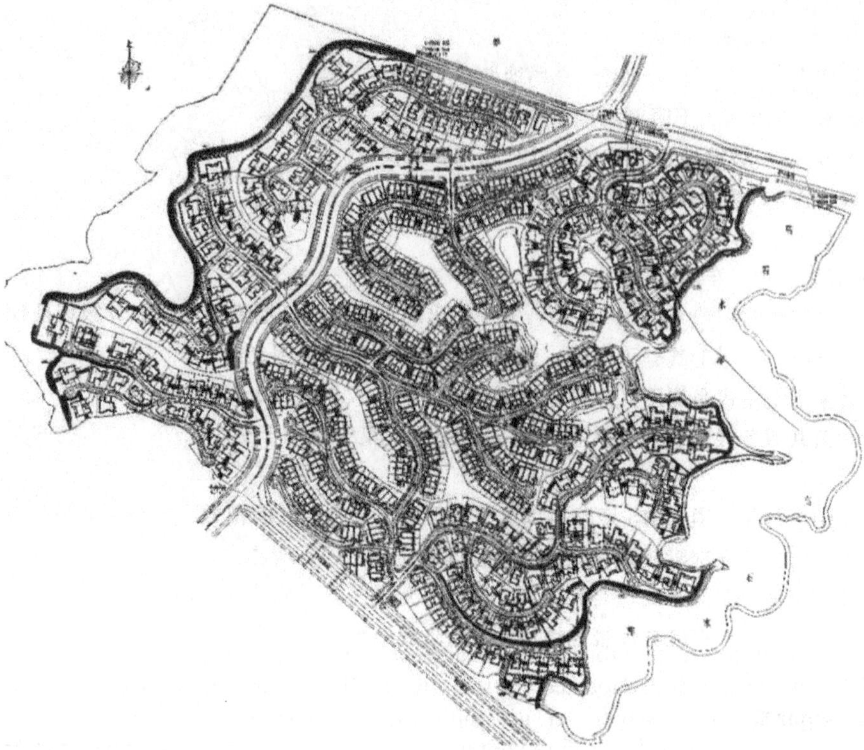

Fig. 4 Road system and building layout of Chongqing Longfor Blue Lake Villa Cluster in Chongqing. *Source* Luo Danxing. *Research Series of Modern Residential Development in Chongqing-Exploration of Theory and Methods for Planning of Low-rise and Low-density Residential Areas in Suburbs.* Master's Thesis of Chongqing University. 2006, p. 96

(2) Organizing traffic according to pedestrian requirements as much as possible to reduce motor vehicle traffic flow

In mountain cities, the urban layout should adopt the combination of comprehensive area layout organization and work and residence as much as possible, in order to make it possible to walk for commuting or daily shopping activities within the walking distance. In addition, the width of the sidewalk should be properly widened to reduce the traffic flow of vehicles and the pollution and interference to the environment (Fig. 5). In the mountainurban road traffic system, it is necessary to create convenient, safe and comfortable conditions for pedestrian traffic. The pedestrian system composed of horizontal walkway and vertical stairway traffic can take a shortcut and achieve a good effect of spatial connection.

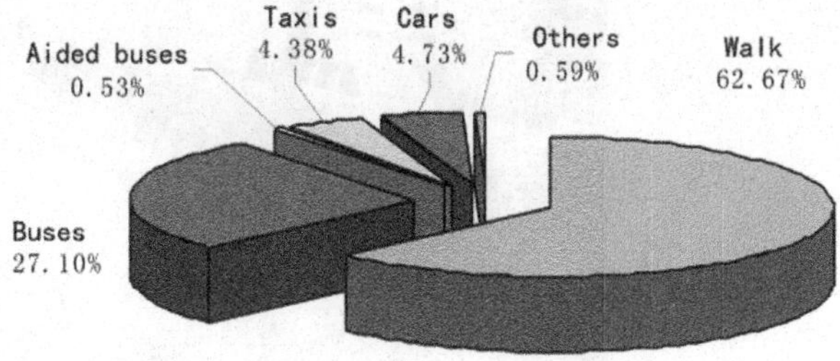

Fig. 5 Proportion of Various Modes of Transportation in Chongqing Municipality in 2002. *Source* Text of Master Plan of Metropolitan Area in Chongqing in 2005

(3) Promoting bicycle traffic

The road traffic system organization and road sectional design should provide convenient conditions for bicycle traffic. In the organization of road sections, the set-up of non-motor vehicle (bicycle) lanes should be considered. The gradient of a road suitable for cycling should generally not exceed 2.5%. Although it is not difficult for motor vehicles to drive on a gradient greater than 2.5%, it has an impact on bicycle traffic. In many mountain cities, the longitudinal gradient of urban trunk roads exceeds 3% due to complex terrain. For example, the maximum gradient of the trunk road on North Street in Yuzhong District, Chongqing Municipality exceeds 7%, and the longitudinal gradient of the trunk road on Yanan Road in Guiyang City exceeds 4%, which are very difficult for cyclists and tricycles. This is also the main reason why the development of bicycles is restricted in mountain cities such as Chongqing.

When the longitudinal slope of a bicycle lane is large, if the slope length increases, the driving speed on an uphill slope will be greatly reduced, and the speed on a downhill slope will be accelerated, which will affect safety. Therefore, the length of road slope should also be controlled and not be too long. A gentle slope section should be set between two long slope sections to slow down the consumption of physical strength, and at the same time, a gentle slope section should be added at the end of the ramp to slow down the vehicle speed.

In order to give full play to the traffic capacity of the trunk roads of motor vehicles, and take into account the driving capacity and safety requirements of non-motor vehicles, the separate set-up of motor and non-motor vehicle lanes can be adopted in the mountainurban road traffic organization (Fig. 6).

(4) Giving priority to the development of public transportation system

We should reasonably organize the motor vehicle transportation system and give priority to the development of public transportation, in order to reduce the consumption of land resources and energy as well as the pressure on the environment (Fig. 7). According to the traffic type and nature, passenger and freight transport distributary should be carried out, and the pedestrian system should also be separated from the vehicle system as much as possible, in order to allow different types of vehicles to go their own way and form a reasonable traffic network.

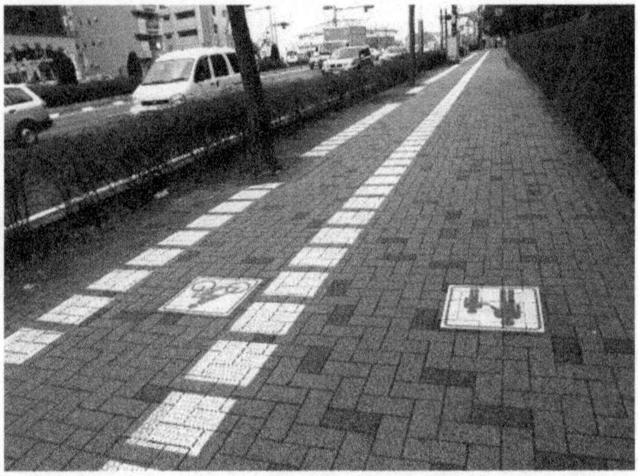

Fig. 6 Bikeway and Walkway Organization in Nagoya, Japan. *Source* Photograph by the author in 2006

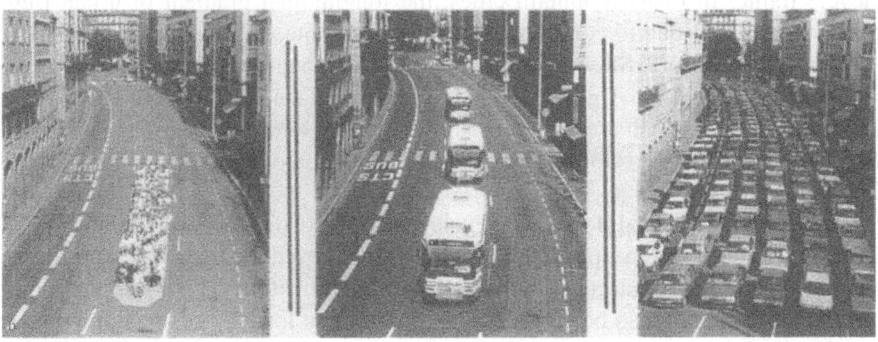

Fig. 7 Comparison of Road Space Required to Transport 300 Passengers. *Source* Jin Fan. "Development of BRT in China" in Urban Planning Overseas. March 2006, p. 30

All kinds of traffic converge at intersections to form interweaving and conflicts between people and vehicles, and among vehicles. Therefore, in the mountainurban road traffic organization, complex and deformed intersections should be avoided as much as possible, in order to reduce the number of intersections, ensure a good channelization design of intersections, and improve traffic capacity and traffic conditions.

According to the data of the urban traffic survey conducted in Chongqing Municipality in 2002, 62.67% of residents in urban clusters traveled on foot, while the road traffic area and road network density of the city were far lower than those of mega cities such as Beijing and Shanghai, and the traffic congestion is not more serious than these cities. This strongly demonstrates the advantages of the "organic, loose, concentrated, zonal, balanced, multi-center and clustered" spatial structure.

The eco-city Curitiba in Brazil has adopted the concept of giving priority to the development of public transportation and achieved good results, which divides its public transportation into four levels, namely, bus rapid transit, direct bus transportation, interval line public transportation and branch line public transportation, thus forming the city's public transportation network system.

Before the 1960s, Curitiba adopted the traditional circle-layer development pattern. In the city's master planning and design competition held in 1964, the Brazilian architect Hall William put forward a new planning scheme which was selected. The planning and design scheme changes the past road traffic development pattern of organizing circular radiation around the old urban center, but guides the expansion of urban space by forming the urban structural axis with bus rapid transit.

In terms of land use, different density distribution grades are formed with the structural axis as the main body. The highest residential and commercial agglomeration areas are distributed along the axis, and the farther away from the axis the lower the density. The bus transportation system of Curitiba has successfully guided and promoted the expansion of urban space, with the urban population growing from only 140,000 in the 1950s to 1.8 million in 1999. Curitiba has a car ownership rate of 3,000 per 10,000 people, which is the highest in Brazil. However, the proportion of public transport trips in the city can reach as high as 68%. For cars, this proportion is 22%, while for taxis, this proportion is 8%.

(5) Flexible and diverse modes of transportation organization

In order to meet the convenient connection of all functional parts of mountain cities, the density of trunk road network should be appropriately increased according to the complexity of mountainurban terrain. For example, in large and medium-sized plain cities, it is generally considered that the distance between trunk roads is 700–1,100 m, and the corresponding trunk road density is 2.8–1.8 km/km^2. In mountain cities, the distance can be appropriately reduced to approximately 500 m. In small cities with complete terrain and small gradient, the distance can be appropriately increased (Huang 2002). In cities with fragmented and complex terrain, the distance can be appropriately reduced, and the corresponding road network density can also be adjusted in a proper way. Refer to the recommended data listed in the table below for the maximum longitudinal slope limit value of roads and motor vehicles in mountain cities (Table 1).

Table 1 Maximum longitudinal slope limit value of motor vehicles on Urban roads

Calculated running speed (km/h)	80	60	50	40	30	20
Maximum longitudinal limit value (%)	6	7	7	8	9	9
Recommended value for maximum longitudinal slope (%)	4	5	5.5	6	7	8

In a plateau area with an altitude of 3,000–4,000 m, the recommended value of the maximum longitudinal slope of the urban road is reduced by 1% according to the listed value, and the recommended value of the maximum longitudinal slope in cold areas with snow cover is not more than 6%. Xu et al. (2005)

(6) Mountainurban landscape environment and street space art creation

The mountainurban road traffic system not only meets the needs of pedestrian and vehicle traffic, but also serves as the main place to display the mountainurban features. First, the urban road system should be organically combined with the landscapes, such as natural environment, mountain water system and green space system, perspectives and ventilation corridors, and considered with street buildings, public activity squares, cultural relics and various public facilities in a unified way, in order to form a rhythmic external space sequence. According to the nature, scale and specific conditions of different streets, an overall arrangement is made for the design of road alignment, the well-spaced treatment and irregular layout of buildings, street furniture, ornament greening, color and lighting, in order to form a coordinated and varied spatial combination (Figs. 8 and 9).

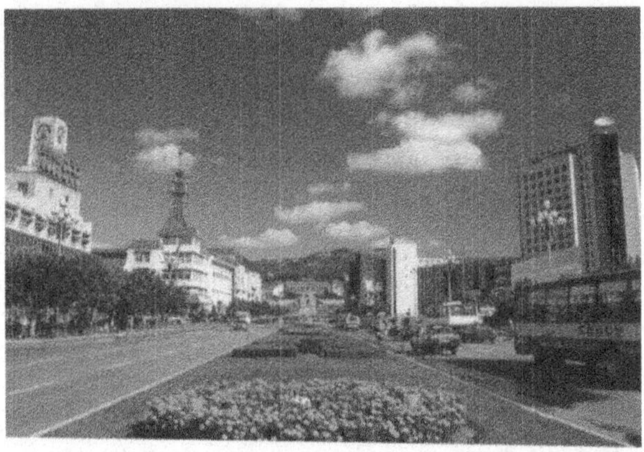

Fig. 8 Urban Garden Street Landscape and Its Organization in Weihai City. *Source* Division of Urban Construction of Ministry of Housing and Urban-Rural Development of the P.R.C. *Landscape Garden Cities in China*. Beijing: China Architecture & Building Press. September 1999, p. 73

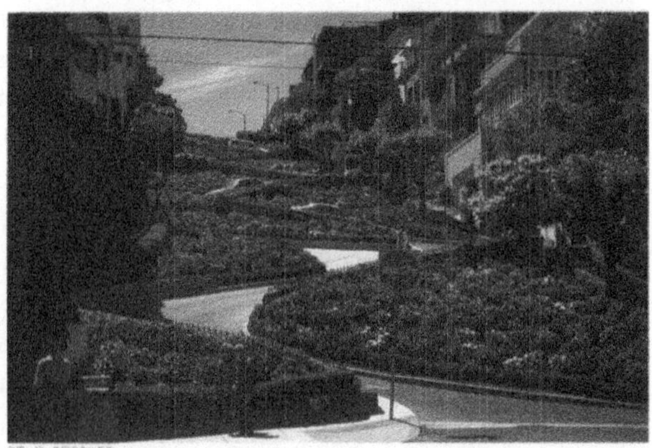

Fig. 9 Lombard Street in San Francisco. *Source* Photograph by the author in 2005

(7) Various types of transportation facilities

In order to overcome the contradiction of complex terrain and inconvenient connection, various types of transportation facilities are adopted in mountain cities, such as light rail (Fig. 10), subway, grand outdoor stairway (Fig. 11), outdoor tunnel (Fig. 12), outdoor vertical elevator (Fig. 13), outdoor escalator (Figs. 14 and 15), telpher (Fig. 16), river cableway (Fig. 17), river crane (Fig. 18), etc.

The outdoor environment of mountain cities often presents complex natural and artificial terrains with multiple steps, fortresses, revetments, streams and gullies where accidents of falls and injuries often occur. Planning designers should pay special attention to humanized treatment, consider the walking safety and convenience of adults, and focus more on that of children, the elderly and people with mobility difficulties. For example, the surface of the ground and steps should not be too slippery, anti-skid treatment should be provided, and facilities such as guardrails should be set up on both sides of the stair climbing. At the same time, it should be

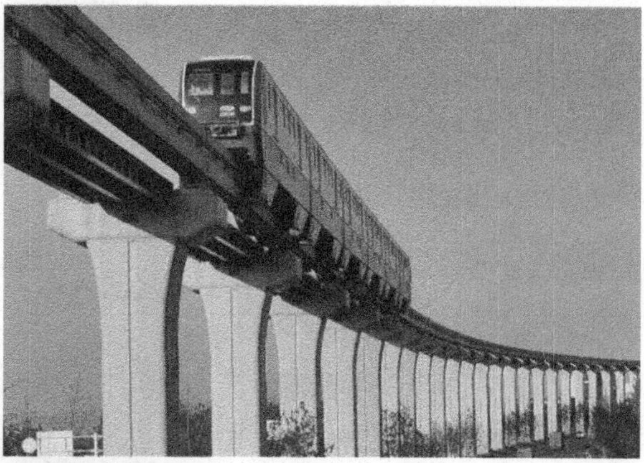

Fig. 10 Light Rail in Chongqing. *Source* Website of Chongqing Rail Transit http://www.cqmetro.cn/

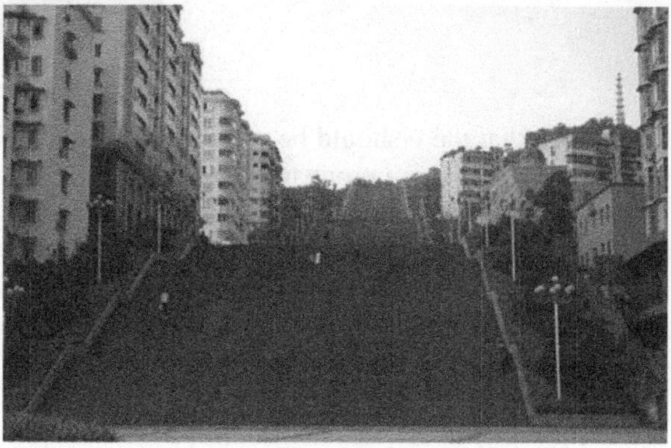

Fig. 11 Grand Stairway in Yunyang New County, Chongqing. *Source* Photograph by the author in 2005

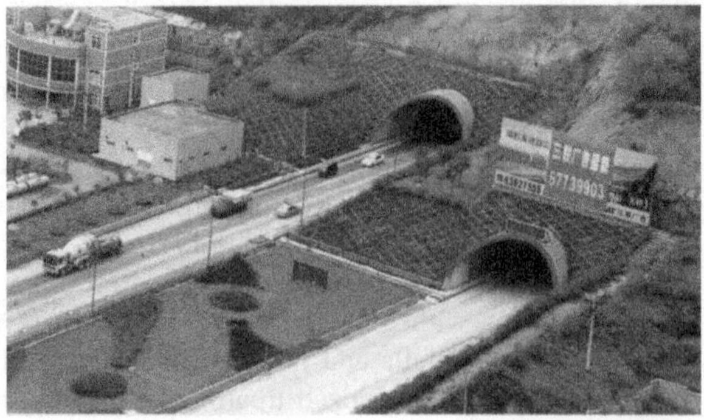

Fig. 12 Tunnel on Zhenwu Mountain in Chongqing. *Source* Chongqing Municipal People's Government. *City · Planning · Art*. Shanghai: Academia Press. October 2005, p. 38

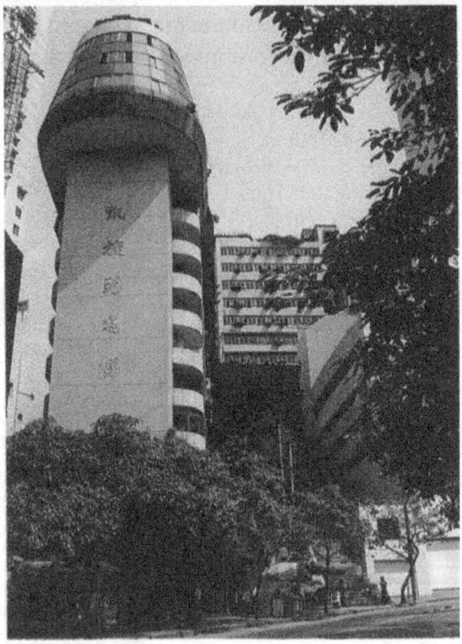

Fig. 13 Outdoor Elevator on Kaixuan Road in Chongqing. *Source* Chongqing Municipal People's Government. *Fifty Years of Construction in Chongqing*. Chongqing: Chongqing Publishing Group. October 1999, p. 76

considered that there should be warning signs in places that are usually not easy to attract attention but are prone to safety accidents. At night in particular, there should be enough lighting to ensure the safety of pedestrians.

In addition, pedestrians should also be careful when walking, just as Du Xunhe, a realist poet of the late Tang Dynasty, said that "The reefs in the river are very dangerous with heaving waves. People are very careful when passing by. Hence, no one has been ever heard of accidentally falling into it and drowning. It is precisely in a place where the current is slow without reefs that people are often heard to be drowned". It demonstrates the truth of never lack of vigilance and nipping hazards in the bud.

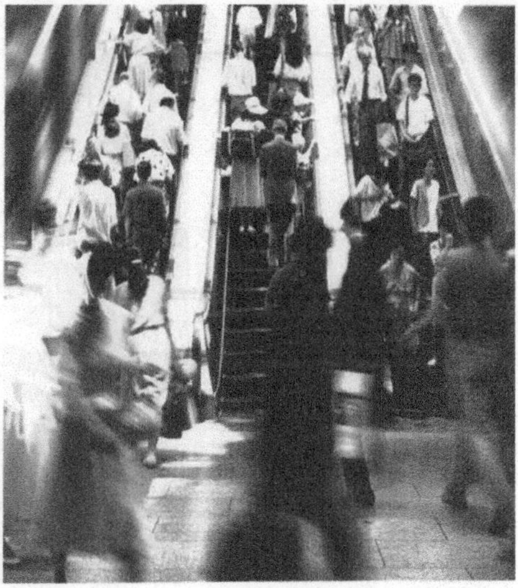

Fig. 14 Escalator on Lianglukou Street in Chongqing Municipality. *Source* Chongqing Municipal People's Government. *Fifty Years of Construction in Chongqing*. Chongqing: Chongqing Publishing Group. October 1999, p. 73

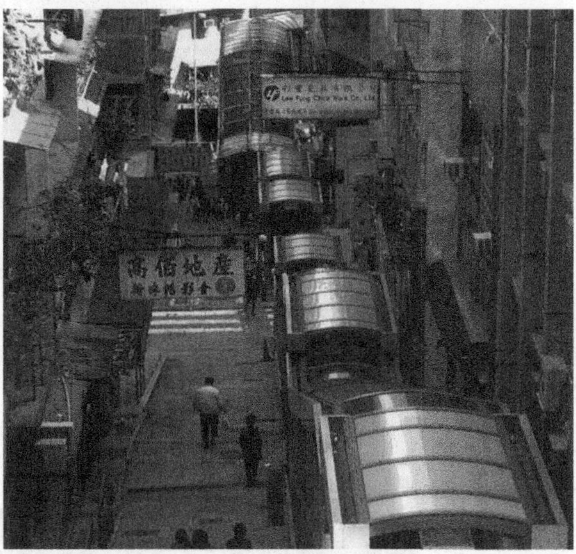

Fig. 15 Escalator in Mid-levels District, Hong Kong. *Source* Lu Jiwei, Wang Haisong. *Mountain Architecture Design*. Beijing: China Architecture & Building Press. February 2001, picture on the inside cover page

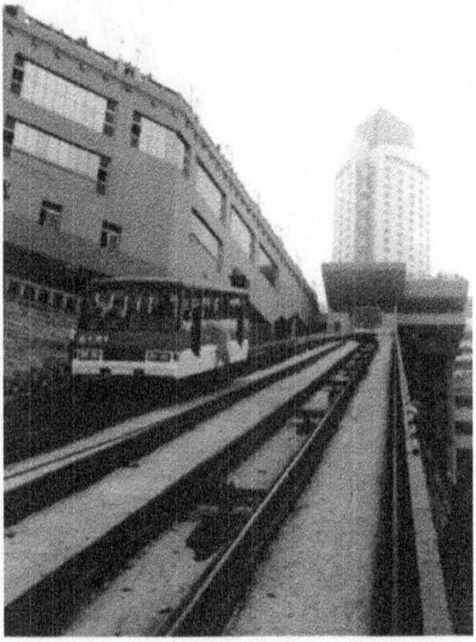

Fig. 16 Telphers in Chaotianmen, Chongqing. *Source* Photograph by the author in 2002

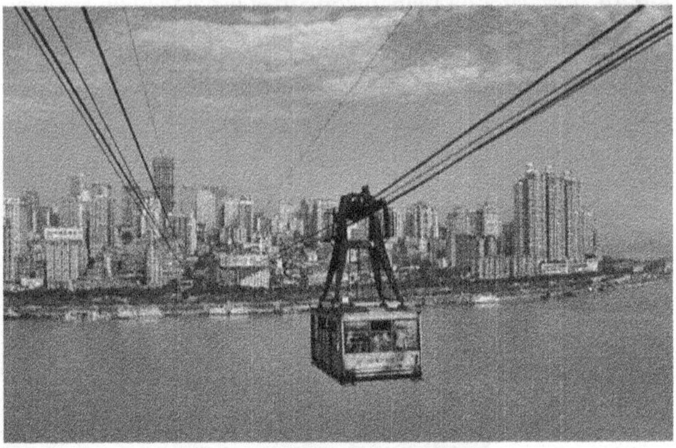

Fig. 17 River Cableway in Chongqing. *Source* Photograph by the author in 2002

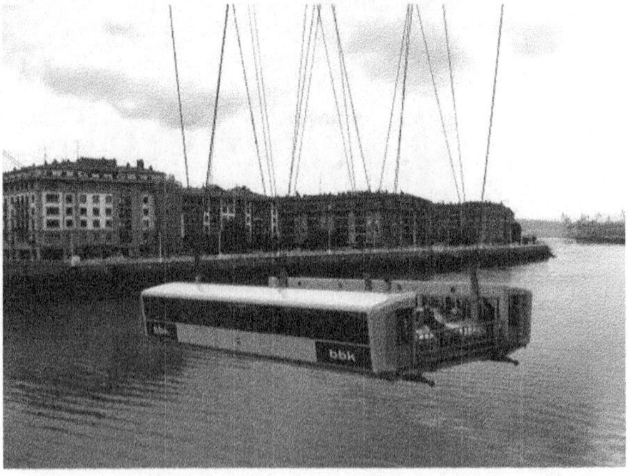

Fig. 18 River Crane in Bilbao, Spain. *Source* Photograph by the author in 2005

References

Huang G (2002) Mountainurbanology. China Architecture & Building Press, Beijing
Huang G (2003) Collection of mountainurban planning and design. Chongqing University Press, Chongqing
Huang G, Chen Y (2002) Ecocity: theory and design approach. Science Press, Beijing
Xu X et al (2005) Urban road and transportation planning. China Architecture & Building Press

Chapter 9
Ecology

1 Mountain Urbanization and Eco-Environmental Planning

(1) Ecological impact of mountain urbanization

Since the 1980s, China's urbanization has begun to accelerate its development, China's social economy has embarked on a period of rapid development, the material and cultural living standards of the people have been significantly improved throughout the country, and the development goal of a moderately prosperous society has been achieved in advance. In July 1997, J. E. Stiglitz, an American economist and Nobel Laureate in economics, listed urbanization in China and the high-tech in the United States as the two key factors influencing the human development in the 21st century. In October 2000, the Fifth Plenary Session of the 15th CPC Central Committee adopted the *Proposal of the CPC Central Committee for Formulating the 10th Five-year Plan for National Economic and Social Development*, which clearly stated the "active yet prudent promotion of urbanization". It is predicted by relevant authorities that by 2030, China's urban population will increase by 400 million, which is equivalent to the total population of Europe today. This will bring an unprecedented impact on China's natural environment, which is mainly reflected as follows:

(1) In the man-made activities of mountain areas and urban construction, such as road construction and real estate development, the undulating natural landforms have been razed to the ground due to the large-scale filling and excavation in spite of the natural terrain, and the vegetation has been damaged, resulting in the man-made landscape of mountain "plainness", which destroys the balance and stability of natural landforms, and accelerates the frequent occurrence of geomorphic disasters such as ground erosion, soil erosion, landslide, mudslide, and collapse.

(2) Urbanization has changed the process of surface hydrological cycle, exacerbated the scouring, denudation and migration of surface materials, accelerated the deposition of rivers and lakes, and increased the hazards of flood

© The Author(s), under exclusive license to Springer Nature Singapore Pte Ltd. 2021
G. Huang, *Theory of Mountainurbanology*,
https://doi.org/10.1007/978-981-16-0819-3_9

disasters. Urbanization has also changed the underlying surface factors of climate formation. The discharge of a large number of harmful gases affect the climate and human health, produces heat island, arid island, humid island, rain island and turbid island effects, forms heat island circulation, and affects regional climate and climate variation, thereby causing the frequent occurrence of various meteorological disasters, such as floods, droughts and rainstorms.

(3) Urbanization has changed the natural habitat conditions. The original natural ecological network system is segmented and fragmented, green spaces are squeezed and shrunk, biological development is restricted, biological species are mutated, animal migration corridors are blocked, species are reduced, and biodiversity disappears, which weakens the buffering capacity of the natural environment, resulting in eco-environmental fragility and affecting the stability of the ecosystem.

(4) Due to the excessive intensity of human activities and over-strengthened unsustainable modes of urban production, land resources are heavily occupied, which makes the contradiction between man and land more prominent, overloads the bearing capacity of resources, greatly weakens the ability of the artificial eco-environment to resist external interference, making it face great difficulties in achieving the strategy of sustainable development. Due to the high concentration of population and industry, many cities and regions with high-speed development have experienced repeated construction, disordered competition, huge consumption of resources and energy, and serious waste, which have caused expanding pressure on the environment, resulting in unlimited urban sprawl and disordered construction (Shang 2002). Accordingly, this has also brought about the deterioration of living environment, business environment, cultural and entertainment environment, public health environment, traffic environment and interpersonal communication environment, which leads to the imbalance of urban ecosystem, produces air pollution, soil pollution, water pollution, noise pollution, solid waste and electromagnetic radiation pollution, and causes the frequent occurrence of various disasters and diseases.

Mountainurban eco-environment is a huge and complex ecosystem of society, economy and environment, which is a unity of interaction, restriction and dependence among urban life organisms with urban population as the main body, natural environment and social environment.

The mountainurban eco-environmental planning is made for the countermeasures, objectives, and measures of urban eco-environment construction in a certain period, which aims to mitigate the negative impact of urbanization on the environment, coordinate the contradiction between urban social and economic development and urban eco-environment, create a healthy, safe and sanitary urban environment for citizens, and realize the sustainable development of the city.

Since China's reform and opening up, the situation of high input, high consumption and low output, which promote rapid economic growth, and has attracted widespread attention at home and abroad. For example, in 2004, China's GDP accounted for about 4% of the world's total, but the consumption of disposable energy accounted for approximately 12% of the world's total, while fresh water is 15%, alumina is 25%, steel is 28%, and cement is 50%. The sharp contradictions of global population, resources, and environment have made China's modernization face severe challenges. Even if the international market can make up for the lack of resources in China's development, the heavy ecological and environmental costs will be unbearable (Li and Zhao 2004). Therefore, it is only possible to achieve the goal of building a harmonious society and a moderately prosperous society in all respects only by transforming the economic growth mode and development pattern, developing circular economy, building a conservation-oriented city, developing an environmentally-friendly society, taking the path of new industrialization, and realizing comprehensive, coordinated and sustainable development.

In October 2005, the Fifth Plenary Session of the 16th CPC Central Committee adopted the *Proposal of the CPC Central Committee for Formulating the 11th Five-year Plan for National Economic and Social Development*, which clearly put forward that the Scientific Outlook on Development should guide the overall economic and social development, focus on independent innovation based on scientific development, and promote the building of a harmonious society. The "11th Five-Year Plan" reflects the overall requirements of the Scientific Outlook on Development in terms of planning concept, system, content and approach.

(2) Tasks and contents of mountainurban eco-environmental planning
(1) Tasks of mountainurban eco-environmental planning

A city is a living and orderly physical environment. The urban spatial structure and form are the skeleton of different urban physical environments, and are the basis of distinguishing different urban features and characteristics. The formation and development of a city, the urban spatial structure and form are inseparable from the natural ecological conditions and social and economic conditions. The urban physical environment is the result of the organic combination and interaction between artificial construction and natural environment. All cities are formed and developed on the basis of a certain natural and geographical environment. Natural ecological conditions such as specific geographical location, climatic characteristics, mountains, rivers, topography and forest vegetation are the external conditions for the formation of urban spatial structure, while the organization of urban functional elements, the infrastructures such as urban industrial structure, market, education, transportation network and energy, and the social and economic factors such as the urban status quo and historical characteristics constitute the internal conditions of urban spatial structure and form that are the result of the interaction between the external and internal urban conditions (Song 2000). Urban society, urban economy and all urban activities must be premised on maintaining the urban natural eco-environment which organisms rely for existence. The relationship between urban structural form and urban natural eco-environment must be established in an adaptive and coordinated way. For example, the urban layout structure, development scale, and road skeleton network cannot be separated from the basic conditions such as geographic location, climatic conditions, topography, and environmental capacity of river and lake systems and land. If the two are not suitable or coordinated, or the artificial construction exceeds the allowable environmental capacity, or the urban road network cannot be combined with local natural conditions, topography and landforms, there will be "urban diseases" such as high population density, traffic congestion, increased urban construction and operation costs, damage of urban landscape, dissipation of features, aggravation of environmental pollution, decline of sanitary conditions, environmental deterioration, and infestation of diseases, which cannot be a healthy, comfortable and sustainable city. In the urban eco-environmental planning, we should not only study the specific local physical environmental planning, but also first seek the reasonability of urban structural form on a whole, and the organicity and integration of urban spatial structure, road network structure, natural conditions, and topography, in order to make the urban spatial structure and natural eco-environment harmonious and adaptive (Liu 2001). Therefore, the task of urban eco-environmental planning is to guide the urban development to the correct direction in coordination with the environment, and combine development, control and guidance well.

(2) Contents of mountainurban eco-environmental planning

Mountainurban eco-environmental planning and protection mainly includes four aspects, namely, optimization of mountainurban physical environment, intensification of mountainurban landscape ecosystem, control and protection of mountainurban physical environment, and evaluation on mountainurban eco-environmental influence.

2 Optimization of Mountainurban Physical Environment

(1) Optimization of Mountainurban spatial structure

The optimization of mountainurban spatial structure refers to the reasonable space-time structure and development pattern formed by the mountainurban socio-economic-natural complex ecosystem on a specific land. A reasonable urban spatial structure should be reflected in different historical stages of urban development. All the urban functional elements can achieve the effect of coordinated development over time. According to the overall requirements of the Scientific Outlook on Development, the main problem for the optimization of mountainurban spatial structure is to establish a spatial structure development pattern in different types of mountain cities that meets its own development requirements according to its natural, economic and social characteristics, thereby achieving the coordinated development between cities and villages, work and residence, production and life, and population development and resources and environment in specific mountain areas (Wu 2001).

(1) **Adjustment of mountainurban industrial layout**

The urban industrial layout not only has an important influence on a city's economic structure and economic development, but also has a direct impact on the overall urban layout, the flow of people, the operation of logistics, the organization of road traffic, and environmental protection. In order to optimize and purify the urban eco-environment, it is necessary to make necessary adjustments to the unreasonable urban industrial layout. For the adjustment of industrial layout, we should first investigate, analyze and evaluate the current situation of urban industrial layout, and put forward suggestions for adjustment according to the results of investigation, analysis and evaluation.

① The adjustment of industrial layout should be conducive to the coordinated development of urban economy and eco-environment.
② The adjustment of industrial layout should be conducive to the industrial transformation and the requirements of ecological development. According to the requirements of industrial ecology and circular economy, we should adjust the internal production process and the coordination among different industries to reduce energy consumption and environmental pollution.
③ According to the degree of environmental pollution in the process of industrial production and the difficulty of treatment, specific measures, such as treatment within a time limit or relocation, closure, shutdown, merger and transfer, should be taken.

(2) **Optimization of mountainurban residential environment**

Living is one of the basic functions of a city. Residential quarters are an important part of a city. The spatial environment and overall image of residential quarters not only have a direct impact on the daily life, mental and physical health of residents, but also reflect the basic features of a city to a large extent. The planning and design of residential quarters should not only meet the basic living needs of residents, but also strive to create a beautiful and comfortable spatial environment to provide them with the needs of outdoor activities such as daily communication, rest, walking, and fitness. According to Maslow, an American psychologist, people have seven basic needs: survival, physiology, safety, love, respect, beauty, and self-actualization. Merely after the former need is satisfied will the latter higher-level need be generated. The environmental design of residential quarters is to meet the needs of various activities of residents and create conditions for the realization of their higher-level needs. Therefore, it is far from enough to complete the design of residential and public buildings themselves in residential quarters, and it is also necessary to pay enough attention to the optimization design of outdoor environment, and elaborately organize and design the layout of adjacent open spaces such as green areas, small gardens and parks, activity spaces for children, teenagers, the elderly, and the disabled, road organization, road and site pavement, as well as the arrangement of architecture, sculpture, ornaments and plants. In addition, the requirements of ecological development should be taken into account in site greening and pavement, in order for the surface water to penetrate into the underground as much as possible, and create a high-quality living space and eco-environment for residents.

(3) **Intensification of mountainurban infrastructure**

There must be modern municipal infrastructure in modern cities. Urban infrastructure planning is to reasonably arrange the infrastructure construction and its time sequence, which is an important content of mountainurban planning and eco-environmental protection.

Urban infrastructure contains a wide range of contents, including comprehensive facilities such as road traffic, water supply and drainage, energy, environment, and disaster prevention, which serves as an important material basis for the survival and development of a city, closely correlated to the guarantee of urban production, life and safety, and urban efficiency and quality.

Urban infrastructure generally consists of six major systems:

① **Transportation system**: External and internal transportation facilities.
② **Water circulation system**: Water resources protection and purification, water supply and drainage facilities, water treatment, recovery, recycling, etc.
③ **Energy system**: Electric power, gas supply, heating facility system.
④ **Communication system**: Information communication network, postal service, television broadcasting system.
⑤ **Environmental system**: Urban environmental protection and public health, sanitation and urban solid waste recovery, treatment and recycling.
⑥ **Disaster prevention system**: Disaster reduction and prevention facility system such as urban earthquake resistance, firefighting, flood control, air defense, typhoon and lightning protection.

The above six systems should be integrated, arranged and optimized in an all-round way, in order for these to form a complex network system that is mutually independent, but functionally interconnected, in the mountainburban plane and spatial layout. Merely in this way can the giant machine of modern city operate efficiently, and perform a comprehensive and coordinated function.

(2) Control of overall mountainurban image

Urban landscape environment is the concrete expression of a city's image, which focuses on showing its characteristic style, architecture, historical and cultural features.

The urban landscape environment control planning is to conduct overall design and control of urban landscape environment under the guidance of the principles of landscape ecology and visual arts, in order to form a diverse and unified urban landscape with different characteristics.

The control of urban landscape environment includes the control of the overall urban image and the landscape environment control of urban scenic spots, scenic spots and historical and cultural relics.

The mountainurban overall spatial structure and landscape image often leave a deep visual impression, such as coastal mountain cities such as Sanya City, Qingdao City, Yantai City and Dalian City, landscape cities such as Guilin City, Hangzhou City, Leshan City, Chongqing Municipality, plateau cities such as Lhasa City and Lijiang City. The overall urban image control is the control of three-dimensional space, which is mainly to optimize the urban spatial structure and the overall landscape image, the three-dimensional space corridor line, the skyline, and the scale and height control requirements of urban and main activity space and square nodes through the visual analysis of the urban natural environment features and elements, such as mountains, rivers, lakes, seashores and topography, the three-dimensional spatial structure of inland cities in sand areas, and the overall architectural image

(1) **Height zoning**: Reasonable height zoning, such as low-rise area, multi-floor area, high-rise area and mixed area, is determined according to the different natures of urban land use and the overall environmental art needs on the basis of the comprehensive analysis and evaluation of the functional structure of urban land and its relationship with the surrounding areas, in order to achieve the coordination between artificial construction and the overall urban natural environment, the diversity and unity of the overall urban appearance, and the unified coordination of social, economic and environmental benefits

(2) **Building height limit control**: According to the height zoning control, the height limit control requirements are put forward for some areas and buildings, such as the clearance limit for aviation and microwave communication, the building height control on both sides of river and street, and the building height control around important urban landscapes and cultural areas.

(3) **Control of viewing corridors and wind corridors**: According to the features of different types of mountain cities and the requirements of improving the thermal environment, the control lines of landscape viewing corridors and wind corridors are established to prevent artificial construction of corridors obstructing the line of sight and blocking ventilation, and form open viewing corridors and ventilation corridors at different spatial levels.

(4) **City axis planning and control**: Through the urban planning and design of main public activity centers and square nodes, the axis of spatial activity and the skeleton of urban landscape art are formed for the collection and distribution of urban main flow of people in an orderly and varied way, in order to show the characteristics and styles of mountainurban culture in a concentrated manner.

(5) **Landscape environment control of mountainurban scenic spots and historic buildings**: According to the law on cultural relic protection and visual analysis, the scope of scenic spot protection areas and cultural relic protection areas are defined. In recent years, in order to unilaterally pursue the so-called "edge effect", real estate agencies in many cities have carried out development and utilization surrounding important historical and cultural sites and scenic spots, such as the Big Wild Goose Pagoda in Xi'an City, the Black Tower and White Towner, Wu Mountain

and Yu Mountain in Fuzhou City, Yuexiu Mountain in Guangzhou City, and Liberation Monument and People's Auditorium in Chongqing Municipality. The construction of high-rise buildings and some extremely uncoordinated commercial buildings have surrounded the historical and cultural buildings in a tightly packed way, severely damaging the overall urban image and the historical and cultural environment. In many mountain cities, "Where there is mountain, no mountain can be seen; where there is water, no water can be seen", the three-dimensional mountainurban landscape image has been seriously damaged.

(6) **Building density and total construction control**: a city or a community must have a reasonable development intensity. The building density and total construction must be within its allowable environmental capacity. If the density is too high, there will be ecological imbalance and destruction.

In the control of urban landscape environment, the construction of high-rise buildings is a problem that cannot be ignored. In many cities, the construction of super-large buildings and high-rise buildings is out of control, which leads to the serious destruction of urban landscape resources. Many mountain cities imitate plain cities and regard elevation buildings as a symbol of modernization, ignoring the existence of natural mountains and rivers, and blindly constructing large and high-rise buildings irrespective of local natural landscape features, locations and occasions, which have a great boldness of vision of comparing with the height of mountains, and do not coordinate with the original natural environment in terms of building scale, volume and color, resulting in the destruction of natural landscape resources or the disappearance of biodiversity and landscape diversity. Due to the development of urban economy, the progress of construction technology, the expansion of urban scale and the shortage of construction land, many large and medium-sized cities in China have witnessed the rapid development of high-rise buildings, which has its inevitability of social and economic development. However, due to the lack of correct planning, design guidance, strong management and regulatory control, the construction of many urban high-rise buildings is in disorder, which has resulted in the disorderly development of the flow of people, logistics and traffic, and damaged the overall urban image. Due to the large investment and volume of high-rise buildings, it will have a great impact on the surrounding environment and transportation. Therefore, the construction of high-rise buildings must be treated with caution. From the perspective of landscape eco-environment, whether to build high-rise buildings or not mainly depends on:

① hether it is conducive to the protection of the city's natural, historical and cultural characteristics, and whether it is in harmony and coordination with the surrounding environment;

② Whether it is conducive to the reasonable layout of the city and intensification of its landscape environment and features;

③ Whether it is conducive to the reasonable distribution and organization of urban traffic flow to provide a good external spatial environment for citizens.

Detailed analysis and evaluation and elaborate design should be made on site selection, location, height, plot ratio, volume, style, color, red line setback, traffic organization, parking space, and surrounding environment of high-rise buildings, as well as the treatment of the surrounding environment and the favorable and adverse effects on the landscape eco-environment.

In addition to consideration from the landscape eco-environment, the construction of high-rise buildings also involves complex factors such as society, economy, engineering technology, and safety, which need to be comprehensively studied and considered. For key high-rise building construction projects with significant impact in some important sections, environmental impact prediction and assessment shall be carried out according to relevant regulations and procedures.

3 Intensification of Mountainurban Landscape Ecosystem

(1) Intensification of the connection of landscape ecological network

In the process of transforming nature, human beings have exerted profound influences and changes on the natural landscape pattern of mountain areas. According to the degree of human influence on the mountain natural landscape pattern, it can be divided into three forms: disturbance, transformation and reconstruction. Disturbance refers to the change of the characteristics of mountain landscape caused by human activities, which makes the mountain natural landscape fragmented, thus affecting the changes of the mountain natural landscape pattern. For example, urban development and construction activities destroy the natural landforms and forest vegetation, exploration of roads cuts off the corridors of biological migration, and a large number of artificial buildings and municipal facilities change the local climate of urban areas. Transformation means that human beings consciously increase or decrease the elements of mountain natural landscape for the purpose of survival and development, to change the mountain natural landscape to a certain extent, such as mountain excavation, reservoir construction, formation of artificial lakes and wetlands, or shelterbelt construction. Reconstruction is the rebuilding of the original mountain natural landscape pattern and structure. For example, due to the construction of the living environment of towns and villages flooded by the construction of large reservoirs, undulating changes and winding natural mountains and forest landscape have become terraced agricultural landscape.

The mountainurban landscape ecosystem is a combination of artificial and natural landscape ecosystems. Due to artificial construction, the mountain natural landscape ecosystem has been seriously disturbed and destroyed. The intensification of mountainurban landscape ecosystem is to strengthen the protection of the stability of natural landscape ecosystem so as to give full play to the service function of its ecosystem.

The establishment of the national land ecological network system by the government of the Netherlands in 1990 is an example. In the Netherlands, where high-density land use and modern transportation networks are developing at a high speed, mutual connection is reduced due to the division of urban living quarters. To this

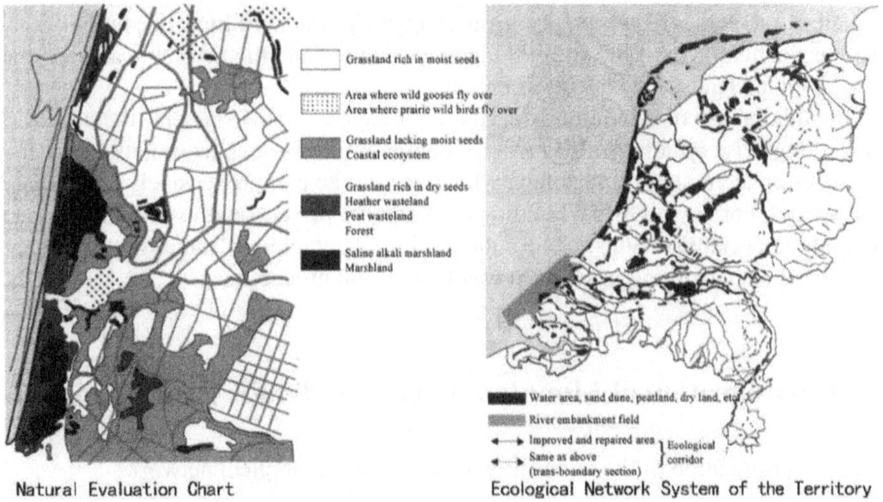

Fig. 1 Land ecological network system of the Netherlands. *Source* Architectural Institute of Japan. (Translator: Li Biyu, et al.). *Handbook of Environmental Design (Comprehensive Edition)*. Beijing: China Architecture & Building Press. February 2003, p. 506

end, the government of the Netherlands formulated the national ecological network system in 1990 to strengthen the connection between the protection of ecosystem and living areas (Fig. 1).

The land ecological network system of the Netherlands is a hierarchical structural network formed on the basis of important ecological elements, such as core part, naturally formed area, ecological corridor, and buffer zone.

(1) Core part: important biological habitats that can preserve and protect the ecosystem;
(2) Naturally formed areas: promote the formation of the ecosystem according to the composition of ecological network;
(3) Ecological corridor: the corridor connecting the core part;
(4) Buffer zone: mitigating the impact of surrounding areas;
According to the different relationship between roads and the construction of different routes, overpass and underground passages should be set up under the condition that each core part is divided to ensure the connection with each residential area.
(2) Optimization of green environment

Green environment plays an important role in urban ecosystem. Humans have an instinctive interest in green plants, which not only have use function and ornamental value, but also have physiological functions. Green plants play an important role in improving eco-environment, adjusting climate, increasing humidity, reducing temperature, alleviating heat island effect, reducing noise, absorbing harmful gas and dust, protecting and increasing biodiversity, giving play to the function of biodegradation and disaster prevention, evacuation, enriching people's spiritual and cultural life, and coordinating the relationship between man and nature. Therefore, the construction of forest plants and green environment is not only directly related to the quality of urban and rural eco-environment and the improvement of residents' quality of life, but also a necessary condition for the economic development of a city and a region, which is the basic guarantee for the sustainable development of urban and rural areas.

The intensification of the landscape ecosystem is to strengthen the protection of the stability and sustainability of the landscape ecosystem, and to adopt compulsory protection for special ecological function areas, such as water conservation areas.

In the past, developed countries such as Europe and the United States have carried out over-intensive exploitation of natural resources as material wealth, so they are constantly punished by nature. As Engels warned in the early days of European industrialization, "Don't be too intoxicated in our victory over nature, and for every such victory, nature retaliates against us". Many countries have learned their lessons from this, changed their course, and strived to protect natural and humanistic civilization resources. The restoration of forest vegetation in Northern European countries such as Sweden, Denmark, Norway and Finland, as well as the development of rural tourism, such as natural tourism, agricultural sightseeing and agritainment are examples. The Finnish law stipulates that if a tree is cut down, three trees must be replanted. In Germany and the United States, even if you cut down a tree in your own home, you must apply to the government. The Luxembourg's parks and nature reserves account for more than 40% of the country's land area. Japan has put forward the slogan "No governance of mountains, no governance of the country". The so-called governance of mountains in Japan, that is, the regulation of river basins in European and American countries, aims to give full play to the ecological benefits of soil and water resources. In the so-called "governing mountains before water governance, afforestation before mountain management", through the protection of mountain forest vegetation, vigorous afforestation, restoration and remediation of sensitive dangerous areas such as wastelands and landslides, we should protect and improve water source forests, and give full play to its function of water conservation, in order to meet the growing demand for water resources in economic and social development, fully satisfy the people's desire for green and nature, and create a good eco-environment on the basis of forests. After the Second World War, Japan's eco-environment and economy were in a severe recession, but instead of logging their own forests, they went to Australia to buy forest wood and China to buy disposable wooden chopsticks. China's current forest coverage rate is only 13.92%, far lower than 60% in Japan. According to statistics, China produces as many as 45 billion pairs of disposable chopsticks a year, which requires 2,700 trees to be cut down every year.

Considering from the requirements of urban ecosystem, special attention should be paid to the construction of the regional level in terms of environmental greening, that is, the ecological greening system of the large environmental circle. In combination with regional planning and regional (municipal) mountain ranges, river systems, farmland water conservancy projects and urban water source protection areas, large, medium and small cities, moderate cities in particular, can build forest parks according to their own topographical and geomorphic characteristics through afforestation, soil and water conservation, vegetation protection, planning and construction of ecological green space systems with various types and functions, and combine the planning of urban environmental greening with the planning of national land improvement, to realize land gardening, and form a ecological green space system network integrating urban and rural areas. At the same time, it is necessary to do a good job in greening at the urban level, residential area level, and household and residential level to form a beautiful and clean green environment.

In order to pursue the agglomeration effect and material civilization, human beings are far away from nature, living in a crowded and narrow reinforced concrete "forest" with ecological imbalance. In order to alleviate the imbalance between mankind and the natural environment and improve the density, the measures to improve the environment by strengthening greening have emerged, because strengthening greening is the fundamental way to improve the environmental quality in cities. The different development patterns of improving urban environment with greening include:

(1) Introducing nature into city—an approach of local improvement
(2) Leading city to nature—increasing the contact between the city and nature through the organic dispersion of spatial structure.
(3) Integration of city and nature - It is the highest form of combination of man and nature to integrate cities with nature by taking advantage of natural conditions such as natural mountains, rivers and lakes, which can only be realized in the urban eco-environment of mountain areas. It is a poetic and painting realm with a blend of mountains, water, fields, forests, and urban landscapes, just as an ancient poem goes, "People can still enjoy the tranquility of mountains and rivers, and the sereneness and seclusion of forests and springs in downtown areas". Therefore, ideal ecological cities will appear in some mountain cities with well-developed landscape conditions in China in the future. Various mountainurban greening has its advantageous conditions, because it can make use of mountains, barren hills, cliffs, steep slopes, gullies, streams, lakes and water surfaces for greening, and combine with streets, squares, fortresses, revetments, building walls and roofs for extensive greening and beautification, constituting a rich and colorful mountainurban wide-range and three-dimensional landscape environment. At the same time, the mountainurban greening network is extended to the peripheral rural areas surrounded by mountains and green water through mountains, rivers and lakes, road traffic corridors, farmlands and forest networks, in order to form a green ecological network system with landscape gardening and urban and rural integration.

In 1972, the *Declaration on the Human Environment* (Stockholm) stated that "Man is the creator and transformer of the environment that provides man with not only material needs, but also opportunities for intellectual, moral and spiritual growth. Man must coordinate with nature and use knowledge to build a better environment".

(3) Green space compensation

Green space compensation is of great significance in urban planning and construction.

Ensuring a certain proportion of green space in urban and rural construction is the strategic work in urban and rural planning, construction and management. However, in the actual construction, when some projects of regional or even national significance have to occupy a certain area of green space, the ecological function of green space must be manifested in other forms through the principle of green space compensation.

Since the energy crisis and environmental movement in the 1960s, countries such as Germany and the Netherlands have attached great importance to the protection of green space, and formulated the green space compensation law to ensure that the ecological service function of green space will not be affected and reduced.

There are three main forms for the green space compensation approach:

(1) **Local ecological compensation**

The green space occupants improve the eco-environment of the occupied area to make up for the loss of its ecological benefits. For example, in the construction of railway or highway, ecological restoration and improvement should be carried out in the ecologically damaged areas through which the line passes.

(2) **Allopatric ecological compensation**

The green space occupants rebuild ecological areas in other areas with similar ecological functions and characteristics as the occupied areas.

(3) **The green space occupants pay the corresponding compensation fee in the form of currency for the loss of the ecological service functions and environmental benefits of the original green space caused thereby**

The first two approaches mainly make up for the loss of the ecological service function and environmental benefits of the original green space by improving the original damaged ecosystem or rebuilding other ecosystems.

The latter monetary compensation is an approach adopted when the first two approaches cannot be implemented. According to Kaldor-Hicks's "compensation test", the essence of the compensation theorem requires that the total gain is greater than the total loss. The benefits of urban green space (such as trees, shrubs, grasslands, etc.) in terms of ecological service functions are calculated, and the various ecological service functions of green space are monetized. Then, the sum of the compensation values of various ecological service functions of green space in the loss area is the compensation fee to be paid by the green space occupants.

(4) Organization of waterfront space

The organization of urban waterfront space plays an important role in improving urban microclimate, enhancing the quality of urban ecological landscape environment, and providing residents with water-enjoyable leisure space. Mountains are beautiful, ethereal and charming because of water. If a city has high-grade and high-quality waterfront space, it will have aura and visitors will stop to enjoy infinite fun, such as the lakeside tour space in Hangzhou City, the waterfront greenbelt in Qianjiang City, tour space surrounding Tai Lake in Wuxi City, waterfront space of Funan River and Living Water Garden in Chengdu City, and waterfront greenbelts in Luzhou City, Jiangjin District and Yunyang County in Chongqing Municipality.

Special attention should be paid to the ecological and water-enjoyable requirements in the organization of waterfront space in mountain cities. The requirement of ecologicalization is to make the urban waterfront space play its ecological service function as much as possible. First of all, waterfront space is regarded as an organic part of urban spatial structure, which is organically connected with the urban trunk road structural system, ventilation corridors, and sight landscape corridors. At the same time, the waterfront and water surface should be kept as natural as possible. For example, the natural coastline can be freely extended without bending or straightening, the rigid engineering treatment can be made, and natural connection can be obtained with the surface water system without cutting off the surface water system.

The water-enjoyable requirement is to create conditions for residents to easily access the water surface and create a water-enjoyable natural space. The waterfront treatment of the Living Water Garden besides the Funan River in Chengdu City, the natural shoreline treatment of the Sibelius Waterfront Park in Finland, the relationship between the greening of Oslo in Norway and roads, and the organization of the waterfront space in Niigata, Japan are all good examples (Fig. 2).

4 Control and Protection of Mountainurban Physical Environment

The control and protection of mountainurban physical environment is to analyze and forecast the development trend of future environmental changes and pollution in combination with the urban social and economic development, put forward prevention measures for main physical environmental factors such as air quality, atmospheric thermal environment, water, noise, soil and solid waste, and formulate the objective of limiting the achievement of control standards in stages.

(1) Control of atmospheric environment

Atmosphere is an indispensable basic material for human survival. In a city, the high-density population and buildings and high-intensity economic and information activities have greatly changed the natural ecological conditions, affected the urban microclimate, produced the urban heat island effect, raised the urban temperature, and greatly increased the harmful ingredients in the atmosphere, which changes the normal components of the atmosphere, generates atmospheric pollution, poses hazard to the industrial and agricultural production, human health and the growth

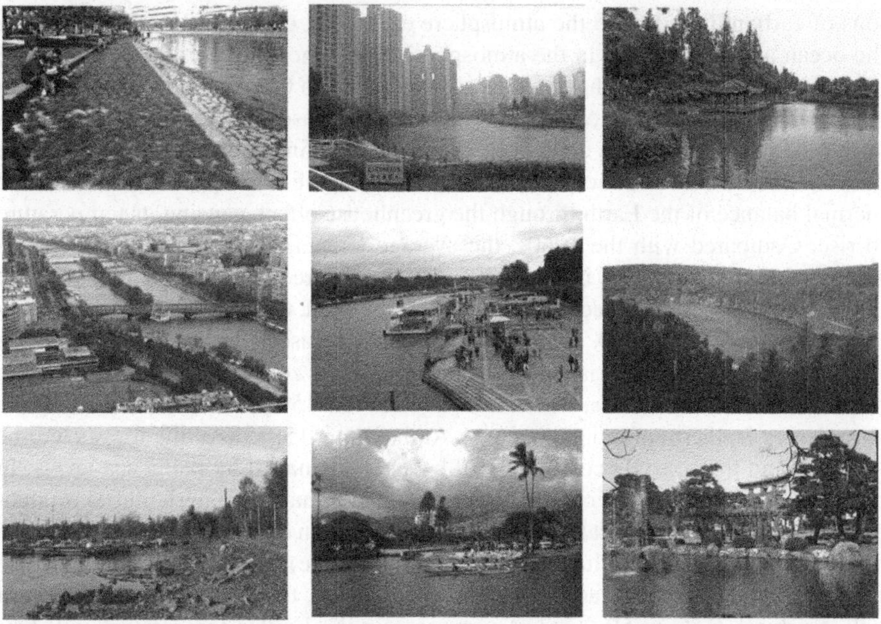

Fig. 2 Cases of various waterfront spatial organizations. *Source* Photographs by the author in 2000, 2004, 2005 and 2006, respectively

of plants and animals, even increases the incidence of cancer, and causes serious problems such as genetic mutations.

The main sources of urban air pollution mainly include industrial pollution sources, domestic pollution sources, and transportation pollution sources. In industrial cities, the most serious sources of atmospheric pollution are often formed due to the large amount, various types and complex components of exhaust gas pollutants discharged from industrial production. The main sources of domestic pollution are the exhaust gas and smoke emitted from kitchen and restaurant cooking. In terms of transportation pollution sources, with the rapid increase of urban automobile ownership and the development of water, land and air transportation, the exhaust emissions of various means of transportation have risen sharply, causing more and more serious pollution to the urban and rural atmospheric environment. The main pollutants that are harmful to human health are sulfur dioxides, carbon monoxide, nitrogen oxides, hydrocarbons, leads, dusts, etc. Carbon dioxide has been in the Earth's atmosphere for at least 2 billion years, with an atmospheric concentration of 0.028% for a long time. However, since the industrial revolution, especially since the 20th century, the global balance of carbon dioxide has been seriously disturbed due to the massive combustion of fossil fuels such as coal, oil and natural gas. From 1850 to 1996, due to human activities (transportation, heating, industrial coal-fired fuel, deforestation, etc.), the carbon dioxide content in the atmosphere had increased by 25%, and the methane content had doubled. At present, humans release approximately 10 billion tons of carbon dioxide into the atmosphere each year, of which 2/3 is absorbed by the ocean and 1/3 remains in the atmosphere. The concentration of carbon dioxide in the Earth's atmosphere has increased from 0.028 to 0.035%. It is estimated that by 2100, 30 billion tons of carbon dioxide will be released into the atmosphere each year, which means that the content of carbon dioxide in the air will be tripled. The increase in the concentration of carbon dioxide in the atmosphere will affect the thermal balance of the Earth through the greenhouse effect, causing its temperature to rise. Compared with the 1960s, the average temperature in the northern hemisphere increased by 0.2 °C in the 1970s, which was predicted to increase by 0.5 °C in the early 21st century. According to the current trend, the Earth's temperature may increase by 3–4 °C by 2100. The influence of this increase in the average temperature of the Earth on the climate is equivalent to the climate zone advancing 150–500 km to the north and south poles, which is equivalent to a 150–500 m upward rise of the zero-degree isotherm in mountain areas. At that time, 15–75% of the forests will be affected, and the coastal ecosystem will be easily damaged by rising sea levels. In addition, there will also be changes in rainfall zones that affect agricultural production. Some scientists advocate that in addition to carbon dioxide, there are at least 15 other trace gases (such as nitrogen dioxide and methane) that also have a greenhouse effect. If these trace gases are also taken into account, the temperature of the Earth will rise by 3–6 °C in 50 years. An increase in the Earth's temperature will first begin to melt the polar ice caps, which account for approximately 2% of the Earth's total water. According to research by the Australian Antarctic scientist Delamarre, the Antarctic sea ice may have melted 25% in the 20 years from the early 1950s to the early 1970s. This change in sea ice area is of global significance because the ice sheet in Antarctica is an important part of the global climate system (Shang 2002).

The Qinghai-Tibet Plateau in China is one of the most sensitive regions to global climate change. As the Science and Technology Daily reported on May 14, 2005, glaciers in the Everest region are in full retreat due to the impact of global warming. According to relevant data, climate change will directly affect the global eco-environment and then the GDP of each country. At present, the cost of weather-related natural disasters around the world is in the tens of billions of dollars a year, which will rise to hundreds of billions of dollars over the next decade with the gradual warming of the global climate.

The main measures for controlling air pollution and improving the quality of air environment are to change the fuel structure, save energy consumption, install dust and smoke abatement facilities, formulate mandatory emission reduction standards, reduce pollution, use clean energy, develop circular economy, and increase the forest green area, as well as to strengthen monitoring and management measures, and strictly implement national standards on atmospheric environmental quality and relevant regulations on environmental protection. In urban and rural planning and construction, the measures should play a positive role in mitigating urban and rural atmospheric environmental pollution, reducing greenhouse gas emissions, slowing down the process of climate change, and reducing its impact on urban and rural life.

(2) Water environment control

Water is the basic material guarantee for human survival. China is a water-scarce country. Therefore, water conservation and water environment control play a particularly important role in urban ecological and environmental planning, which is of great significance to the implementation of sustainable urban development.

The water environment control planning includes both comprehensive utilization and protection planning of water resources and comprehensive improvement planning of water pollution.

(1) **Comprehensive utilization and protection planning of water resources**

① According to the prediction of urban water consumption, we should analyze the balance of supply and demand of water resources, and formulate plans for comprehensive development, utilization and protection of water resources;

② We should formulate different water source utilization and protection planning measures according to different drinking water sources (such as surface water, ground water, etc.) in each city, and actively explore the possibilities and paths of water-saving industries and the ecological development of the water industry. Strengthen management of different water source protection areas to prevent pollution. On the basis of making a comprehensive survey of the reserves of groundwater sources, we should also realize rational exploitation, strictly control the exploitation and ensure planned exploitation;

③ We should strengthen the protection of water resources in coastal cities, formulate protection plans for the protection of coastlines and water areas according to the natural ecological characteristics of coastlines, and strictly control the discharge of land-based pollutants;

④ We should formulate a reasonable distribution scheme of water resources and the countermeasures and measures for water conservation and return water utilization;

⑤ We should improve the urban water supply and drainage system;

⑥ We should explore new ways and approaches of rainwater utilization;

(2) **Comprehensive control planning for urban water pollution**

Water pollution is caused by a large number of pollutants discharged into water body, the content of which exceeds the self-purification capacity of its background content, resulting in deterioration of water quality, thereby destroying the normal function of water body. Urban water pollution is the pollution of water body caused by pollutants produced by urban production and living activities, including industrial pollution, domestic pollution and agricultural pollution sources. Industrial wastewater is the most important source of water pollution with the characteristics of large volume, wide range, complex composition, high toxicity, difficult purification and treatment. Domestic sewage is mostly non-toxic inorganic salts, aerobic organic salts, pathogenic microorganisms and detergent pollution, featuring high levels of nitrogen, phosphorus and sulfur, many bacteria, and seasonal variation of water consumption; agricultural pollution sources include livestock manure, pesticides and chemical fertilizers, which is characterized by high levels of organic substances, phytonutrients, pathogenic microorganisms, pesticides and chemical fertilizers. The comprehensive control planning for urban water pollution mainly includes the following contents:

① Predict the discharge of urban sewage according to the urban development plan;
② Correctly determine the urban drainage system and sewage treatment scheme, develop ecological treatment approaches, popularize recycling technology, and reduce the amount of sewage treatment;
③ Strictly control the urban land development plan to reduce soil erosion and pollution sources;
④ Strengthen the discharge control of pollution sources such as industrial wastewater and domestic sewage.

(3) Acoustic environment control

Noise has a great impact on the health of urban residents. With the development of urban economic and cultural construction and the increase in traffic volume, urban noise has become an important source of urban pollution, which mainly includes traffic noise, capital construction noise, factory production noise and activity noise. Among them, traffic noise has become the most common source of pollution that has the greatest impact on cities and residential areas. About 70% of urban environmental noise comes from vehicles, such as cars, trains, airplanes, and ships, which constitute a cubic noise source of urban space. At present, the noise on both sides of a highway in China is approximately 70–80 dB. When the speed of a motor vehicle is doubled, the noise can increase by 6–10 dB. In urban areas with developed air transportation, aircraft noise is the most harmful. When an aircraft flies at an altitude of 15,000 m, the pressure waves can spread to the ground within a range of 30–50 km on both sides of the flight track, affecting many residents. At present, approximately half of the world's people live in a noise-polluted environment, and approximately 40% of the urban residents in China live in an environment that exceed noise standards.

The noise control and reduction planning measures mainly include:

(1) Adjust the layout of urban structure, and correctly deal with the relationship between the railway entry route, highway entry route, airport site selection, noise-polluted factories and urban development and functional areas;
(2) Optimize the industrial structure, adjust the industrial layout, and take measures to control or relocate the factories disturbing residents, in order to fundamentally eliminate large noise sources;
(3) Carry out acoustic environmental impact and assessment on newly developed urban land, and reasonable planning and design according to national standards for environmental noise in urban area and other functional requirements;
(4) Reasonably organize the urban traffic network to avoid the interference of traffic noise on residential areas as much as possible;
(5) Reduce the impact of noise by using greening treatment and setup of isolation belts;

(6) Prevent the spread of noise and reduce the impact of noise by using the change of terrain height;
(7) Restrict the flow of vehicles on major traffic arteries or restrict certain vehicles from entering or leaving specific areas.
(4) Prevention and control of soil pollution

(1) **Characteristics of soil pollution**

Soil pollution mainly refers to the phenomenon that pollutants enter the soil and form accumulation, which deteriorates the soil quality. When the amount of "three wastes" (waste gas, waste water, industrial residue) invading the soil exceeds the environmental capacity and transfer and transformation ability of the soil, and destroy its original balance, the soil composition, structure and function will be changed to cause soil pollution. Soil pollution mainly comes from industrial and urban wastewater and sewage, solid waste, chemical fertilizers and pesticides, atmospheric pollution sediments and biological pollution.

The remarkable characteristic of soil pollution is its sustainability and long-term nature. Pollutants that enter the soil move slowly and are not easily removed by large-scale measures. For example, it takes more than 10 years for organochlorine pollutants to decompose naturally in the soil, while the natural degradation of foamed plastic materials and plastic bags known as "white pollution" in the soil takes more than hundreds of years. In some cases, after the soil stops polluting, the pollution still exists even if these are reused after three or five years. When soil is contaminated, it will harm human health through the food chain and groundwater. When the soil of a city is contaminated, the natural ecological balance of the microbial system in the soil is destroyed, which makes a large number of bacteria propagate and spread, resulting in the spread of diseases in densely populated urban areas.

(2) **Causes of soil pollution**

The pollution of industrial wastewater and urban sewage to soil mostly comes from the unscientific use of sewage to irrigate farmland. Sewage and wastewater contain many nutrients needed by plants, nitrogen, phosphorus and potassium in particular. The reasonable use of seeds can achieve the effect of increasing production, but if the sewage is arbitrarily irrigated without necessary treatment, it will cause serious consequences because sewage and wastewater also contain various toxic and harmful substances, including organic and inorganic compounds such as heavy metals, phenols, cyanide, mercury and arsenic. These toxic and harmful substances are deposited in the soil and will remain in the roots, stems, leaves, flowers, fruits and seeds of crops after being absorbed by crops, and finally reach the human body and cause serious harm. It often occurs all over the world to cause soil pollution and endanger the health of humans and animals by unscientific use of sewage for farmland irrigation. In many areas of China, the soil pollution caused by unreasonable use of sewage irrigation is very serious. For example, an irrigation area in Lanzhou City, Gansu Province used sewage for irrigation, resulting in 17,700 μ of farmland polluted, in which 7,700 μ was forced to abandon cultivation due to excessive salt content. From 1977 to 1978, 19,000 μ of farmland in the suburbs of Jilin City, Jilin Province was too polluted to be used for farming due to sewage irrigation.

Excessive use of chemical fertilizers and pesticides is another important cause of soil pollution. Long-term heavy use of nitrogen fertilizer may destroy the soil granular structure and reduce crop yields. Excessive application of phosphate fertilizer may cause the lack of elements such as iron and zinc in the soil, and phosphate fertilizer is fixed into an ineffective state, thereby reducing crop yields. In addition, heavy use of calcium cyanide may also cause soil pollution, which will produce toxic substances such as dicyandiamide, cyanic acid and hydrogen cyanide in the soil, leading to inhibition of nitrification and reduced crop production. Chemical pesticides have a significant effect on the prevention and control of diseases and insect pests, but improper use may cause soil pollution, which may cause water and air pollution, and also enter the human body through the food chain. According to statistics, chemical pesticides have experienced rapid development in the world in recent decades, which presently have more than 500 varieties, with an output of 1.5–2 million tons. The degree of soil pollution by pesticides and the duration of their retention in soil depend on the stability of pesticides themselves, the degradation of microorganisms in soil, meteorological conditions, and the physical and chemical properties of soil. At present, pesticides that pollute the soil and plants heavily are mainly organochlorine pesticides and pesticides containing heavy metals such as lead, arsenic and mercury, as well as certain specific herbicides.

Solid waste pollution is another important factor of soil pollution. Solid waste refers to solid and muddy substances discharged from human production and life, which can be divided, as follows: industrial solid waste, mining solid waste, agricultural waste, radioactive solid waste and urban waste according to its source; organic waste and inorganic waste according to its chemical property; hazardous waste and general waste according to its hazard status. There are many kinds of solid wastes with different toxic elements. Industrial solid waste and mining solid waste often contain multiple toxic elements, and urban domestic waste contains a variety of pathogens, parasites and a large number of toxic substances. If these solid wastes are discharged arbitrarily, these will cause serious pollution to the soil and water body after being leached by rainwater.

(3) **Prevention and control of soil pollution**

Due to the relatively fixed location and obvious regional characteristics, soil resources are susceptible to pollution from urban development and construction, industrial and domestic wastewater, sewage, waste gas and solid waste in mountain areas and cities. Once the soil is contaminated, it is difficult to recover in a short period of time. Therefore, strict and effective preventive measures must be taken against the pollution sources of urban production and life, so as to eliminate all kinds of pollution sources that may be produced by urban and suburban soil.

In urbanized areas, except for suburban agricultural zones and park green spaces, the soil layer of urban land has been removed and stripped due to ground buildings. Instead, it is replaced by hardened artificial ground, forming a layer of impermeable ground pavement between ground and underground, which isolates the ecological channel for the interpenetration of surface water and groundwater, and causes new ecological problems. Therefore, in the planning and construction, it is necessary to expand the pavement scope of permeable ground as much as possible, and reduce the proportion of impermeable ground.

(5) Control of urban solid waste

Solid waste, including residential refuse, construction waste, factory waste and commercial waste, is an important source of urban pollution. With a staggering domestic waste output in China, each urban resident generates about 300 kg of waste on average a year, which is increasing by 10% each year. Due to the large amount, wide range, great varieties and complex composition of urban solid waste, secondary pollution is often easy to occur in the process of collection, accumulation and treatment, which brings great difficulties to control and treatment.

The control and treatment of urban solid waste is of great significance for comprehensive utilization and recovery, purification of urban environment, improvement of environmental quality and economic benefits. In terms of the control of solid waste, we must first reduce the generation of solid waste from the source as much as possible, for example:

(1) Actively develop green industries, advocate green consumption, minimize the generation or non-generation of solid waste pollution sources, strictly control "white pollution" and develop degradable commodities;
(2) Improve the environmental awareness and civilization of the whole people, develop good hygiene habits, consciously maintain the cleanliness of the environment, improve the recovery and comprehensive utilization of solid waste, turn waste into treasure, and realize the recycling and commercialization of solid waste;
The treatment of solid waste is to use physical, chemical, biological and other treatment approaches to carry out the final treatment of the waste. The treatment approaches generally include sanitary landfill, biotechnological composting and efficient incineration. Regardless of the approach adopted, sufficient site area is required. Hence, this is a critical link for the reasonable selection of solid waste disposal sites and treatment approaches.

① The treatment site should be selected according to different treatment approaches. The above three treatment approaches have different requirements for site selection;
② In terms of the area of treatment site, we should consider not only its own capacity and occupation requirements of production facilities, but also the needs of supporting facilities and development;
③ The economic efficiency and reasonability of transportation distance should be considered;
④ Consideration should also be given to the ecological and environmental conditions of the site, such as climate, soil, geology, hydrology, and other natural conditions and the impact on the surrounding environment.

As the construction of centralized urban solid waste treatment plant (factory) covers a large area, with large investment, complex treatment and possible adverse impact on the surrounding environment, environmental impact prediction and assessment should be conducted for the determination of its location, site selection, treatment approach and planning and design, including physical, chemical, eco-environmental, social, economic and cultural impacts on the environment.

5 Mountainurban Environmental Impact Assessment

(1) Purpose of environmental impact assessment

The purpose of environmental impact assessment is to prevent and mitigate the negative impacts and adverse consequences of development activities on the eco-environment, including the pre-development construction process and the process of completion, operation and utilization.

(2) Items of environmental impact assessment

The items of environmental and cultural impact assessment that should be carried out in mountainurban planning include land use, transportation and facilities.

① **Land use**

a. Industrial and mining development or construction and expansion of industrial zones;
b. Construction and expansion of amusement areas and parks (zoos);
c. Construction of new areas;
d. Renewal of old districts;
e. Development of comprehensive industrial and commercial areas;
f. Important public buildings and super high-rise buildings (above 100 m);
g. Development of underground space, construction and expansion of underground commercial streets and underground facilities.

② **Transportation**

a. Construction of highways and expressways;
b. Bridges, viaducts, flyover crossing;
c. Underground tunnel engineering or rail transit engineering;
d. Construction and expansion of railway or highway transportation hubs;
e. Construction and expansion of port terminals.

③ **Facilities**

a. Sports fields;
b. Construction and expansion of various cultural and educational facilities;
c. Development of medical buildings;
d. Water supply facilities;
e. Sewage treatment facilities and sewer systems;
f. Construction and expansion of substations of 100,000 kV and above;

g. Construction and expansion of waste treatment facilities and resource recycling projects;
h. Construction and expansion of waste transfer station;
i. Construction and expansion of graves and crematoria;
j. Construction and expansion of slaughterhouses.

(3) Main contents of environmental impact assessment
(1) Whether the statutory documents comply with the relevant provisions of government examination and approval;
(2) Environmental impact assessment report and instructions;
(3) Whether the amount of environmental impact complies with relevant legal provisions;
(4) Whether the development behavior is located in an ecologically sensitive location;
(5) Whether the development behavior is in line with environmental receptivity, such as land supply capacity, allowable limit of landscape influence, supply capacity of water resources, pollution bearing capacity of water body, etc.; The determination of whether the environmental impact prediction and assessment approaches are reasonable includes physical and chemical impacts on the physical environmental, eco-environmental impacts, socio-economic impacts and cultural impacts.

References

Liu N (2001) Natural geography. Science Press
Li W, Zhao J (2004) Review and prospect of ecology research. China Meteorological Press, Beijing
Shang Y (2002) General ecology, 2nd edn. Peking University Press, Beijing
Song Y (2000) Urban ecology. Shanghai Normal University Press, Shanghai
Wu L (2001) Introduction to sciences of human settlements. China Architecture & Building Press, Beijing

Chapter 10
Disasters

Environmental problems have existed since human beings began to use natural resources to meet their needs. However, with the process of urbanization, the development of industrialization and the increase in population, environmental problems have become increasingly prominent. More and more attention has been given to environmental pollution and urban disasters. Climate warming, water and soil erosion, anabatic land desertification, increased frequency of flood disasters, urban traffic problems, public health problems, pollution, diseases, fires and increased social crime rates have aroused people's attention to urban disasters and prevention.

In the past, insufficient attention has been given to disaster prevention in urban planning. Hence, cities often suffered from huge losses due to natural or man-made disasters. However, regardless of natural or man-made disasters, it is possible to minimize the losses caused by disasters if we can do a good job of prevention in advance, make a quick response and have a full range of contingency measures in case of disasters, and deal with the aftermath quickly and provide places of refuge and rescue after disasters. In addition, in order to achieve the level of a safe city, urban social and economic security is also very important in addition to environmental safety, allowing urban residents to really feel safe and at ease.

There are two main types of disasters in mountain cities: natural disasters, such as geological disasters (earthquakes, landslides, faults, etc.), mountain torrents and floods, and man-made disasters, such as fires.

1 Mountain Torrents and Flood Disasters

Mountain torrents are the main manifestation of flood disasters in mountain areas and cities, which are characterized by short action time, sudden and sharp rise and fall, high flow velocity, and great surface erosion and destructive power to mountain areas and cities. The outbreak of mountain torrents is often accompanied by geological disasters such as landslides, collapses, and mudslides.

© The Author(s), under exclusive license to Springer Nature Singapore Pte Ltd. 2021

G. Huang, *Theory of Mountainurbanology*,
https://doi.org/10.1007/978-981-16-0819-3_10

(1) Causes of mountain torrents and flood disasters

(1) **Geomorphic conditions**

Geomorphic characteristics play an important role in the formation of floods. Steep ground line gradient in the basin accelerates the confluence speed, resulting in the rapid fluctuation of floods, large peak discharge and high water level. Mountain and hilly areas cover 60–80% of the drainage area of China's seven major rivers (Yangtze River, Yellow River, Huai River, Hai River, Liao River, Songhua River and Pearl River), where the mountain torrents caused by heavy rains are fierce, causing the river water to rise and fall steeply and severe damage and loss to infrastructures such as towns, roads, and bridges, and crops in the upper reaches of the basin. After the floods with large volumes and high peaks gathered in these areas enter the plain area, the river channel in the plain area has gentle slope, low flow velocity, which cannot be accommodated by the riverbed, resulting in greater flood disasters. For example, in 1987, a catastrophic flood occurred in the lower reaches of the Three Gorges of the Yangtze River, with the peak flow reaching 110,000 m^3/s, while the riverbed of the Jingjiang River could only accommodate 6000 m^3/s, causing serious floods (Xianghao 2000).

(2) **Climatic conditions**

Climatic conditions are another important factor in the formation of floods and cause of flood disasters. Most rivers in China are of rainfall pattern, and the runoff is mainly supplied by rainfall which is directly affected by the Pacific subtropical high. During April to early June in general years, the ridge line of the subtropical high is at 150–200° north latitude. Hence, heavy rain and floods occur in the Pearl River Basin and coastal areas. From mid-June to early July, the ridge line of the subtropical high moves to 200–250° north latitude, causing plum rain in the Jianghuai area and the rise of river water level. From late July to mid-August, the ridge line of the subtropical high moves to 300° north latitude, and the zone of rainfall moves northward to Haihe River Basin, Hetao area and northeast area, which becomes the main flood season of river channels in this area. At this time, tropical storms and typhoons constantly land, causing rainstorms and floods in southern China; in late August, the ridge line of subtropical high moves southward. Hence, the rainy season in northern and central China ends. It can be observed that floods in river basins in China are closely correlated to climatic conditions. Therefore, whenever rainstorm occurs in a river basin, a flood will be formed, whose magnitude depends on the intensity of rainstorm, runoff area and rainfall time in each basin.

(3) **Geological conditions**

In the vast loess region of western China, the soil is of uniform quality, high permeability and extremely poor impact resistance. In case of rainstorm, there will be very large surface erosion and very high erosion modulus of surface soil, which leads to the erosion of a large amount of sediment and the collapse of hillsides, forming deep gullies. The middle reaches of the Yellow River flows through the Loess Plateau, with an area of water and soil erosion of 430,000 km^2. A large amount of sediment flows into the river channel along with surface runoff, making the sediment concentration of the Yellow River rank first in the world. As a result, the riverbed in the middle and lower reaches is seriously silted up, and is 5–10 m above the ground, becoming a famous perched river. At the same time, the riverbed erosion and deposition of this kind of sediment-laden river drastically changes, and the riverbed is extremely unstable. Hence, the river bank can easily overflow and break up the levees in case of a catastrophic flood.

(4) **Vegetation conditions**

Vegetation can preserve soil moisture and reduce rainwater erosion and surface runoff. Plant roots can also increase soil organic matter, improve soil structure and soil fertility. According to the case data of the Yongding River Basin, when the local vegetation coverage rate is 20%, the soil loss per mu is 7.4 m^3/year. When the local vegetation coverage rate is 40%, the soil loss per μ is 3.6 m^3/year. When the coverage rate is 60%, the soil loss rate per μ is 1.3 m^3/year. According to observation data in northern Shaanxi from 1971 to 1974, vegetation coverage can reduce surface runoff by 60–80%, and soil erosion by 70%.

(5) **Human factors**

Improper human activities have a direct impact on the formation of floods and flood disasters in the river basin, which mainly include:

① Deforestation and destruction of vegetation;
② Unreasonable farming and overgrazing;
③ Fill-up of rivers and lakes, aquaculture in lakes and destruction of wetlands lead to a large reduction in lake area and the decrease of flood storage capacity;
④ The construction of roads, buildings and structures on flood land, and even the development of cities and towns affect the normal flood discharge of river channels;
⑤ Dumping garbage to river channels, or encroaching on the river section to build embankments and hardening river banks, etc.

The above unreasonable human activities will greatly increase the probability of the occurrence of floods or aggravate the losses caused by floods.

(2) Prevention of mountain torrents and flood disasters

The great Dujiangyan water conservancy project takes an important position in the history of world water conservancy construction. Before its implementation, the Minjiang River often suffered from disaster caused by flooding water. In 256 B.C., Li Bing, the Prefecture Chief of Shu State in the Qin Dynasty, presided over the construction of the Dujiangyan water conservancy project. According to the *Records*

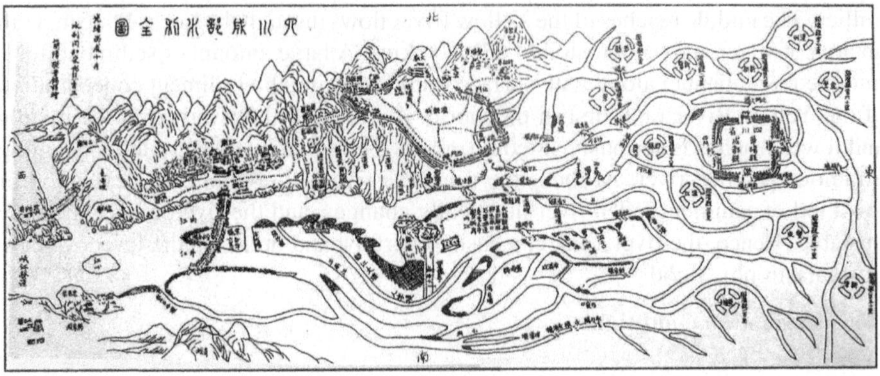

Fig. 1 Panorama of Dujiangyan water conservancy irrigation. *Source* Ying Jinhua, Fan Bingeng. *Famous Historic and Cultural Cities in Sichuan Province*. Chengdu: Sichuan People's Publishing House. October 2000, p. 131

of the Grand Historian of China • Annals of Rivers and Canals, "Li Bing, the Prefecture Chief of Shu State, chiseled the Yulei Mountain, and built the Dujiangyan irrigation system, eliminating the water damage of the Min River. Two rivers were dug in Chengdu, which were navigable, and the surplus water was used for irrigation. Local people enjoyed great benefits from this." It is recorded in the *Annals of Huayang State • Annals of Shu State* that "The Shu State had a vast and fertile land, known for its abundant natural resources and wealth. The water was introduced when it was arid, and the water gate was blocked when it rained. Therefore, it was said that wind and rain came on time, so the people did not know what it was like to be hungry. People all over the world called this place enviously as the land of abundance (Fig. 10.1).

The Dujiangyan water conservancy project is divided into three main parts, namely Yuzui, Feisha Weir and Baopingkou (Fig. 10.2). The Min River is divided into Inner River and Outer River by Yuzui, of which the Outer River is its main stream, and the Inner River flows to Chengdu and the West Sichuan Plain through Baopingkou, playing the role of navigation, irrigation and flood diversion; Baopingkou is the throat that controls the flow of the Inner River; Feisha Weir is a project of flood diversion and siltation reduction of the Inner River into the Outer River, which is a low weir about 180 m long, made of bamboo cages filled with stones. In times of flood, water from the Inner River overflows from the top of the weir into the Outer River; the weir can be washed away by the surging floodwater, and the water will pour directly into the Outer River. When it is used in conjunction with Baopingkou, it can ensure that the Inner River irrigation area has no water shortage when the water volume is small, and no flood when the water volume is large.

The Dujiangyan water conservancy project has been playing an important role for thousands of years, whose irrigation area has continued to expand to one million hectares up to now. Figure 10.3 is the valuable experience of flood channeling summed up by Chinese ancestors in their long-term struggle against floods. It is also

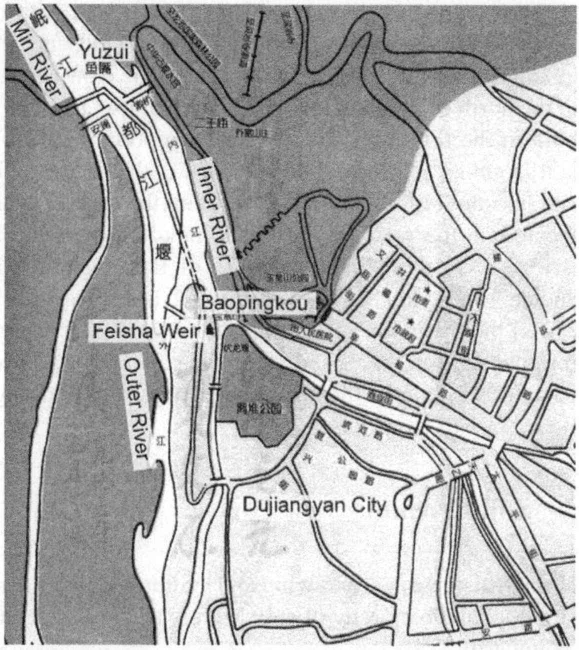

Fig. 2 Dujiangyan sketch map. *Source* Wang Chunwu. *Eternal Magic Achievement Dujiangyan*. Beijing: China Three Gorges Publishing House. April 2001, p. 9

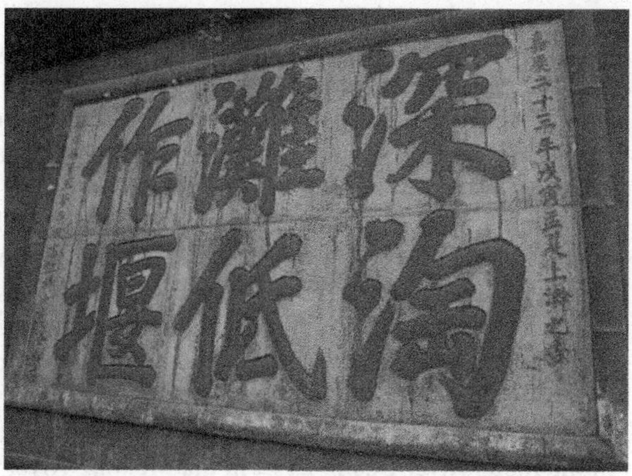

Fig. 3 Deep clean beach, low weir. *Source* Photograph by the author in 2005

the crystallization of wisdom reflected in the history of water conservancy construction over thousands of years of China's development, which is truly a great pioneering work that excels nature and benefits all generations. The great beauty and ingenuity of its engineering design conception is a model of the perfect combination of natural landscape, water conservancy project and urban planning in ancient China.

The layout of Dujiangyan City is integrated with the Dujiangyan water conservancy project and the surrounding natural landscapes. The north of the city is close to Jingui Mountain on the east bank of the Min River. There is a vital road leading to the north of Sichuan Province. For defense, two city walls and two gateways were built in the north, with strategically important terrain and beautiful scenery. There is a Douping platform, below which is the Erwang Temple in memory of Li Bing and his son. Fulong Temple is built on the central island of the Min River, which is located on the west bank of Baopingkou. There is a bamboo cable bridge, Anjian Bridge, over the Min River north of Baopingkou, which was a famous bridge in ancient times, and has been rebuilt into a steel cable bridge. There is a gallery bridge across the diversion channel in the west of the city, with benches on both sides for rest and sightseeing. Standing on the gallery bridge, we can see the water of the Min River flowing under the bridge straightly from Baopingkou to the downstream plain.

2 Soil Erosion

Human disturbance, and wind and hydraulic effects can seriously damage the natural structure and hierarchy of soil. Merely the soil protected by vegetation can maintain its integrity.

Soil is very sensitive to various human activities of construction and vulnerable to erosion once it loses the protection of vegetation and litter layer. Wind and hydraulic forces can blow or wash away soil particles. The surface layer of soil is rich in humus, and has a granular structure and strong absorption capacity. If the surface layers are lost, there will be deficiency of soil humus on the lower layer, with poor absorption capacity and stability. Once exposed to the Earth's surface, it is extremely susceptible to erosion. If the lower layer of soil is clay, its absorption capacity is even worse. In case of flood, it will form a rapid surface runoff, which is extremely erosive to soil.

Soil erosion is now the biggest threat to the continents in the world. Due to the interference of human activities, land reclamation has greatly increased the rate of soil erosion. According to the data, under the same rainfall conditions, the erosion rate of cultivated soil is 100 times higher than that of forest soil. The annual soil loss per km^2 in the world is 180 t on average, compared with 2804 t in China's Yellow River Basin, 257 t in the Yangtze River Basin, 1518 t in the Ganges River Basin, 449 t in the Indus River Basin, 214 t in the Mekong River Basin, 97 t in the Mississippi River Basin, 63 t in the Amazon River Basin, and 37 t in the Nile River Basin.

The amount of water and soil erosion in China has reached 5 billion t a year, and the two major rivers, the Yangtze River and the Yellow River, alone transport 2 billion t of sediment into the sea each year. At present, China's soil erosion area has accounted for one sixth of the national land area, about 1.5 million km^2, with a tendency of aggravation. Compared with the soil erosion rate, the soil regeneration rate is quite low. Under normal circumstances, the soil can only be supplemented by the weathering of parent rock, which takes approximately 200–1000 years to be thickened by 1 mm. If the soil is thickened by 3 mm by natural weathering, it may take 10,000 years. In recent years, the sandstorms in China have also intensified. By 1997, China's desertification area had reached 2.62 million km^2, accounting for 27% of the country's land area, with a trend of further expansion.

(1) Wind erosion

Loose and bare soil is very susceptible to wind erosion. Wind can blow soil particles to high and far places to form dust weather, and sandstorm and black storm may be formed in severe cases. For example, in the 1930s, due to blind reclamation in the western region of the United States, grassland vegetation was seriously damaged, and three "black storms" occurred in succession. The most serious one occurred in May 9–11, 1934, which swept from the west coast to the east coast at a speed of more than 100 km per h, rolling up more than 300 million t of soil and destroying tens of millions of hectares of fertile farmland. In North and Northwest China, the lack of effective protection of vegetation is the main cause of soil wind erosion, dust weather and sandstorm due to drought, extensive soil exposure and overgrazing. When dust is blown to high altitude by the wind, it can be transported horizontally for hundreds of kilometers or even tens of thousands of kilometers. Wind erosion and sandstorm can not only expose the roots of plants or bury vegetation, but also have a serious impact on the normal urban production and life. In many areas, wind erosion is more harmful than water erosion.

(2) Water erosion

Rainwater has an impact effect on the soil surface, just like an idiom says, "Constant dripping wears away a stone". Rainfall can damage the surface of soil which is lost in the form of surface runoff, especially in mountain areas with undulating terrain. There are three main different forms of rain erosion on soil.

Sheet erosion, that is, the surface of eroded area is evenly impacted and brings up the surface soil.

Rill erosion, that is, rainwater falls down rapidly along small ditches and rills, resulting in long-term cutting of soil.

Gully erosion, that is, a rill is deepened and widened by the scouring of rainwater to form a true gully. In the gully area of the Loess Plateau, gully erosion usually starts from a rut left by a vehicle off the road, which forms a deep gully after long period of rainwater scouring.

In the development of mountain areas and urban planning and construction, serious wind and water erosion accidents often occur due to the violation of strict site reconditioning procedures on large exposed construction sites. For example, in 1997, Taipingxi Town was among the first batch of new construction and relocation in the Three Gorges Reservoir area. However, in the process of relocation and construction, no detailed planning for mountainurban construction or the preparation for site drainage in the early stage of construction were made in strict combination with the terrain. Under the scouring of several rainstorms, a large amount of earth and rock excavated in the construction site poured into the river, causing the site to be severely damaged and having to choose a new site.

Under the severe reality that the contradiction between man and land is becoming increasingly prominent, the vegetation and agricultural land should be strictly protected. The development and utilization of land should be strictly based on natural conditions and characteristics of natural resources to develop farming, forestry, animal husbandry, side-line production and fishery industries according to local conditions. All land not suitable for cultivation should be returned to forests or grasslands, overgrazing is strictly prohibited, the land damaged by mining must be ecologically rehabilitated, and the erroneous practice of focusing only on mining and transformation without considering protection and management of environmental construction must be corrected.

3 Geological Disasters

(1) Earthquake

(1) **Concept of earthquake**

Earthquake, as a frequent natural phenomenon, is crustal vibration caused by the intense release of the Earth's internal energy, which has a major impact on the natural environment and human life. The ground damage from a strong earthquake is caused by strong shock waves. The magnitude of an earthquake is a level that indicates the amount of energy released by the seismic focus, which is usually measured on the Richter scale proposed by Richter. Earthquake intensity refers to the damage degree of earthquake to the ground and buildings, which is both related to and different from the magnitude. An earthquake has only one magnitude, but the same earthquake will have different intensities due to different epicentral distances. The closer to the epicenter, the greater the intensity, and vice versa.

(2) **Classification of earthquakes**

① **According to the geological origin**, it can be divided into tectonic earthquake, volcanic earthquake, collapse earthquake and induced earthquake.

Tectonic earthquake: The earthquake caused mainly by fault changes due to the tectonic changes in the deep crust is called tectonic earthquake, which often occurs in areas with greater activity on the Earth's crust, and generally in the active zones of folds and faults in modern times. This is characterized by frequent activities, long duration, wide range of influence and strong destructiveness. Furthermore, 90% of earthquakes on the Earth's crust are of this type.

Volcanic earthquake: An earthquake triggered by volcanic activity is called a volcanic earthquake, whose seismic focus is usually limited to the volcanic activity zone, with generally not large impact range and a depth not more than 10 km. Such earthquake accounts for approximately 70% of the total.

Subsidence earthquake: It is caused by the collapse and subsidence of rock stratums, which mainly occurs in karst developed areas or steep mountain areas, with a low incidence and little harm.

Induced earthquake: Due to human activities, factors such as reservoir impoundment and underground nuclear explosions can also trigger earthquakes, which are collectively called induced earthquake.

② **According to the depth of seismic focus**, it can be divided into shallow-focus earthquake (depth of 0–70 km), intermediate-focus earthquake (depth of 70–300 km) and deep-focus earthquake (depth of –300 km), in which the number of occurrence decreases in turn, accounting for about 72.5, 23.5 and 4% of the total number of earthquakes, respectively.

③ **According to the surface structure of earthquake occurrence**, it can be divided into marine earthquake and continental earthquake, accounting for approximately 85% and 15% of the total number of earthquakes, respectively.

(3) **Earthquake resistance and disaster prevention**

Earthquake, continental earthquake in particular, is one of the most serious natural disasters suffered by mankind. China is a country with high intensity of seismic activities and severe earthquake disasters in the world. According to the statistics recorded by instruments since the twentieth century, the continental earthquakes in China accounts for 33% of the world's total, which are not only of high frequency but also of great intensity. Once it occurs, it not only strongly causes ground subsidence and the destruction of buildings, but also often leads to a series of secondary disasters. Therefore, it is very important to do well in earthquake resistance and disaster prevention, which requires not only to strengthen the knowledge of earthquake prevention and disaster reduction, but also to reinforce buildings and build quakeproof facilities in hazardous areas with high intensity. In terms of urban layout, we should not only consider improving the comprehensive capability of earthquake resistance and disaster prevention, but also do a good job in earthquake prediction. Based on China's national conditions and combined with domestic and foreign experience and lessons, relevant Chinese departments have proposed the policy of "prevention first, comprehensive defense" for earthquake prevention and disaster reduction, which mainly includes four links: earthquake monitoring and prediction, earthquake disaster prevention, earthquake emergency, earthquake relief and reconstruction. The implementation of each link must rely on scientific and technological progress.

(2) Landslide

Landslide is a common geological disaster in mountain areas and cities, which is a phenomenon that the rock and soil mass on a slope slides slowly along a certain sliding surface under the action of gravity. In addition to factors such as soil, rock nature, strength, slope and stability, the occurrence of landslides is also directly correlated to soil vegetation, rainfall intensity, rain erosion and interference of human activities.

Due to its complex geological conditions and frequent urban construction activities, Chongqing is a city and region prone to landslides in China. For example, as reported by Chongqing Daily on October 10, 1998, there were 29,896 geological disasters such as dangerous rocks and landslides in the city, with a slipping mass of 50–105 m^3, causing harm to the urban social economy and the safety of people's lives and properties, which requires frequent strengthening of governance. For example, the old landslide mass of Jipazi was partially resurrected in Yunyang County due to the rainstorm in July 17–18, 1982, which destroyed 1730 houses and caused 15 million m^3 of landslide mass to slide into the Yangtze River, with a maximum slipping distance of 200 m. It also resulted in a siltation of the Yangtze River bed of more than 30 m, forming a 600 m-long rapids and dangerous shoals, which caused great difficulties to the shipping of the Yangtze River and caused great economic loss.

In addition to the main triggering factor of the torrential rain, the occurrence of the Jipazi landslide is also caused by river water scouring, artificial soil and stone quarrying at the front edge of the old landslide mass, forming a 40 m-high scarp along the river. The resurrection of the old landslide mass is the result of the superposition of natural and human factors. In addition, the section of the old landslide mass that was not damaged by humans survived the rainstorm unscathed.

(3) Collapse

(1) **Causes of collapse**

Collapse is a phenomenon that rock and soil mass on a slope falls suddenly under the action of gravity. Collapse is a kind of sudden disaster that occurs very quickly and is very destructive. For example, the great collapse of Jiguanling on the left bank of the Wu River on July 3, 1994 caused a rock mass of 5.3 million m^3 falling into the river, forming a rock-fill dam with a length of 110 m, which cut off the flow of the Wu River for half an hour and the river water to surge by 10 m. Three coal mines were destroyed, the docks were damaged, five ships were sunk, six people died, and the economic loss amounted to more than 9 million ¥.

Collapse usually occurs on steep hillsides or rivers, lakes and coasts, and is more likely to occur when the rock fissures are developed and the structure is broken, especially when the rock stratum and fracture surface are consistent with the slope inclination. Collapse mainly occurs in the season of heavy rain and melting of snow and ice. Due to the large amount of water absorption by the rock and soil, the load increases sharply, and the friction within the rock and soil mass decreases. Hence, this is most prone to collapse. In addition, when large-scale artificial blasting is carried out in the construction of roads and tunnels in mountain areas, the rock mass structure is easy to be destroyed and collapse.

(2) **Prevention and control of collapse**

① The terrain that may collapse should be avoided.

② In the development and construction of mountain areas, it should be avoided as much as possible to excavate a large number of steep mountains or slopes, and destroy the stability of the natural structure of rock and soil.

③ Reinforcement engineering measures such as excavation, anchoring or blocking should be adopted for the rock and soil mass with stable surface.

(4) Debris flow

(1) **Causes and classification of debris flow**

Debris flow is a kind of sudden natural disaster in mountain areas, which often causes severe disasters to mountain cities in an instant.

Debris flow is a mudflow that carries a large number of mudstone masses flowing in a river and gully after rainstorm or snowmelt, which is a kind of mud-like substance formed by the mixture of sediment accumulation on both sides of the bank and rainstorm runoff after the collapse of rock slope during the rainstorm, with the characteristics of fierce outbreak, short duration and extremely destructive power. The destructive force of debris flow lies not only in the rapid increase of its flow with the collapse and admixture of the deposits on both sides, but also in the impact force of the rocks contained in the debris flow.

According to its fluidity, debris flow can be divided into turbulent debris flow, viscous debris flow and comprehensive debris flow.

Turbulent debris flow is a mixture of water and solid materials such as fine-grained soil and stones. In the process of debris flow, the fine-grained soil is in a turbulent state, and solid materials move forward by the drag force at the bottom. When the debris flow stops, the impact cone formed by solid materials has a certain sorting type.

The water, fine-grained soil and stones in viscous debris flow have a certain viscosity and structure that remains almost unchanged in the process of flow. Solid materials such as stones and branches flow slowly in a straight line in one direction on a plane in suspension, but when the debris flow stops moving, the slurry seems to be in a solidified state, keeping its original structure unchanged.

Comprehensive debris flow is a kind of debris flow between turbulent debris flow and viscous debris flow.

According to the different particle composition and physical properties, debris flow can be classified into three categories: debris flow, mudflow and water-rock flow.

There are many factors for the formation of debris flow, mainly including topography, geology, vegetation in basin areas, surface weathering substances and precipitation.

In terms of the formation of debris flow, debris flow basin can be divided into three areas. The upper part of the basin is the formation area or supply area of debris flow, which water and solid substances such as sediment and stones in the debris flow mainly come from, the middle part is the passage area and local recharge area of debris flow, and the lower part is the deposition area of debris flow, that is, the area where the impact cone is formed.

Debris flow usually occurs on rivers and gullies in mountain areas with complex landforms. A steep slope is the condition for violent flash floods and scour after the

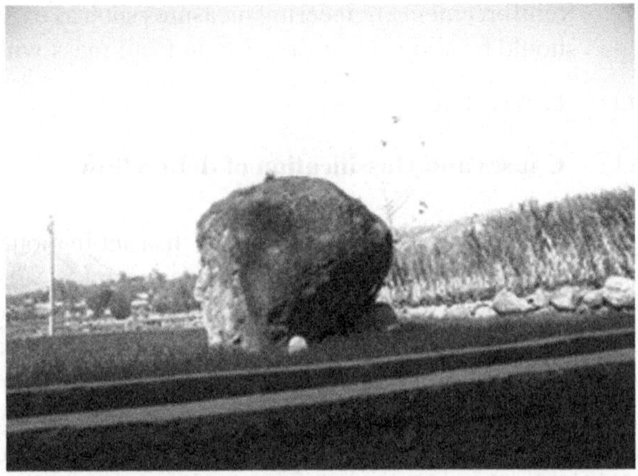

Fig. 4 Treatment measures for check dam project in Qiaojia county. *Source* Photograph by the author in 2006

sudden melting of precipitation or large amounts of snow. The collapse and landslide in geological tectonic movement and strong weathering of rocks and other natural processes cause rocks to disintegrate, fall off, and break, which provides the solid material supply basis for the formation of debris flow. These are mixed with water to form a certain concentration of mud, which causes huge stones to move forward in a suspended state in mud. In the semi-arid or glacial mountain areas in Southwest, Northwest and North China, debris flow is often prone to occur. For example, on May 27, 1984, a catastrophic debris flow broke out in Heishangou, Dongchuan Mining Bureau, Yunnan Province, which caused 155 deaths, destroyed a large number of industrial and civil buildings, and resulted in a 14-day shutdown of the entire mine and a direct economic loss of 11 million yuan; in 1753 (the 18th year of the reign of Emperor Qianlong in the Qing Dynasty), a large-scale debris flow suddenly broke out after a rainstorm in Bainigou, 4 km southeast of Qiaojia County, Yunnan Province. The entire county was buried, killing hundreds of people and inundating more than 1000 μ of farmland. Figure 10.4 presents the remnant of a large stone taken by the author from the scene (the largest one has been quarried for construction work).

Vegetation in the watershed can block the flow velocity of surface runoff, and make the soil infiltrate more water. The root system of plants can consolidate soil particles on the surface to prevent soil erosion. Therefore, the destruction of vegetation in the watershed is one of the important factor for the formation of debris flow.

(2) **Prevention and control of debris flow**

The prevention and control of debris flow should be based on prevention. In many mountain areas in West China, many cities are built in the accumulation area downstream of debris flow gullies. These low-frequency debris flow gullies have fewer floods in ordinary years, which are not enough to cause disasters. Most of these have not occurred debris flow for decades or even hundreds of years. However, there should be no fluke in these debris flow gullies. Effective preventive measures must be taken to protect the eco-environment in the gullies and reduce soil erosion. Hazard

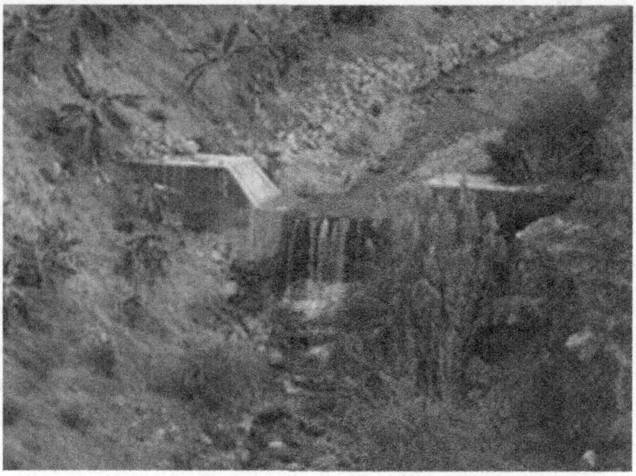

Fig. 5 Boulders left by debris flow. *Source* Photograph by the author in 2006

areas should be marked out in the lower reaches of the gullies, where the construction of any permanent buildings is forbidden. As long as human activities leave the hazard areas of debris flow, there will be no major disaster losses. Therefore, it is the most effective measure for the prevention of debris flow disaster to avoid the hazard areas of debris flow in the site selection of mountain new cities and expansion of old cities.

① **Engineering measures**: It includes the overall layout of engineering, individual projects such as sediment storage dams, check dams (Fig. 10.5), gully regulation and debris flow drainage channels, structural types and selection of engineering materials.

② **Biological measures**: It includes the zoning of biological governance, ecological restoration measures, the arrangement of dominant tree species and the restoration and protection of vegetation, etc.

③ **Comprehensive measures**: The prevention and control of geological disasters should be combined with the integrated management of Mountainurban planning and environment. In determining the Mountainurban population and construction scale, it is also necessary to consider the allowable bearing capacity of the geological environment in addition to considering the allowable bearing capacity of the ground environment, such as the susceptibility evaluation of land geological disasters, and the possibility of inducing earthquakes, landslides, collapses, and ground subsidence. The construction control and management of complex geological zones should be strengthened to prevent the possibility of inducing geological disasters in engineering construction. In addition, a forecast and early warning system, emergency measures for disaster occurrence, and a prevention and management system should also be established.

4 Fire

(1) Causes of fire

The occurrence of fire is closely related to social production and life. The experience of developed countries shows that the number of fires and the economic losses caused by fires are generally positively correlated with the activity degree of economic activities and the increase in the economic aggregate of a country. China is in a period of rapid economic growth. However, as the factors leading to fires increase, the economic losses caused by fires and the loss of people's lives and property are also on the rise. For example, in 2003, a fire broke out in the merchandise warehouse on the ground floor of a commercial building in Hengyang City, Hunan Province. The building collapsed suddenly while firefighters were fighting the fire, causing more than 20 firefighters to die, and more than 10 firefighters and journalists to be seriously injured.

The causes of fire are various, mainly including:

(1) Buildings are not strictly designed and constructed in accordance with the fire protection standards;
(2) Fire safety evacuation passages are not strictly reserved during use according to the regulations;
(3) Poor management;
(4) Weak awareness of fire control among the masses;
(5) Social crimes and arson.

(2) Fire prevention

(1) To meet the needs of the development of the socialist market economy, we should strengthen the building of fire protection laws and regulations and the fire early warning and monitoring system;
(2) Urban planning and fire planning should be combined;
(3) The publicity and education on fire control should be carried out to raise people's awareness of fire safety.

5 Biological Invasion Disaster

(1) Ecological disaster of biological invasion

"For thousands of years, oceans, mountains, rivers and deserts have provided natural barriers for the evolution of rare species and ecosystems. In recent centuries, these barriers have become ineffective due to the impact of global changes. Many species have travelled across oceans to reach new habitats and become alien invasive species" (*Guidelines for the Prevention of Biodiversity Loss due to Biological Invasion* adopted by the International Union for Conservation of Nature (IUCN) in Switzerland on February 2000).

Due to the process of globalization, natural isolation has been greatly reduced. Alien species enter other places beyond their original habitats through a variety of channels, resulting in a series of ecological security and environmental problems, which have a serious impact on society and economy.

Biological invasion, under the suitable climate, soil, water and transmission conditions, damages the environment and poses a long-term and lasting threat to the ecosystem, which is prone to spread to form a large area of single superior community, and endanger the survival of local plants and animals, resulting in loss of biodiversity. The direct economic losses caused by biological invasion to China's agriculture, forestry, animal husbandry and aquatic products amount to hundreds of billions of yuan each year.

Some infectious diseases (such as smallpox, poliomyelitis, AIDS, etc.) are typical examples of alien species invasion. Globalization will further expand the scope of the impact of new viruses (such as avian influenza, SARS, etc.) that are harmful to humans and organisms.

The State Environmental Protection Administration of China has announced 16 invasive alien species: Eupatorium adenophorum, Mikania micrantha, Alternanthera philoxeroides, Ambrosia artemisiifolia, Lolium temulentum, Spartina alterniflora Loisel, fragrant eupatorium herb, water hyacinth (Eichhornia crassipes), Johnsongrass, Opogona sacchari, Oracella acuta Lobdell, Dendroctonus valens LeConte, Hyphantria cunea, Achatina fulica, Pomacea canaliculata and Rana catesbiana. For example, Eupatorium adenophorum contains toxins, which can easily cause asthma in horses, resulting in a large number of deaths; in the flowering season of Ambrosia artemisiifolia, its pollen can cause symptoms such as asthma, sneezing and runny nose in allergic people, and even other complications and death of frail people; water hyacinth spreads all over the country and becomes a "crazy" disaster, which can quickly block the water surface of rivers and affect the waterways. After the death of a large number of water hyacinth plants, they are mixed with sediment and deposits on the bottom, swamping many lakes and rivers. As water hyacinth plants can absorb a variety of toxic substances such as heavy metals, these may exacerbate the secondary pollution of water quality after sinking into the bottom, resulting in huge indirect economic losses.

According to the statistics of the Dalian Forestry Department, there are currently more than 400 kinds of alien organisms invading China. Over the past decade, there have been at least 20 kinds of new alien organisms, one to two each year on average. The number, frequency, spreading range, harm and economic loss of alien biological invasion are all increasing.

(2) Prevention and control of biological invasion

(1) The legal and regulatory system of biological invasion should be established and improved to strengthen its overall management, and avoid its risk as much as possible;

(2) The risk assessment and early warning system of biological invasion should be established and improved;
(3) The research work on biological invasion should be strengthened, and sound control countermeasures and technical measures formulated for biological invasion species that cause great harms;
(4) The knowledge education of biological invasion should be strengthened to popularize the relevant knowledge, and raise people's awareness of ecological security.

6 Radiation Disaster and Prevention

(1) Causes of electromagnetic radiation disaster

The development of science and technology has promoted the progress of society and the improvement of urban modernization, but it has also brought a series of environmental problems and unsafe factors, among which is electromagnetic radiation. The facilities for artificially generated electromagnetic radiation include high-voltage power transmission and transformation systems (such as high-voltage lines and substations), electrified railways, radio communication transmission systems (such as mobile phone base stations), television transmission towers, broadcasting and relay stations, industrial, scientific, and medical electromagnetic energy facilities, high-frequency smelting furnace, etc. Due to the rapid development of urbanization, many radio communication transmitting stations and substations are surrounded by newly developed residential areas. Some high-rise buildings in urban areas are installed with mobile communication transmitting base stations, which enhances the electromagnetic field strength at high altitude in a city. The increase of automobiles, trams, subways, light rails, and electrified railways has also caused increased urban electromagnetic noise, and the rise of personal radio communications and household appliances has also increased electronic energy density in homes, all of which can emit electromagnetic waves. Nuclear power, as a new energy source, is becoming more and more important in developed countries, which is a relatively clean and safe energy. However, if it is not well managed and any accident occurs, it will cause a great disaster. For example, the accident at the Chernobyl nuclear power plant in the former Soviet Union on April 26, 1986 caused the death of tens of thousands of people, and its impact has not been eliminated to date.

Electromagnetic radiation pollution has become the fourth largest pollution after air pollution, water pollution and noise pollution.

However, not all electromagnetic radiations constitute electromagnetic radiation pollution. The electromagnetic radiation pollution that people worry about refers to the interference intensity of various electromagnetic waves exceeding a certain limit value, which acts on the human body and generates biological effects, causing dizziness, headache, birth defect, leukemia and cancer. As electromagnetic radiation is colorless, odorless, and invisible, it is easy to ignore its hazards. If used improperly, it may cause harm to human health.

(2) Prevention and control of electromagnetic radiation disaster

(1) Safety protection distance should be set in strict accordance with the legal standards and relevant safety regulations stipulated by the state. For example, nuclear power plant, high-voltage transformer and distribution station, high-voltage power line and microwave transmitter tower should be far away from densely populated urban areas, with the setup of adequate safety protective isolation belts. High-voltage power and telecommunication lines should be provided with high-voltage corridors in accordance with regulations. Any building is forbidden to be built under and within a certain distance from both sides of a high-voltage corridor, in order to keep away from the pollution sources of electromagnetic radiation.
(2) Household appliances and electronic communication equipment should be used in strict accordance with relevant operating procedures.
(3) We should strengthen publicity, popularize knowledge about electromagnetic radiation prevention, and raise people's awareness of self-protection and safety.

7 Planning and Management of a Comprehensive Disaster Prevention System

Due to the different types and causes of various disasters, the prevention and control plans are also different. Therefore, it is necessary to formulate disaster prevention plans separately for different disasters. At the same time, various disasters are often interconnected and affect each other, which are also easy to induce secondary disasters. For example, earthquake disaster is easy to induce fire and other secondary disasters, and serious flood is easy to induce infectious diseases and plagues. With the acceleration of globalization and urbanization, the incidence of some emergencies and sudden disasters tends to rise. Therefore, each city must formulate a comprehensive disaster prevention planning and management system to deal with various disasters, which mainly includes the following three aspects.

(1) Establishment of a forecast and early warning system

First, each city should establish a forecast and early warning system that can respond quickly to various disasters before their occurrence. This is the key to disaster prevention and control and reduction of casualties and economic losses. For example, Japan, a country prone to earthquakes, has established a well-developed rapid forecast and early warning system, which can release information nationwide within 10 s before an earthquake.

(2) Establishment of rescue facilities

When a disaster occurs, rescue facilities can be organized to deal with various disasters in an orderly manner, which includes the setting of rescue and evacuation shelters, ensuring the smooth channel of emergency rescue and transport and the unimpeded broadcasting and communication, the supply and backup of lifeline projects such as water, electricity and food, and the distribution of medical and ambulance stations.

(3) Establishment of a post-disaster feedback and monitoring mechanism

It includes rapid handling after disasters, establishment of monitoring mechanism, effective organization of self-help production of residents and reconstruction of their homes, to quickly restore the normal order of urban production and life, and reduce the losses caused by disasters to the minimum.

Reference

Xianghao Z (2000) Introduction to mountain science and mountain research in China. Sichuan Science and Technology Press, Chengdu

Chapter 11
Aesthetics

1 Urban Aesthetics and Evaluation Standards

Aesthetics is a topic widespread in human society, which is both common and profound, and both simple and complex. The aesthetic concept has existed since the existence of mankind, and urban aesthetics has existed since the existence of a city. For example, we have visited many cities, some of which give us a good impression, but some a bad impression. This is a preliminary impression of the city intuitively, including environment, buildings, traffic, streets, squares, as well as people's style of conversation, costumes and dressing. Therefore, urban aesthetics may contain a lot of contents, which is manifested in many aspects. This can be reflected, as follows: in nature, such as urban mountains, rivers and landscape; in urban society, such as people of all ages and both genders, daily life and clothing and apparel; in urban economy, such as industry and commerce, market, and financial situation; in urban culture, such as advertising, television, culture, and education; in urban architecture and engineering, such as residences, shops, restaurants, and bridges; and also in cultural landscape and style, such as cultural ethos and spiritual outlook (Guangyu 2002). Some cities have a very tense rhythm of life, while others are more relaxed, as observed from the people walking on the road. Therefore, urban aesthetics can be reflected in all aspects of society, economy, humanities, environment, etc.

We can decompose a beautiful city into several single elements for analysis and evaluation, including:

(1) Comfort: It means a psychological and physical comfort;
(2) Clarity: It is reflected in the urban layout, structure, function and traffic system;
(3) Accessibility: It means easy access and proximity, which not only has perfect traffic system structure, but also provides convenience for residents to participate in various social activities;
(4) Diversity: Life itself is rich and colorful, so people always hope to get rich and colorful life satisfaction, but do not want to live in a very simple and monotonous city;

© The Author(s), under exclusive license to Springer Nature Singapore Pte Ltd. 2021

G. Huang, *Theory of Mountainurbanology*,
https://doi.org/10.1007/978-981-16-0819-3_11

(5) Selectivity: Only with diversity can there be selectivity that also includes a variety of choices such as employment, transportation, etc.;
(6) Flexibility, i.e. adaptability: It means the flexibility, adaptability, elasticity and scalability should be reflected from single residence to the whole city;
(7) Sanitation: It is the basic requirement of a clean, culturally advanced and healthy city;
(8) Sense of security: It is the basic condition for the development of a culturally advanced city, which gives people a sense of belonging from residential rooms to families, communities, neighborhoods and to the whole city.
The above eight elements are all necessary for a beautiful city, of course, it is far more than these. In a city with beautiful environment and pleasant scenery, people will feel comfortable and willing to stop for appreciation, while a dirty, disorderly and bad city is certainly impossible to make people feel good or even depressed.
(9) Sense of neighborhood, that is, the atmosphere of community life: This is a cordial feeling of neighborhood harmony, respecting the elderly and caring for children, and solidarity and mutual assistance;
(10) Sense of native land: Native land is the soil on which the family depends, and is a characteristic formed by the long-term cultural accumulation of a city. The soil of our hometown has cultivated our "roots", and our childhood life will be deeply imprinted in everyone's mind. No matter how long a person has lived in his or her hometown, this kind of nostalgia is difficult to be indifferent and forgotten. It is the spiritual pillar for the inheritance and continuance of culture. In order to inherit and continue its own culture and cultivate the spirit of native soil, each city proposes the idea of "loving our own city" and "knowing our own city". Exploring each city's own historical and cultural accumulation is a good way to motivate citizens and inspire future generations;
(11) Sense of prosperity: A city should maintain a sustained social and economic development for a long time, with prosperity and vitality.

The above 11 aspects are the common requirements for all cities and the common evaluation standards for a beautiful city.

2 Aesthetic Characteristics of Mountain Cities

As a mountain urban, there is also a special requirement on its aesthetic composition in addition to the common and general requirements of the above 11 aspects, that is, the creation of the mountainurban characteristic beauty. In general, it should be seized and embodied in three aspects.

(1) Overall mountainurban harmonious beauty

China's mountain cities have always been nurtured in the embrace of natural mountains and rivers, making the urban artificial environment and natural landscape organically integrated. Traditional Chinese architecture, gardens and urban planning are all deeply influenced by the aesthetic concept of "landscape culture" and the philosophical concept of "unity of man and nature". The "man" here is not above nature, but a part of nature. They are people who blend in and coexist with nature, who consciously incorporate all activities of transforming nature into a virtuous cycle of natural ecology, instead of challenging and destroying nature. As an old saying goes, "Mountains are full of vitality for the existence of water, water is magnificent for the existence of mountains, and a city will be more beautiful with the existence of mountains and rivers." A city is particularly beautiful against the landscape of mountains, rivers, lakes, trees and terraces. From many traditional mountainurban forms, we can see the harmonious relationship between mountains, water, city and people, and the reciprocal symbiosis between artificial and natural systems, which makes people deeply feel the organic and harmonious aesthetic feeling between the city and the natural environment, just as an ancient poem goes, "People can still enjoy the tranquility of mountains and rivers, and the sereneness and seclusion of forests and springs in downtown areas."

(2) Rich and varied mountainurban beauty

The earth, mountains and rivers are important material spaces for human survival. The rich and varied topography and landscape environment are precious natural resources of mountain cities. "People should live in a place with mountains and rivers." and "Mountain take water as its blood, plants as its hair, and mist and rosy clouds as its charm." (Guo Xi, Song Dynasty), which provide an important material basis for the creation of dynamic mountainurban spatial structure and landscape characteristics, resulting in a development form and overall image different from other cities. With the changing skylines and street traffic, stacked and staggered buildings and multi-dimensional spatial organization, a unique mountainurban three-dimensional art takes shape.

(3) Dynamic mountainurban beauty

In the changes of four seasons in spring, summer, autumn and winter, day and night, and wind, flower, snow and moon as well as morning sunlight and sunset glow, people may feel the dynamic changes of time and space of a mountainurban in a particularly profound way. As Guo Xi said in the same book as above, "The sceneries of real landscape are different in the four seasons. The shadows of the mountains in spring are as light as the smiles of beauties, the mountains in the summer are fresh and green, and the mountains in the autumn are as bright as the facial makeup of beauties, and the mountains in winter are bleak and colorless, as if falling asleep." The rich and varied architectural cultural environment of a mountainurban is also deeply integrated into the dynamic natural environment of the changing seasons, forming a time and space beauty that is like a smile, a drop, a makeup and falling asleep, and weaving a colorful dynamic beauty of artificial and natural integration.

"The supreme state of oriental aesthetics is a tacit understanding between man and nature" (Qiuyu 1995). As the only one of the five major ancient civilizations in the world, the Chinese civilization is still being carried forward. One of its fundamental reasons lies in the profound ecological wisdom of the Chinese nation. In the long process of understanding and transforming nature, we have continuously summed up experience and lessons, and formed the ecological consciousness of the mutual dependence and coordination between man and nature. "If mankind is to survive in the twenty-first century, we must look back at 2500 years and learn from the wisdom of Confucius" (a paragraph in the declaration at the gathering of Nobel Prize Winners organized by the UNESCO in Paris in 1989). An important aspect of Confucius' wisdom is that it represents the ecological wisdom of harmony between man and nature, which has been nurtured and grown in China for more than 2000 years.

References

Huang G (2002) Mountainurbanology. China Architecture & Building Press, Beijing
Yu Q (1995) Fragments of civilization. Chunfeng Literature and Art Publishing House, Shenyang

Chapter 12
Governance

With the rapid development of China's social economy, various contradictions between urban social and economic development and environment will be further highlighted. With the gradual improvement of the socialist market economy, the regulatory role of urban planning in social and economic development will be further strengthened. Furthermore, the role of urban planning regulation and control can only be realized by effectively implementing and strengthening urban governance through urban planning.

The implementation of urban planning is different from the implementation of a certain project. The implementation of urban planning is not only a huge and complex systematic project of society, economy and environment, but also a long-term project, which is comprehensive, policy-based and highly local. The implementation of urban planning is both a basic function of municipal government, but also a common function of citizen organization and the common obligation of all citizens (Guangyu 1994). Through the administrative functions of the municipal government and the formulation of a sound legal system for administration according to law, and through the organization of enterprises, public institutions, non-governmental organizations and the active participation, coordination and interaction of all citizens, the objectives of urban planning are gradually implemented.

1 Improving the Legal System of Urban Governance

By means of legislation, the development of laws and regulations on urban governance should be strengthened, in order for all urban and regional development and construction activities and eco-environment maintenance to be brought into the track of the legal system, and allow the macro decision-making and objectives of urban planning to be gradually implemented.

© The Author(s), under exclusive license to Springer Nature Singapore Pte Ltd. 2021

G. Huang, *Theory of Mountainurbanology*,
https://doi.org/10.1007/978-981-16-0819-3_12

2 Scientific Implementation of Decision-Making

The implementation of urban planning is comprehensive, involving various aspects of social, economic and environmental construction and protection of regions and cities. The implementation of urban planning is also a dynamic process of long-term development. Therefore, it must be carried out step by step through scientific decision-making from regional planning to overall planning, detailed planning, construction design and each planning and design link and level.

Scientific decision-making is the essence of modern management, which involves the analysis, comparison and demonstration of multiple factors, objectives and programs, and the selection and optimization of schemes. In this process, a series of public policies and conflicts and coordination of various interests must be considered, including land supply policy, hillside development and ecological maintenance policy, real estate development and housing supply policy, industrial layout policy, circular economy policy, regional coordination policy, policy of coordinated urban and rural development, policy of historical and cultural tangible and intangible heritage protection, and environmental protection policy. At the same time, consideration should also be given to how to establish a balance and coordination mechanism in the implementation process, correctly handle the national interests and local interests, long-term interests and local interests, economic development and environmental protection, the coordination between the interests of different groups, as well as the relationship between efficiency and fairness. The implementation of these public policies fully embodies the guidelines of the Scientific Outlook on Development and the building of a harmonious society, and reflects the development goals that the country and municipal governments must achieve in a certain period of time.

3 Implementation of the Principle of Decision-Making Democratization and Mass Participation

The core content of urban governance is the implementation of public participation in urban planning and decision-making. Urban residents can exercise citizens' right to actively participate in strengthening management in urban planning and implementation in accordance with the Constitution. Urban planning and implementation are related to the vital interests of every citizen. Therefore, a mechanism for the participation of all people in supervising its implementation should be established throughout the planning formulation and implementation. In the past few years, it is a good form for the establishment of the planning committee system, the expert consultation and evaluation system for urban planning projects, the inspector system for the implementation of urban planning, and the hearing system for the urban planning and design process in cities to improve the quality of planning, public participation in urban planning and transparency in its implementation. This demonstrates that

China is gradually exploring experience and effective approaches in implementing the democratization of decision-making and in the direction of urban governance. Indeed, there is a process for the establishment and improvement of these systems, and there should be legal and institutional guarantees to prevent the abuse of power by some government decision-makers.

4 Improvement of the Functions of Community Organizations

Community is the most basic autonomous unit of a city. Strengthening the function construction of community organizations is an important embodiment of enhancing citizens' participation in democratic management, improving their autonomy and interaction, and enhancing the capability of urban governance. If a city has a sound, civilized and vibrant community organization, it will greatly improve the level of civilization construction of the whole city.

Reference

Huang G (1994) Mountain-city planning and construction and environmental development. Science Press, Beijing

The manufacturer's authorised representative in the EU is Springer Nature Customer Service Centre GmbH, Europaplatz 3, 69115 Heidelberg, Germany. If you have any concerns regarding our products, please contact ProductSafety@springernature.com

Printed and bound by CPI Group (UK) Ltd, Croydon, CR0 4YY
30/09/2026
02225330-0008